Water Quality, Pollution and Management

Water Quality, Pollution and Management

Edited by **Raven Spoon**

☐ **Syrawood**
PUBLISHING HOUSE

New York

Published by Syrawood Publishing House,
750 Third Avenue, 9th Floor,
New York, NY 10017, USA
www.syrawoodpublishinghouse.com

Water Quality, Pollution and Management
Edited by Raven Spoon

International Standard Book Number: 978-1-68286-187-5 (Hardback)

Printed in the United States of America.

Contents

Preface

This book aims to highlight the current researches and provides a platform to further the scope of innovations in this area. This book is a product of the combined efforts of many researchers and scientists from different parts of the world. The objective of this book is to provide the readers with the latest information in the field.

Sustainable water resources for consumption are constantly depleting due to pollution, climate change and other environmental factors. Therefore, water quality monitoring and management of water resources has become essential. The book aims to shed light on some of the unexplored aspects of water quality, pollution and management. It encompasses some of the most important topics and concepts like waste water treatment, water recycling and reuse, emerging water management practices, water contamination and pollution, etc. This book includes researches and case studies from various parts of the world. Students, researchers and readers in general will find this text an invaluable source of reference.

I would like to express my sincere thanks to the authors for their dedicated efforts in the completion of this book. I acknowledge the efforts of the publisher for providing constant support. Lastly, I would like to thank my family for their support in all academic endeavors.

Editor

Spatial Quantification of Non-Point Source Pollution in a Meso-Scale Catchment for an Assessment of Buffer Zones Efficiency

Mikołaj Piniewski [1,†,*], Paweł Marcinkowski [1], Ignacy Kardel [1], Marek Giełczewski [1], Katarzyna Izydorczyk [2] and Wojciech Frątczak [2,3]

[1] Department of Hydraulic Engineering, Warsaw University of Life Sciences (WULS-SGGW), Nowoursynowska str. 159, Warszawa 02-776, Poland;
E-Mails: p.marcinkowski@levis.sggw.pl (P.M.); i.kardel@levis.sggw.pl (I.K.); m.gielczewski@levis.sggw.pl (M.G.)

[2] European Regional Centre for Ecohydrology of the Polish Academy of Sciences, Tylna str. 3, Łódź 90-364, Poland; E-Mails: kizyd@biol.uni.lodz.pl (K.I.); wfratczak@wp.pl (W.F.)

[3] Regional Water Management Authority in Warsaw, 13B Zarzecze, Warszawa 03-194, Poland

[†] Current address: Potsdam Institute for Climate Impact Research (PIK), P.O. Box 60 12 03, Potsdam 14412, Germany.

[*] Author to whom correspondence should be addressed; E-Mail: m.piniewski@levis.sggw.pl;

Academic Editors: Lutz Breuer and Philipp Kraft

Abstract: The objective of this paper was to spatially quantify diffuse pollution sources and estimate the potential efficiency of applying riparian buffer zones as a conservation practice for mitigating chemical pollutant losses. This study was conducted using a semi-distributed Soil and Water Assessment Tool (SWAT) model that underwent extensive calibration and validation in the Sulejów Reservoir catchment (SRC), which occupies 4900 km^2 in central Poland. The model was calibrated and validated against daily discharges (10 gauges), NO$_3$-N and TP loads (7 gauges). Overall, the model generally performed well during the calibration period but not during the validation period for simulating discharge and loading of NO$_3$-N and TP. Diffuse agricultural sources appeared to be the main contributors to the elevated NO$_3$-N and TP loads in the streams. The existing, default representation of buffer zones in SWAT uses a VFS sub-model that only affects the contaminants present in surface runoff.

The results of an extensive monitoring program carried out in 2011–2013 in the SRC suggest that buffer zones are highly efficient for reducing NO_3-N and TP concentrations in shallow groundwater. On average, reductions of 56% and 76% were observed, respectively. An improved simulation of buffer zones in SWAT was achieved through empirical upscaling of the measurement results. The mean values of the sub-basin level reductions are 0.16 kg NO_3/ha (5.9%) and 0.03 kg TP/ha (19.4%). The buffer zones simulated using this approach contributed 24% for NO_3-N and 54% for TP to the total achieved mean reduction at the sub-basin level. This result suggests that additional measures are needed to achieve acceptable water quality status in all water bodies of the SRC, despite the fact that the buffer zones have a high potential for reducing contaminant emissions.

Keywords: water quality; nutrient; ecotones; vegetative filter strips; buffer strips; hydrological model; Pilica; diffuse pollution

---

1. Introduction

1.1. Water Management Context

Fulfilling the requirements of the European Water Framework Directive [1] and the Nitrates Directive [2] by reducing pollution emissions to water ecosystems remains a major challenge faced in water management. Particularly, the main issue is the reduction of non-point pollution that originates from agricultural land. The contributions of agriculture to the pool of nitrogen and phosphorus compounds in water ecosystems are high.

In Poland, the large share of farmland consisting of highly fragmented arable land and strongly dispersed developments has resulted in major pressure from pollution emission sources, including (1) pressure from agriculture related to the use of inappropriate farming practices (transport of organic and mineral nitrogen and phosphorus compounds from fertilizers to the environment) and (2) pressure from scattered households that are not connected to sewage systems.

Thus, the development of N and P reduction strategies is a major task for water authorities throughout Europe. One example of activities that are undertaken to achieve sustainable water management goals in agricultural catchments is the EU-funded EKOROB project (Ecotones for reducing diffuse pollution). The main objective of this project is to develop an Action Plan for reductions of diffuse pollution in the Pilica River catchment (Poland) and will help achieve a good ecological status for the water in the Sulejów Reservoir, particularly by reducing eutrophication and decreasing the frequency and intensity of cyanobacterial blooms.

The Action Plan is based on the ecohydrology concept [3–5], which assumes that the basis for integrated river basin management is the quantification of catchment-scale processes that are part of the hydrological cycle. The concept of ecohydrology involves quantifying hydrological processes at the basin scale and the entire hydrological cycle to quantify ecological processes. This quantification includes the patterns of hydrological pulses along the river continuum and the identification of various human impacts on point and non-point sources of pollution [6]. Thus, this quantification should be the first step when developing regulatory processes for sustainable water use and ecosystem protection. Although many mathematical tools are available for this task, the Soil and Water Assessment Tool (SWAT) [7] is one of the most widely used and comprehensive tools.

1.2. The Use of SWAT for Quantifying Emission Sources

SWAT is a comprehensive hydrological/water quality model that is increasingly being used to address a wide variety of water resource problems across the globe [8]. Several studies have investigated the spatial variability and distribution of various pollutant emissions/losses in catchments of different sizes [9–12]. Niraula et al. [9] calibrated SWAT (and a less complex GWLF model) for a small catchment in Alabama and used it to identify Critical Source Areas (CSAs) for sediment, TN and TP based on the loadings per unit area (yield or emission or losses) at the sub-basin level. Another application of SWAT in a medium-sized Greek catchment resulted in similar findings, but with a finer level of Hydrological Response Units (HRUs) [10]. Wu and Chen [11] investigated the influences of point source and diffuse pollution on the water quality of a relatively large catchment in south China by using SWAT. These authors concluded that diffuse pollution overwhelmingly surpasses point source pollution for all constituents except TP. In addition, these authors identified CSAs at the HRU and sub-basin levels. Finally Wu and Liu [12] calibrated SWAT for a large catchment in Iowa and showed a relationship between the shares of agricultural areas with sediment and NO_3-N emissions by using the calibrated model.

1.3. Riparian Buffer Zones and Their Modeling in SWAT

Riparian buffer zones (ecotones, vegetative filter strips) are an effective Best Management Practice (BMP) for buffering aquatic ecosystems against nutrient losses from the agricultural landscapes. Buffer zones are strips of permanent vegetation (including herbs, grasses, shrubs or trees) that are adjacent to aquatic ecosystems and used to maintain or improve water quality by trapping and removing various non-point source pollutants from overland and shallow sub-surface flow [13–16].

For pollutants transported in surface runoff, the process of sediment and nutrient trapping by buffer zones is reasonably well understood, particularly for grass filter strips (cf. review [17]). Reductions in the surface flow velocities due to the increased hydraulic roughness of the vegetation in the buffer enhanced particle deposition. Vought et al. [18] reported that a buffer strip with a width of 10 m can reduce phosphorus loads, which are typically bound to sediments, by as much as 95%. Buffer zone are effective for removing sediments and other suspended solids contained in surface runoff; however, soluble forms of nitrogen and phosphorus are not removed as effectively as sediments [19]. Results collected from 44 fields (row crops with slopes range from 1%–14%) showed that a 10-m buffer zone reduced the total suspended solids, soluble phosphorus and nitrate-nitrogen contents by 64%, 34% and 38%, respectively [20]. The efficiencies of narrow buffer zones (5–10 m) in Norway varied from 81%–91%, 60%–89% and 37%–81% for particles, total phosphorus and total nitrogen, respectively [21].

Buffer strips are normally less efficient for removing nitrate than orthophosphates from surface runoff. In contrast with orthophosphate, nitrogen is very labile and is not largely adsorbed within the soil [18].

However, the impacts of the sub-surface flow efficiency of the buffer zones on reducing nitrogen are well described in the literature and often reach a concentration reduction of 90% [17]. A meta-analysis of nitrogen removal in riparian buffers based on data from 65 individual riparian buffers from published studies indicated a mean removal effectiveness of 76.7% [22]. However, the efficiency of buffer zones for reducing phosphorus in shallow groundwater is not well documented, and some studies suggest that riparian zones are ineffective for removing dissolved phosphorus or could even release phosphorus to the water [23,24].

Buffer zones along small streams that are more exposed to pressure from agriculture are more efficient than buffer zones along larger rivers. A key factor for determining the efficiency of buffer zones is their continuity. Continuous and narrow riparian buffers are more efficient than wider and intermittent buffers [25]. Hence, an important issue for effective water management is the selection of priority areas that have the highest emissions of diffuse pollution. Next, concentrating measures, such as buffer zones, should be applied in these areas. Catchment-scale water quality modeling is one possible solution for quickly identifying priority areas.

Numerous examples are available regarding the application of SWAT for simulating the affects of buffer zones on diffuse pollution [26–29]. Older versions of SWAT used a very simplistic equation that was only based on filter width for calculating the HRU-level reduction rate of buffer zones. This equation was based on empirical data from the US regarding buffer strip efficiency [27,28]. Since then, SWAT has undergone certain modifications to address variable source areas within watersheds and vegetated buffers adjacent to streams [26,30]. The new VFS sub-model currently used in SWAT reduces the sediment, nitrate and phosphorus loading in streams as a function of estimated reductions in runoff. Hence, the new VFS sub-model only affects contaminants present in surface runoff and neglects nutrient trapping in shallow groundwater. As mentioned previously for nitrogen in sub-surface flow, buffer zones are very efficient measures [17]. However, little consensus has been reached for phosphorus [23,24]. This result suggests that the buffer zone efficiency is case-specific and depends on local conditions. Hence it is equally as important to apply existing models as it is to measure the efficiency of existing buffer zones in the field to gain more confidence regarding their behavior.

1.4. Objective

Two objectives of this paper are:

1. Spatial quantification of NO_3-N and TP emissions from major pollution sources in a meso-scale catchment using SWAT.
2. Simulation of buffer zone effects on the mitigation of pollution losses when applied in Critical Source Areas through the combined use of the default SWAT VFS sub-model and local field monitoring data.

The term "meso-scale catchment" refers to catchments with an order of magnitude between 10 and 10^3 km^2 [31]. A part of the Pilica catchment selected as the case study in this paper, a demonstration catchment of the EKOROB project, satisfies this condition.

2. Materials and Methods

2.1. Study Area

The study was conducted in the Sulejów Reservoir catchment (hereafter referred to as the SRC). Sulejów is a shallow and eutrophic artificial reservoir that was built in 1974 and is situated in the middle course of the Pilica River in central Poland. Two main tributaries supply water to the Sulejów Reservoir: the Pilica and Luciąża Rivers. At its full capacity, this reservoir covers an area of 22 km^2, with a mean depth of 3.3 m and a volume of 75×10^6 m^3. The Sulejów Reservoir was used as a drinking water reservoir for Łódź agglomeration until 2004 and is currently an important recreational site that has been extensively studied (*cf.* review [32]). Microcystis aeruginosa is the dominant species of bloom-forming cyanobacteria in the reservoir and produces microcystin-LR, microcystin-YR, and microcystin-RR [33–35].

The SWAT model is used in this study for the entire SRC area upstream of the dam, which occupies 4933 km^2 (Figure 1). This area consists of the Pilica (contributing 79.8% of area) and the Luciąża (15.3%) River catchments and a direct reservoir sub-catchment with several smaller streams (4.9%). The elevation of the SRC varies from 154 m in the lowland areas in the north to 499 m in the highland areas in the south. The distribution of land cover in the SRC area is as follows: 44.4% arable land, 38.6% forest areas, 12.3% grasslands, and 4.7% urban areas (mainly low-density residential areas), with the remaining land occupied by other types of land cover (data according to Corine Land Cover 2006). The predominate soil types in this area are loamy sands and sands. The climate of this area is typical for central Poland, with a mean annual temperature of 7.5 °C and mean January and July temperatures of −4 °C and 18 °C, respectively. The mean annual precipitation is 600 mm. The highest amounts of precipitation occur in June/July, and the lowest amounts of precipitation occur in January. The flow regime is characterized by early spring snow-melt induced floods and summer low flows with occasional summer floods. The quantitative pressure on surface water resources is relatively low. The fish farming industry scattered around the catchment and the Cieszanowice Reservoir constructed in 1998 on the Luciąża River (volume of 7.3×10 m^3 at full capacity) are the only considerable sources of flow alteration.

In contrast, multiple point and non-point pollution sources in the area result in elevated N and P loads flowing into the Sulejów Reservoir, which eventually contribute to toxic cyanobacterial blooms in its waters. These different sources will be described systematically in terms of SWAT model inputs in Sub-Section 2.4.

2.2. SWAT Model

2.2.1. General Features

SWAT is a physically based, semi-distributed, continuous-time model that simulates the movement of water, sediment, and nutrients on a catchment scale with a daily time step. The basic calculation unit, referred to as a "hydrological response unit" (HRU) is a combination of land use, soil, and slope overlay. Both water balance components, which is a driving force behind affect all processes that occur in a watershed, and water quality, output parameters are computed separately for each HRU. Water, nutrients and sediment leaving HRUs are aggregated at the sub-basin level and routed through the stream network to the main outlet to obtain the total flows and loadings for the river basin.

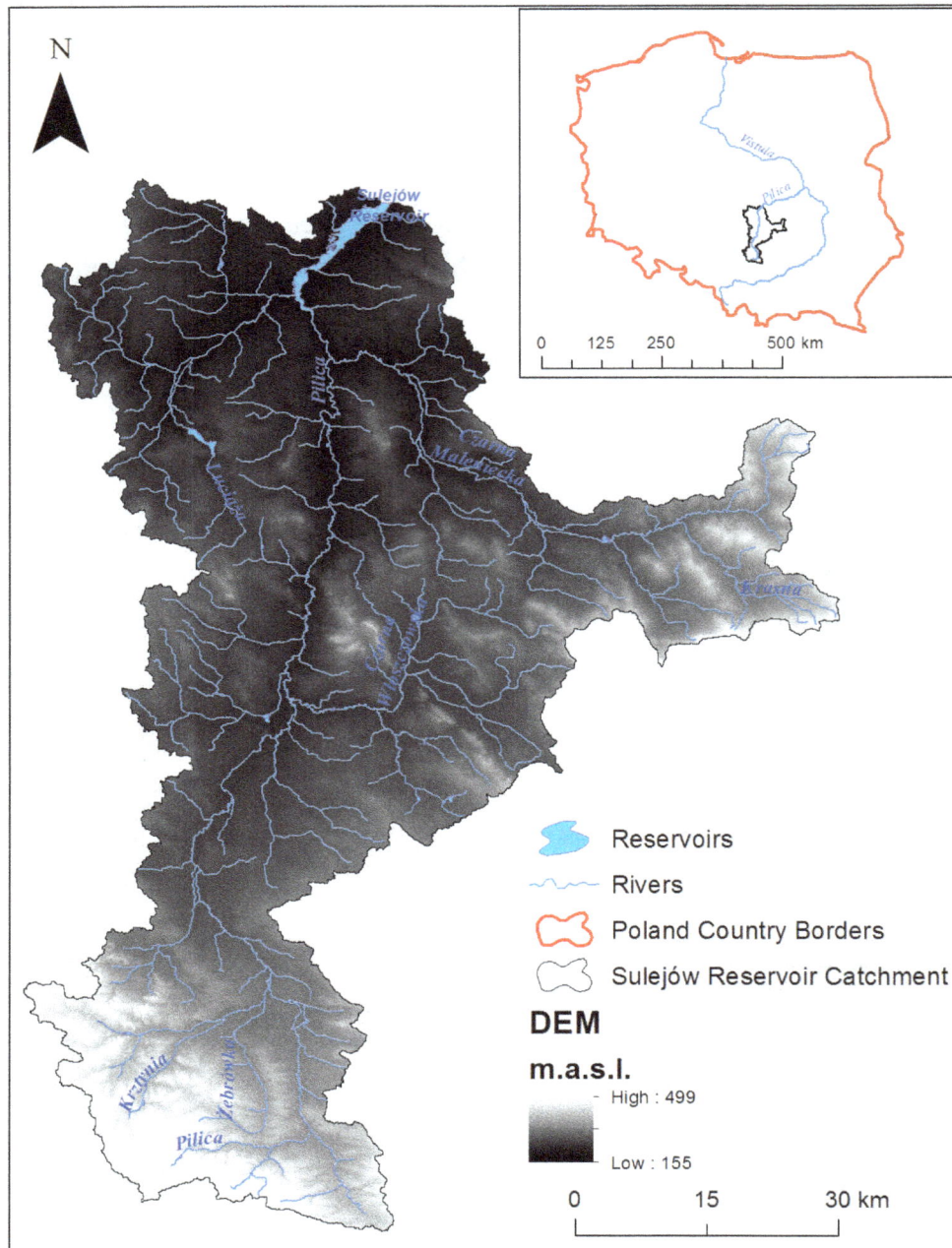

Figure 1. Study area.

In this study, channel routing was modeled using a variable storage coefficient approach. The modified USDA Soil Conservation Service (SCS) curve number method for calculating surface runoff and the Penman-Monteith method for estimating potential evapotranspiration (PET) were selected. In the model, snow-melt estimations are based on the degree-day method. The SWAT adapted plant growth model, which is used to assess the removal of water and nutrients from the root zone, transpiration, and biomass/yield production, is based on EPIC [36]. The in-stream water quality component allows us to control nutrient transformations in the stream. The in-stream kinetics used in SWAT for nutrient routing are adapted from QUAL2E [37].

SWAT simulates the movement and transformation of several forms of nitrogen and phosphorus in the watershed. In the nitrogen cycle, the main processes are denitrification, nitrification, mineralization, plant uptake, decay, fertilization, and volatilization. In the phosphorus cycle, the main processes are mineralization, fertilization, decay, and plant uptake. The nutrient transport pathways from upland areas to stream networks correspond to the following hydrological transport pathways: surface runoff, lateral subsurface flow and groundwater flow. Additionally point source discharges of water and contaminants can be defined that are directly input into the water routed through the stream network.

From the point of view of modeling buffer zones in SWAT, it is important to note that HRUs are lumped and non-contiguous geographic units within each sub-basin. A SWAT model setup may consist of thousands of such units, and each of them may represent one field, a portion of a field, or, more likely, portions of many fields [38].

2.2.2. Runoff-Reduction-Based Buffer Zone Sub-Model

A key characteristic of the buffer zone sub-model implemented in SWAT is that it works at the HRU level and reduces the loads of sediment, nitrate and phosphorus that enter the stream as a function of estimated reductions in runoff. Hence, the sub-model only affects contaminants that are present in surface runoff and neglects the potential affects of buffer zones on shallow groundwater. This sub-model was developed and evaluated using measured data derived from the literature and included data that were collected using differing experimental protocols and under conditions with different soils, slopes and rainfall intensities. When measured data were unavailable, predictions from the process-based Vegetative Filter Strip MODel (VFSMOD) [39] and its companion program, UH, were used. The UH (upland hydrology) utility uses the curve number approach (USDA-SCS, 1972), unit hydrograph and the Modified Universal Soil Loss Equation (MUSLE) [40] and allowed us to generate a database of sediment and runoff loads that enter the VFS. VFSMOD simulations were used to evaluate the sensitivities of various parameters and correlations between the model inputs and predictions. Consequently, an empirical model for runoff reduction by VFSs was developed, as described by the following equation:

$$R_R = 75.8 - 10.8 \ln(R_L) + 25.9 \ln(K_{sat}) \tag{1}$$

where R_R is the runoff reduction (%); R_L is the runoff loading to the buffer zone (mm); and K_{sat} is the saturated hydraulic conductivity (mm $\cdot$ h^{-1}). An important consideration is that SWAT conceptually partitions VFS sections within an HRU into two parts: a short part that occupies 10% of its length and receives flow from the 0.25–0.75 of the field area and a part that occupies 90% of its length and receives the remaining amount of flow. A fraction of the flow through the most heavily loaded 10% that is fully channelized is not subject to the VFS sub-model. Although the buffer zone width is an essential and intuitive characteristic that influences its trapping efficiency, it is not implemented as a parameter of the VFS sub-model in SWAT. Instead, the drainage area to buffer zone area ratio ($DAFS_{ratio}$) that is negatively correlated with the buffer zone width is combined with the HRU-level predicted runoff to estimate R_L.

The sediment reduction model based on VFS data removes sediment by reducing runoff velocity and enhancing infiltration in the VFS area, which is described by the following equation:

$$S_R = 79.0 - 1.04S_L + 0.213R_R \tag{2}$$

where S_R is the predicted sediment reduction (%) and S_L is the sediment loading (kg $\cdot$ m^{-2}).

The nitrate reduction model was only based on runoff reduction, as described by the following equation:

$$NN_R = 39.4 + 0.584R_R \tag{3}$$

where NN_R is the reduction of nitrate nitrogen (%).

The model for total phosphorus reduction was based on sediment reduction, which is described by the following equation:

$$TP_R = 0.9S_R \tag{4}$$

where TP_R is the total phosphorus reduction (%); and S_R is the sediment reduction (%).

2.3. SWAT Setup of the SRC

Table 1 lists all major data items and sources used to create the SWAT model setup of the SRC. The specific applications of this data at different stages of model development are described below.

Table 1. Data items and sources used to create the SWAT model setup of the SRC.

Data	Source
Digital Elevation Model	CODGiK (Central Agency for Geodetic and Cartographic Documentation)
Water cadastre GIS layers	RZGW (Regional Water Management Authority in Warsaw)
Corine Land Cover 2006	GDOS (General Directorate of the Environmental Protection)
Orthophotomap	GUGiK (Head Office of Geodesy and Cartography: geoportal.gov.pl)
Agricultural statistics	The Local Data Bank of GUS (Central Statistical Office)
Agricultural soil map 1:100,000	IUNG-PIB (Institute of Soil Science and Plant Cultivation - National Research Institute)
Forest soil maps 1:25,000	RDLP (Regional Directorate of State Forests in Łódź, Radom and Katowice)
Atmospheric deposition of nitrogen	GIOS (Chief Inspectorate of Environmental Protection
Climate data	IMGW-PIB (Institute of Meteorology and Water Management - National Research Institute)

The automatic watershed delineation of the SRC was based on the Digital Elevation Model (DEM) and stream network GIS layer. A 5 m resolution DEM characterized by a mean elevation error of 0.8–2.0 m was created from the ESRI TIN DEM available from CODGiK (Polish Central Geodetic and Cartographic Agency). This DEM resulted in the division of the catchment into 272 sub-basins with average areas of 18.1 km^2 (Figure 2). The Corine Land Cover (CLC) 2006 layer was used as the primary data source for the land use/land cover map. However, this layer was enhanced by several supplementary datasets and analyses:

- The (open) drainage ditch layer was used to sub-divide the CLC grasslands class into those under (code: FES2) and beyond (code: FESC; *cf.* Figure 3) the influences of drainage. It was assumed that the influences of the drainage ditches occurred within a 100 m buffer around the ditches.
- The orthophotomap was used to identify which SWAT crop database classes should be assigned to the "Heterogeneous agricultural areas" CLC class (code 2.4). Based on a manual, case-by-case investigation, the following three classes were most frequently assigned: "Agricultural land generic" (AGRL), "Urban low density" (URLD) and "Mixed forests" (FRST).
- The commune-level (39 units) agricultural census statistical data from 2010 were used to sub-divide the "Non-irrigated arable land" CLC class (code 2.1.1) into classes that represented particular crops that were cultivated in the SRC. This subdivision was done using a set of GIS techniques, including the "Create Random Raster" tool in ArcGIS. Thus, a 100 m resolution raster dataset that represented the random (yet preserving the commune level of crop distribution) locations of 6 major crops was created and combined with the final land cover map used as SWAT input. Although it may seem risky to generate random crop locations, we believe that this approach does not significantly impact the modeling result due to the lumped nature of the HRUs within each sub-basin.

Overall, the following five crops were distinguished as well as a fallow/abandoned land class (BERM): spring barley (BARL), rye (RYE), potato (POTA), corn silage (CSIL) and head lettuce (LETT). Figure 3 shows the final distributions of all land cover classes in the SWAT model setup.

The numerical soil map (scale 1:100,000) from the Institute of Soil Science and Plant Cultivation (IUNG) and numerical soil maps (scale 1:25,000) from the Regional Directorate of State Forests were used to create a user soil input map with 27 soil classes. By overlaying the land use and soil maps, 3401 HRUs were delineated in the catchment. The following area thresholds were used in the HRU delineation: 30 ha for land cover and 50 ha for soils. Thus, when using this method, all land cover types below the first threshold in each sub-basin were removed and aggregated into the remaining classes.

The meteorological data required by SWAT (precipitation, solar radiation, relative humidity, wind speed, and maximum and minimum temperatures) were acquired from the Institute of Meteorology and Water Management-National Research Institute (IMGW-PIB) for 1982–2011. Precipitation data were obtained from 49 stations, whereas data for other variables from 17 stations. To improve the spatial representation of climate inputs, spatial interpolation of all variables (apart from solar radiation, which was only available for one station) was performed before reading the SWAT input files. For precipitation, the Ordinary Kriging (OK) method was applied. However, the Inverse Distance Weighted method was used for the other variables. Szcześniak and Piniewski [41] showed that the OK method outperforms the SWAT default method for precipitation. For other weather variables, the interpolation process did not significantly affect the modeling results.

2.4. Pollution Sources in the Model Setup

Parameterization of point and non-point source pollution plays a critical role in water quality modeling and has attracted considerable attention in this paper. The following anthropogenic pollution sources were identified: (1) diffuse pollution from agricultural areas; (2) sewage treatment plants and septic tanks and (3) fish ponds. Atmospheric deposition of nitrogen was also considered.

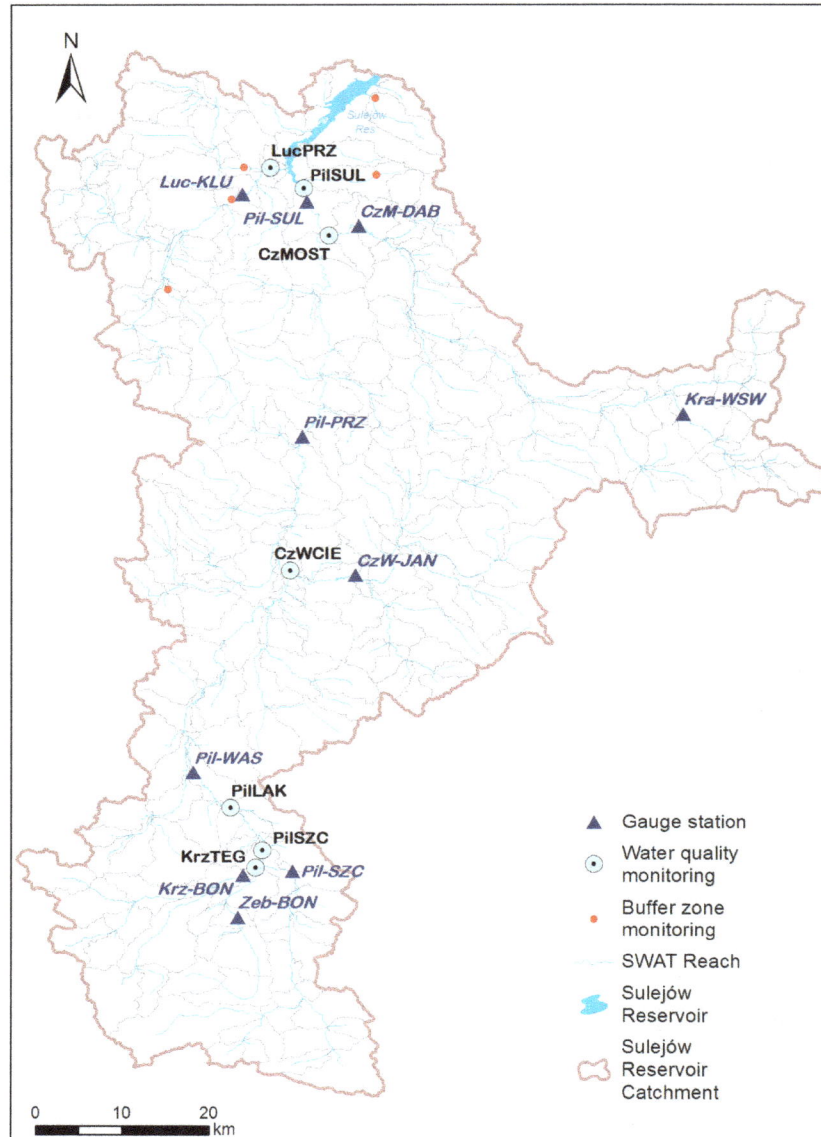

Figure 2. Delineation of the SRC into sub-basins and the gauging station locations used for calibration and validation.

2.4.1. Atmospheric Deposition

The mean concentration of nitrogen in precipitation measured in Sulejów near the inlet of the Sulejów Reservoir was obtained from the Chief Inspectorate of Environmental Protection (GIOS). The monthly results covered the time period of 1999–2010. Precipitation samples were collected daily, and an integrated sample was created and measured in the laboratory each month. Between 1999 and 2010, the mean annual concentrations varied significantly between 1.39 and 2.2 mg N $\cdot$ L^{-1}. The final value input into the model was subject to calibration and equaled 1.48 mg N $\cdot$ L^{-1}.

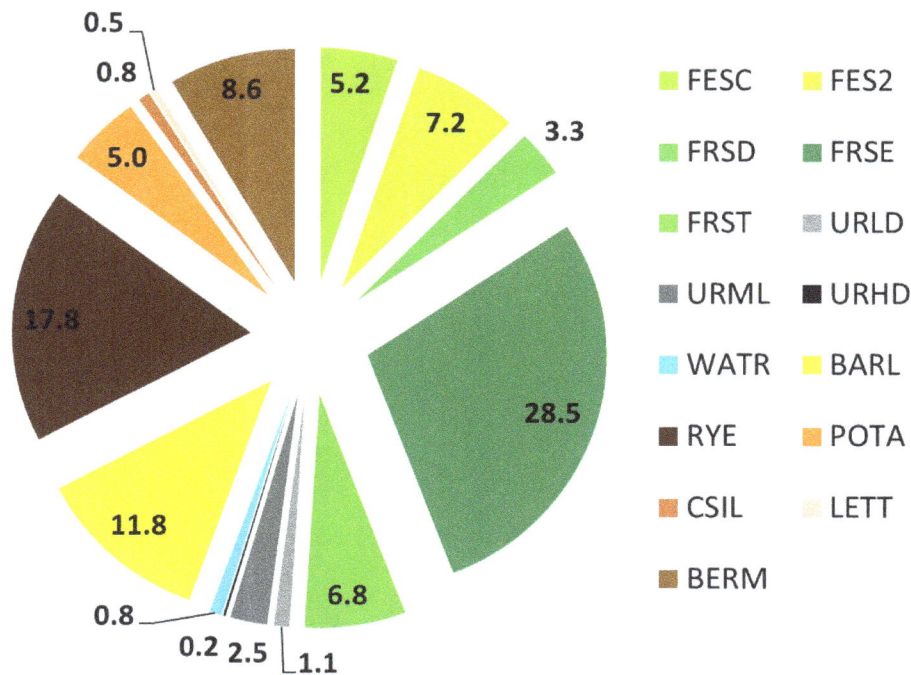

Figure 3. Land cover classes used as input for the SWAT model of the SRC. All codes as in the default SWAT plant database with one exception: FES2 has the same parameters as FESC, but is under the influence of open drainage ditches.

2.4.2. Fertilizers

Diffuse N and P pollution from agricultural fields mainly results from fertilizer use. SWAT enables us to define crop- and soil-specific management practices scheduled by date for each agricultural HRU. Typical management practice schedules, including the dates, types and amounts of fertilization, were obtained by consulting local extension service experts. First, derived management schedules were assigned to agricultural HRUs by using the ArcSWAT interface. However, this approach typically leads to bias in the total amounts of spatially-averaged fertilizer when compared with data from external sources. In this study, we used commune-level data from the Central Statistical Office (as for 2010) to determine mineral fertilizer use and livestock population. The livestock population was used to calculate the amount of available organic fertilizer (manure or slurry). In the final step conducted using GIS software, correction factors for fertilizer rates were defined for the sub-basins that overlapped with different communes. The commune layer intersected the sub-basin layer so that the total amounts of fertilizer used annually in different communes (expressed in tons of N and P) could be distributed over the SWAT sub-basins proportionally to the area of agricultural land in each sub-basin. Simultaneously, we aggregated the total fertilizer use per sub-basin from the model output based on initially implemented management schedules. Next, correction factors were calculated for each sub-basin as the ratios of total fertilizer use at the sub-basin level from census data to the total fertilizer use obtained from the SWAT output files. In the final step, each HRU fertilizer rate in the operation schedules was adjusted using the calculated correction factors. After this adjustment, the bias in the spatially averaged amounts of fertilizer largely decreased. Figure 4 shows the final, sub-basin-averaged rates of mineral and organic fertilizers applied in the SWAT model of the SRC.

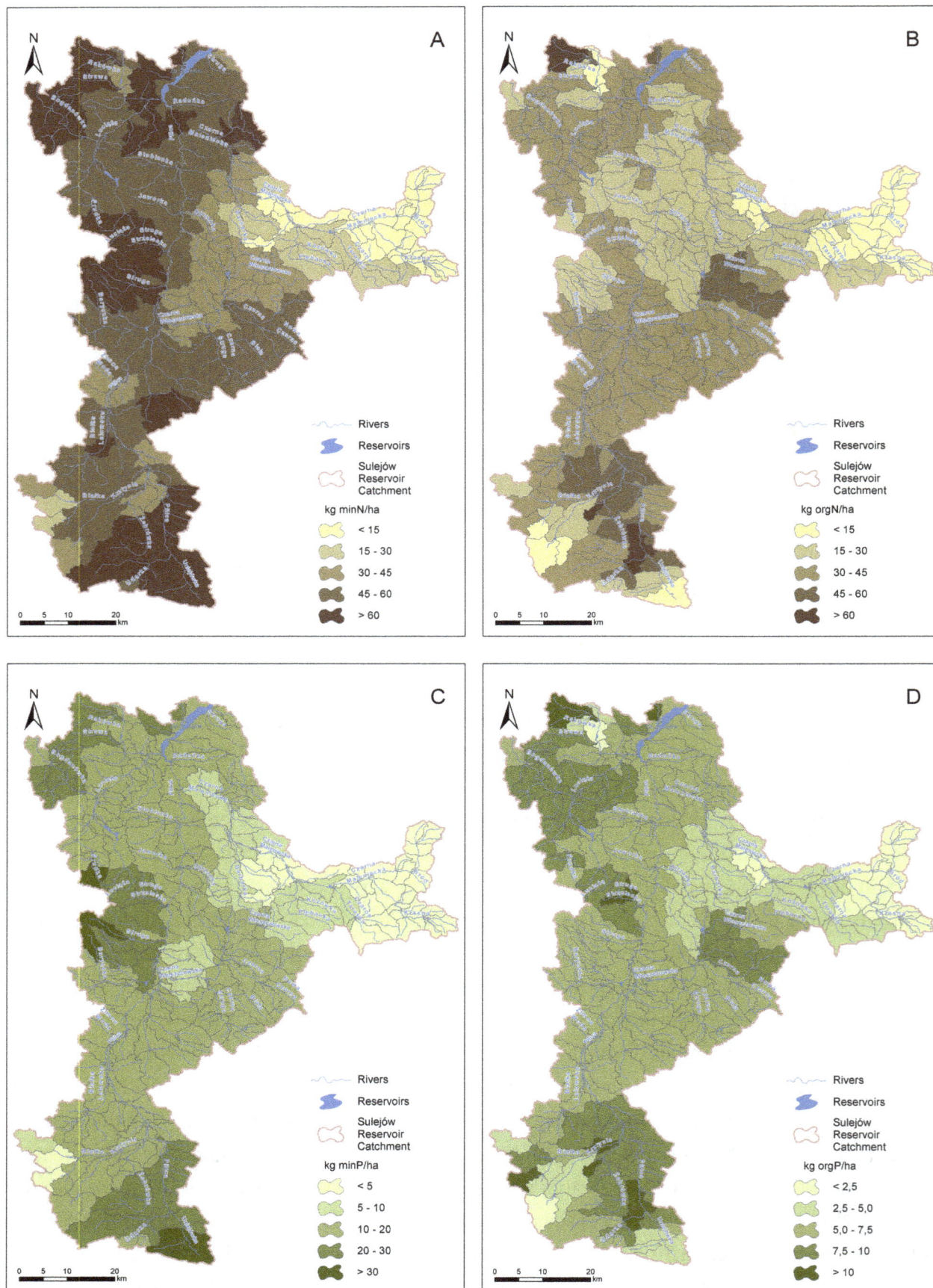

Figure 4. Sub-basin-averaged N and P fertilizer rates: (**A**) mineral nitrogen; (**B**) organic nitrogen; (**C**) mineral phosphorus; (**D**) organic phosphorus.

2.4.3. Sewage

Twenty three sewage treatment plants discharging an average of more than $2 \text{ L} \cdot \text{s}^{-1}$ of treated sewage water annually were identified in the SRC and used in the model setup (Figure 5A). The largest plant of the SRC situated in Piotrków Trybunalski discharges its sewage water downstream of the Sulejów Reservoir and was therefore neglected during model setup. For each sewage treatment plant, discharge and nutrient loads were expressed as constant or mean monthly values depending on the available data. These values were obtained directly from plant operators in most cases by using a telephone/electronic survey.

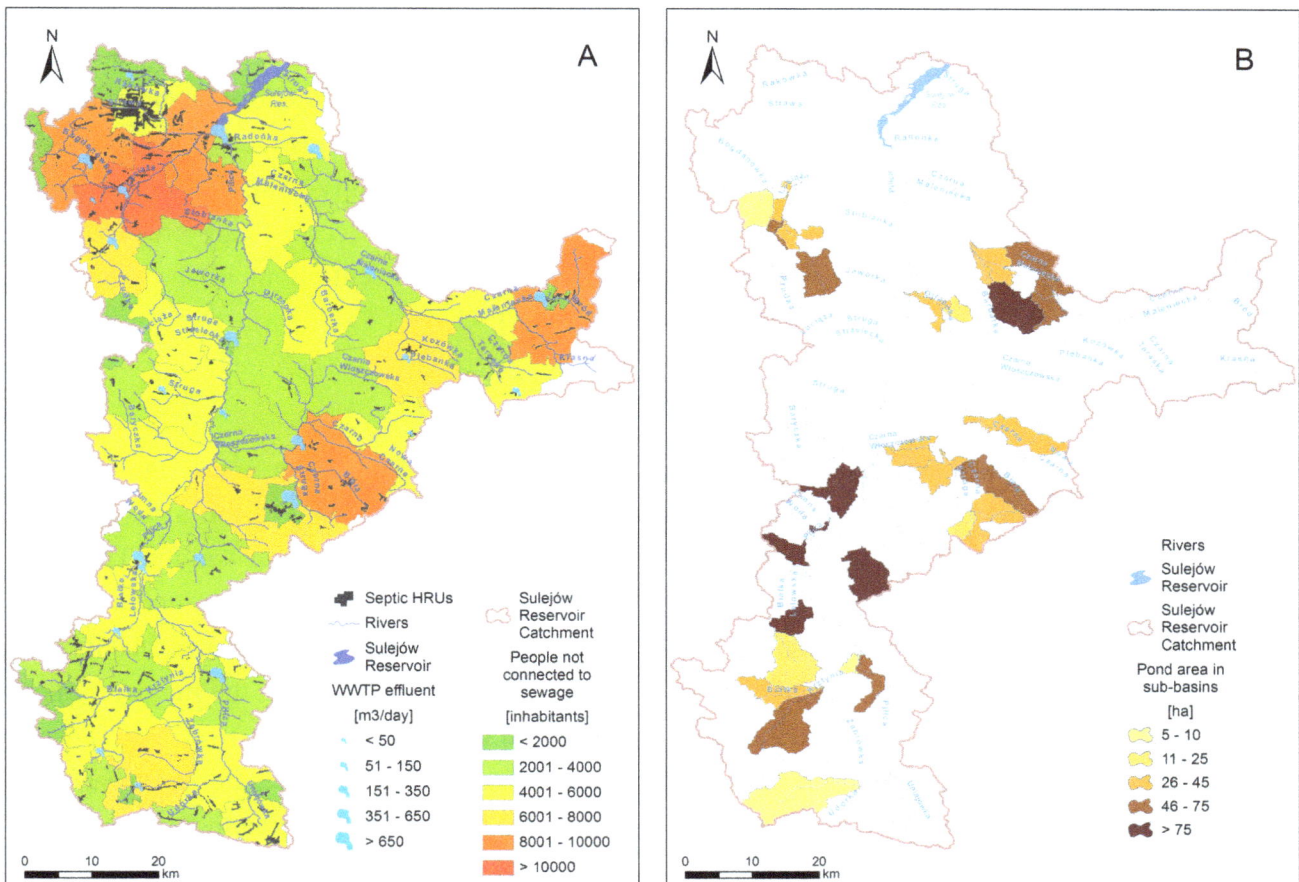

Figure 5. Sewage treatment plants (**A**) and fish ponds (**B**).

Even though water management in Poland is undergoing rapid modernization, which is manifested, for example, by investments in treatment plants, many rural and suburban areas remain disconnected from sewer systems. In such cases, domestic septic systems are usually used for sewage treatment. However, one common problem associated with domestic septic systems in Poland is leaking septic tanks [42]. The SWAT model uses a biozone algorithm [43] to simulate the effects of on-site wastewater systems. The type of septic tank widely used in Poland can be approximately represented by the so-called "failing systems" in SWAT. To identify approximate locations of septic systems in the SRC, commune-level data for the number of people disconnected from wastewater treatment plants were used, which were obtained from the Local Data Bank of the Central Statistical Office (GUS). This number was estimated as 229,000 people for the entire SRC. Spatial analysis of these data made it possible to identify 202 HRUs with the land cover classes "URLD" (urban low density) or "URML" (urban medium-low density), in which the septic function was initiated (Figure 5A). The water quality parameters of the sewage effluents were specified based on the available literature [44]. For example, the TN and TP concentrations were 60 mg · L^{-1} and 20 mg · L^{-1}, respectively.

2.4.4. Fish Ponds

An important feature of the SRC is carp breeding in earth ponds at traditional land-based farms adjacent to the river channels. The area of ponds identified in the 32 sub-basins was significant (Figure 5B). The ponds were represented in SWAT by defining monthly water use parameters (water withdrawn from the reaches of the river for filling the ponds in the spring and maintaining the desired water level until late summer) and point source discharges (representing water release to adjacent reaches of the river in October to empty the ponds before winter). The quantities of the abstracted and released water were calculated based on the estimated pond volume. The water quality characteristics of the discharged water remain largely unknown. Thus, a literature review of the effects of carp breeding on water quality in Central Europe [45,46] was used to define the mean concentrations of the different constituents in the released water: 2.96 mg TN · L^{-1} and 0.7 mg TP · L^{-1}.

2.4.5. Summary

Table 2 lists three main anthropogenic point pollution sources and quantifies the mean annual TN and TP loads that originated from these sources and entered the stream network of the SRC. These estimates are very uncertain for each pollution source. The quantity of released water and the TN and TP concentrations both vary temporally and spatially. Table 2 shows that the order of magnitude is the same for all variables. For TN, the loads from the sewage treatment plants are slightly greater than those from the septic tank effluent and fish pond releases. For TP, the fish pond releases are the major source and the treatment plants and septic tanks are the second and third sources, respectively. However, in our opinion, the TP load from septic tank effluents is underestimated because SWAT does not simulate the downward movement of P to the groundwater. In addition, while the loads from treatment plants (in SWAT) are usually constant with time, the loads from the fish ponds only occur in October. In contrast, the loads from septic tank effluents are variable with time because the travel time between the bottom of the tank to the nearest river depends on the soil physical properties and hydrological conditions.

Table 2. Mean annual TN and TP loads entering the stream network in the SRC that originated from different pollution sources.

Pollution Source	TN (kg/year)	TP (kg/year)
Sewage treatment plants	81,989	6,913
Septic tank effluent	65,572	1,462
Fish pond releases	53,313	9,273
Total	200,874	17,648

Table 2 does not include the loads from the two remaining sources of atmospheric deposition and agricultural production. Because these sources are land-based sources, the load that enters the soil profile is generally known (e.g., Figure 4) but the load that enters the stream network is not. The load that enters the stream network is definitely smaller than the load entering the soil profile due to soil retention, plant uptake etc., and its direct estimation would require modeling.

2.5. Spatial Calibration Approach

SWAT-CUP is a program that allows to use a number of different algorithms to optimize the SWAT model. In addition, SWAT-CUP can be used for sensitivity analysis, calibration, validation and uncertainty analysis [47]. In this paper, we applied SWAT-CUP version 2009 4.3 and selected the optimization algorithm SUFI-2 (Sequential Uncertainty Fitting Procedure Version 2), which is an inverse modeling program that contains elements of calibration and uncertainty analysis [48]. Although SUFI-2 is a stochastic procedure, it does not converge with any "best simulation" and quantifies standard goodness-of-fit measures, such as the Nash-Sutcliffe Efficiency (NSE) or R^2 for each model run. Hence, SUFI-2 indicates the "best simulation" among all of the performed runs, which corresponds to the run with the highest/lowest value of the earlier defined objective function. In this study, we used the widely used NSE as an objective function. The NSE can range from $-\infty$ to 1, where 1 is optimal. Moriasi *et al.* [49] recommended the value of 0.5 as the threshold for satisfactory model performance for a monthly time step, mentioning that under certain circumstances (e.g., daily time step, high uncertainty of observations) this requirement could be made less stringent. We also tracked other goodness-of-fit values, such as R^2 and percent bias (PBIAS). The PBIAS measures the average tendency of the modeled data to be larger or smaller than their observed counterparts. Positive values indicate model underestimation bias, and negative values indicate model overestimation bias.

Calibration was performed in three steps, beginning with continuous daily discharge, continuing with irregular (approximately one measurement per month) and daily NO_3-N loads and ending with TP loads. The calibration period was from 2006 to 2011, and the validation period was from 2000 to 2005. Figure 2 presents the locations of the 10 flow gauging stations (data acquired from IMGW-PIB) and 7 water quality monitoring stations (concentration data acquired from the General Inspectorate of Environmental Protection), from which the time series were used for calibration and validation. The average daily loads (kg $\cdot$ day^{-1}) on the sampling dates were calculated based on observed daily discharge data (m^3 $\cdot$ day^{-1}) at the closest flow gauging station. If the flow gauging station was situated at another location than the water quality station, discharge data were scaled using catchment area ratios. We evaluated the relationships between the NO_3-N and TP concentrations and discharge for all studied gauges and concluded that the correlations were too low (median R^2 equal to 0.2 for NO_3-N and 0.03 for TP) to use any regression-based methods for continuous load estimation.

Three parameter sets, one for discharge, one for NO_3-N, and one for TP, and their initial ranges applied in SUFI-2 (Electronic Supplement, Tables S1–S3) were chosen based on the previous applications of the SWAT model under Polish conditions [41,50,51], and on the sensitivity analysis performed in the SRC.

In most SWAT studies, calibration is restricted to the catchment outlet. In some cases, especially in small (*i.e.*, <100 km^2) catchments, this approach is justified and sometimes inevitable due to data scarcity.

However, wide variations occur in the runoff that is produced in different sub-areas of large river basins due to variations in the physical catchment properties and the associated hydrological processes [52]. Variations in water quality may be even higher due to natural and anthropogenic factors. One of the most effective methods used to account for this type of variation is to perform spatially distributed calibration (*i.e.*, multi-site or multiple gauges, hereafter referred to as "spatial calibration"), as performed by [52–54].

Spatial calibration is a much more complex task than single-gauge calibration, and its complexity depends on the number of gauges used and the spatial dependencies between them. We used the widely applied approach (e.g., [54,55]) of the "regionalization" of parameter values sequentially from upstream to downstream nested catchments. This approach was applied in three steps: discharge, NO_3-N and TP.

After successful calibration and validation, the optimal parameter values were written into the SWAT project and the model was executed for the joint calibration and validation period from 2000 to 2011. Hereafter, this simulation is referred to as the "Baseline" scenario.

2.6. Buffer Zone Efficiency Monitoring in Shallow Groundwater

The monitoring program of the buffer zone efficiency for reducing nitrate and phosphate pollution in shallow groundwater was conducted in 12 transects located in 6 different areas within the SRC. All investigated buffer zones were located between arable fields and stream channels and had variable widths (ranging from 10 to 50 m) and hydrogeological structures (from high to low permeability). The predominant type of land cover of the buffer zones were cultivated meadows, with narrow tall herb fringes and common reed bed communities adjacent to the stream channels.

The groundwater well network was installed in January 2011. Two wells were installed for each transect, one at the edge of the arable land (inlet) and one at the edge of the buffer zone of the stream bank. The wells consisted of HDPE pipes (ϕ 50 mm; Eijkelkamp) that were installed in hand-drilled or machine-drilled holes. The bottom 1 m of each HDPE pipe was perforated. The lithology (granulometric estimation and thickness) was determined through visual inspection of the cores that were collected with the auger during installation.

Groundwater samples were collected monthly from February 2011 until February 2014. Once the water level was measured, the water filling the well bottom were pumped out. Next, the groundwater was sampled by using submersible pumps (Eijkelkamp). During each sampling, temperature, conductivity, and pH were measured in situ. The nitrate, nitrite, ammonium, and phosphate levels were measured using ion chromatography (Dionex ICS-1000, Sunnyvale, CA, USA).

The percent effectiveness of the riparian buffer zones (RR for reduction rate) was calculated by assessing the degree by which the NO_3-N and PO_4-P levels were reduced along the buffer zone.

$$RR_X = \frac{c_{in} - c_{out}}{c_{in}} \cdot 100\% \tag{5}$$

where X denotes a measurement variable, NO_3-N or PO_4-P, c_{in} denotes the inlet concentration and c_{out} denotes the outlet concentration. The values of RR_X were calculated separately in the first step for each year and transect.

The goal was to derive one reduction rate value per variable based on the entire set of sampling results for application in the buffer zone scenario model in SWAT. The mean annual c_{in} across all investigated transects ranged from 0.08 to 31.4 mg NO_3-N $\cdot$ L^{-1} and from 0.05 to 1.49 mg PO_4-P $\cdot$ L^{-1}. Because it was observed that positive RR_X values mainly occur if the inlet concentration exceeds a certain threshold (which usually corresponds to high diffuse pollution in a neighboring field), all measurements with mean annual c_{in} values below 5.65 mg NO_3-N $\cdot$ L^{-1} and 0.166 mg PO_4-P $\cdot$ L^{-1} were removed before conducting further calculations. The thresholds were set according to Polish water legislation. Concentrations above these thresholds are in third or higher classes of groundwater quality (where first and second classes denote very good and good quality, respectively). Consequently, only nine transects located in five different areas (Figure 2) were retained for analysis. Thus, the following values of c_{in}, c_{out} and RR were obtained (mean values across all transects and years):

- For NO_3-N: $c_{in} = 17.6$, $c_{out} = 7.91$ and $RR_{NO_3-N} = 56\%$;
- For PO_4-P: $c_{in} = 0.76$, $c_{out} = 0.18$ and $RR_{PO_4-P} = 76\%$

2.7. Buffer Zone Scenario Assumptions

The VFS SWAT sub-model simulates reductions in sediment and nutrient contents in surface runoff and neglects the role of lateral and groundwater flow in nutrients that contribute to the stream. The field measurements described in Section 2.6 clearly indicated the efficiency of VFS in the reduction of nitrates and soluble phosphorus concentrations in shallow groundwater. Thus, the buffer zone scenario implemented in SWAT in this study consisted the following two items:

- The application of the default SWAT VFS function mimicking reduction of nutrients in surface runoff using the default values of all parameters describing the VFS action; and
- The adjustment of groundwater quality parameters related to nutrient concentrations mimicking reduction of nutrients in shallow groundwater.

At the model parameterization stage, the soluble phosphorus concentrations in the groundwater $GWSOLP$ were specified at the HRU level based on the available field measurements in the wells situated on arable land fields. SWAT does not dynamically model the pool of P in the groundwater. Thus, the concentration remained constant throughout the simulation period. To reflect the role of the buffer zone, the $GWSOLP$ values were multiplied by the estimated phosphorus reduction rate RR_{PO_4-P}.

Unlike phosphorus, the groundwater nitrate pool was modeled in SWAT, which allowed for fluctuations in nitrate loadings in the groundwater over time. To reflect the reduction of nitrate in the buffer zone, the values of the $HLIFE\_NGW$ parameter (the half-life of nitrate in the shallow aquifer) were adjusted. This parameter accounts for nitrate losses due to biological and chemical processes; thus, this parameter can be manipulated to approximate reductions of nitrate due to the acting buffer zone. HRU-specific values of $HLIFE\_NGW$ were decreased by empirical factors, and the nitrate concentrations in the groundwater were reduced by a value of RR_{NO_3-N} relative to the concentrations before the change.

The buffer zone scenario was only implemented in the HRUs that used arable land as a type of land cover and were characterized by high N and P emissions to surface waters. The arable land HRUs accounting for the top 20% of nitrogen and phosphorus emissions were selected. Hence, buffer zones were only tested in Critical Source Areas (CSAs) (i.e., areas with disproportionately high pollutant losses). As proven by the field monitoring results described in Section 2.6, the buffer zone efficiency rapidly decreases when the input concentrations are low (*i.e.*, when the upland field is extensively cultivated). This finding suggests that applying buffer zones in low-emission areas is not efficient. Thus, the application of buffer zones was restricted to the CSAs.

3. Results and Discussion

3.1. Calibration and Validation

3.1.1. Discharge

Table 3 presents the model performance measures for the calibration and validation periods of the three modeled variables. Figure 6 shows simulated *vs.* observed hydrographs for the two main gauges (the last stations on the Pilica and Luciąża before the Sulejów Reservoir). The hydrographs for the remaining 8 gauges are shown in Figure S1 of the Electronic Supplement. The goodness-of-fit values and a visual inspection of the hydrographs both demonstrate good model performance for simulating daily flows in the SRC. However, a few deficiencies were noted.

- During the validation period, the model generally underestimates discharge across the entire range of flow variability. The median value of PBIAS is 0.21;
- The peaks of the largest floods are generally slightly underestimated by most gauges;
- The timing of the flood peaks is sometimes advanced by 1–3 days compared with the timing of the peaks identified in the observed data;
- For the three upstream gauges with relatively small catchment areas (less than 360 km^2) the values of NSE were smaller than 0.5 for either the calibration or validation period.

Table 3. Median values of selected goodness-of-fit measures for discharge, NO$_3$-N and TP for calibration and validation periods.

Variable	NSE cal.	NSE val.	R^2 cal.	R^2 val.	PBIAS cal.	PBIAS val.
Discharge	0.64	0.61	0.70	0.72	0.07	0.21
NO$_3$-N loads	0.56	−0.04	0.69	0.28	0.01	0.26
TP loads	0.48	0.08	0.71	0.25	0.05	0.47

As observed in previous SWAT applications in Poland [51,54], we observed a clear relationship between the model performance indicators and the area upstream of the calibration gauge, at least for NSE and R^2 (Figure 7). The larger catchment size, the higher values of NSE and R^2. No relationship of this type can be identified for the absolute value of PBIAS.

The hydrological conditions for the validation period were much wetter than those during the calibration period, which potentially resulted in the observed differences, particularly the high positive value of PBIAS. The mean discharge at the main outlet in Sulejów was higher. Snow melt floods were dominant during the validation period and storm floods were dominant during the calibration period.

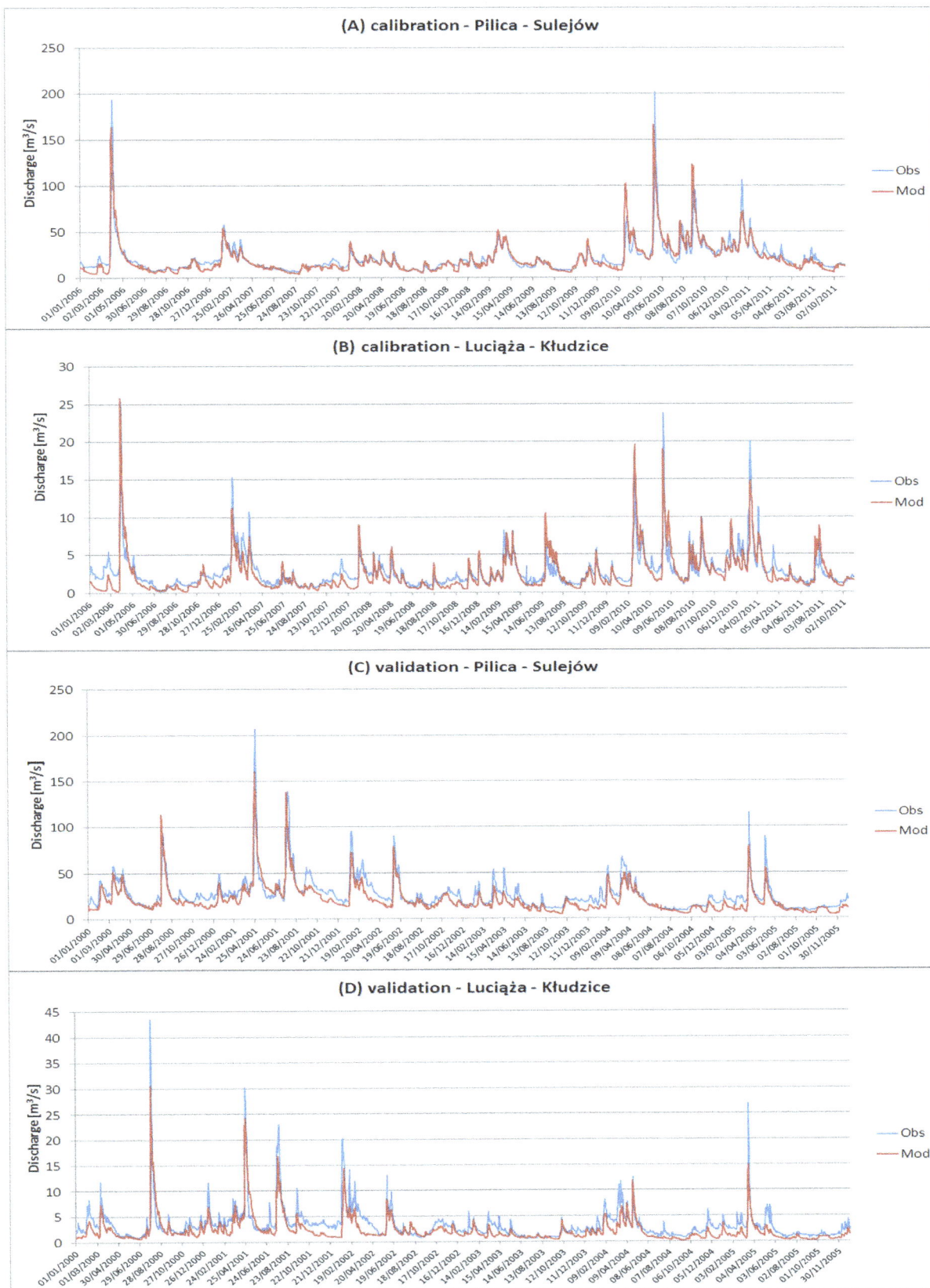

Figure 6. Calibration and validation plots for discharge at the Sulejów gauging station (the Pilica River) and the Kłudzice gauging station (the Luciąża River).

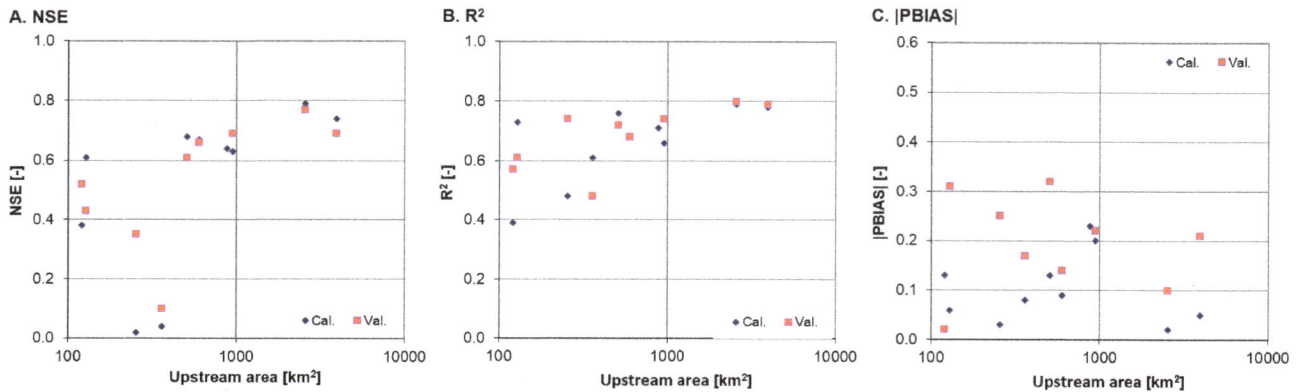

Figure 7. Relationship between the area upstream of a gauge and the model performance measures for discharge for the calibration and validation periods: (**A**) NSE; (**B**) R^2; (**C**) |PBIAS|.

3.1.2. NO_3-N Loads

The goodness-of-fit statistics for NO_3-N loads are not as good as those for discharge (Table 3). During the calibration period, the results are highly variable depending on the gauge. The three problematic gauges with low NSE and R^2 values are situated in the headwater highland part of the SRC. It is very likely that these values are affected by the low performance measure values for discharge simulations in this part of the studied catchment. By contrast, the results are very good for the Czarna Maleniecka and Czarna Włoszczowska Rivers (*cf.* Figure S2 and Table S4 of the Electronic Supplement). In addition, a reasonable fit was observed between the simulations and observations of the two main rivers entering the Sulejów Reservoir (Figure 8A,B).

The model performance during the validation period is slightly worse than during the calibration period. As shown in Figure 8F, the model failed to capture one very large peak. However, a more detailed analysis shows that the modeled peak lagged by 5 days. This lag resulted from the lag in the flood peak from snow melt. Another issue that is visible during the validation period is the considerable bias for most gauges (with a median of 0.26). This bias also reflects the bias in the modeled discharges. Overall, some of the problems identified during hydrology calibration and validation were transposed to the calibration of NO_3-N loads. However, it should be noted that the model preserves several important aspects of the NO_3-N loads and concentration dynamics (e.g., the highest values during the winter and spring and the lowest values during the summer and autumn).

3.1.3. TP Loads

As with NO_3-N, the goodness-of-fit statistics for the TP loads are not as good as those for discharge (Table 3; Figure S3 of the Electronic Supplement). For the calibration period, the comments mentioned in Section 3.1.2 are largely valid for TP. Particularly, lower performance measure values were also noted in the small headwater sub-catchments of the SRC in the south. The very good fit between the simulations and observations is shown in Figure 8C,D.

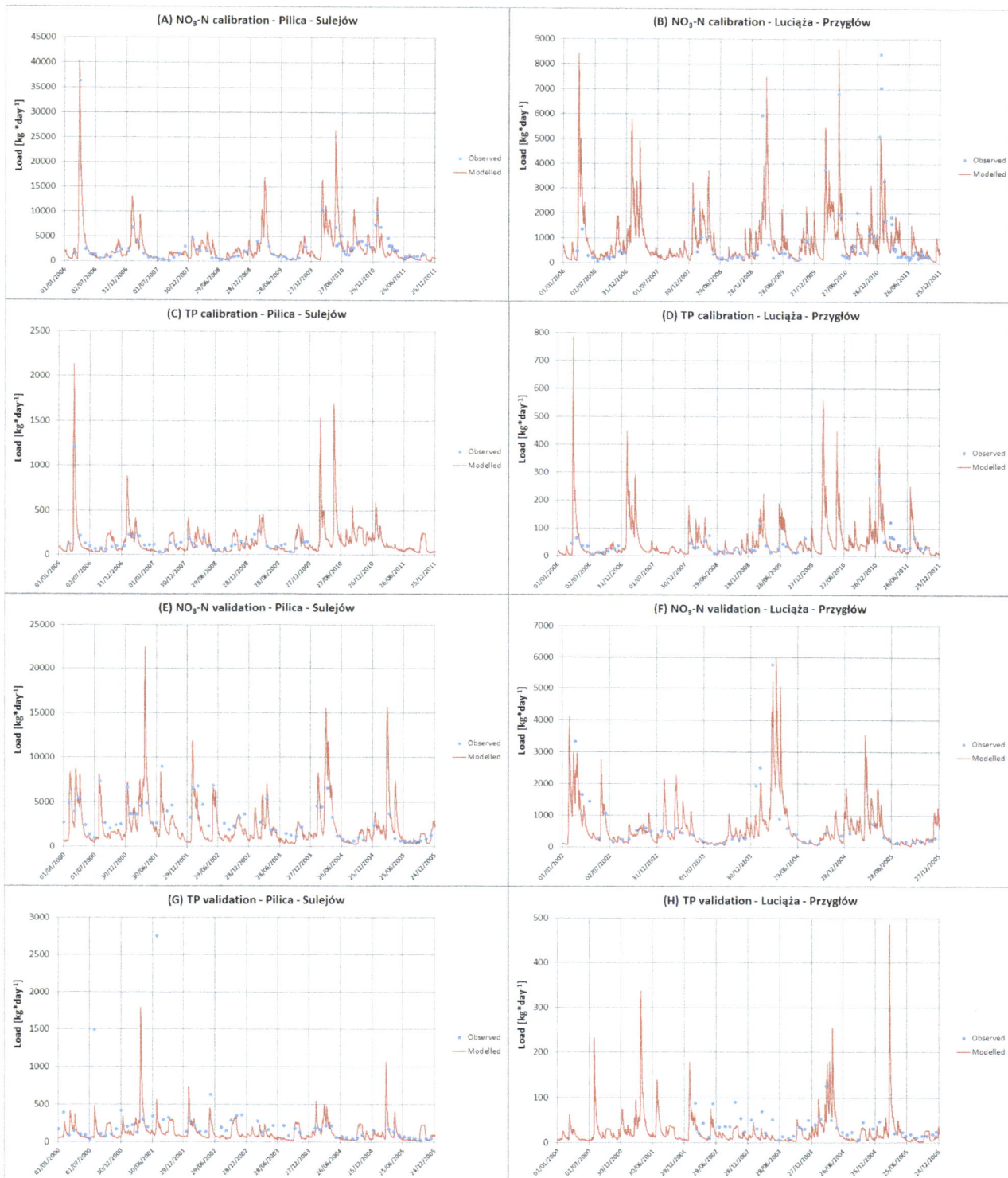

Figure 8. Calibration and validation plots for the NO_3-N and TP loads at the Sulejów (the Pilica River) and Przygłów gauging stations (the Luciąża River).

However, the model significantly underestimates the observed TP loads in most of the gauges, with a median PBIAS value of 0.47. This high bias cannot be explained by the underestimation of discharge alone. In some cases (e.g., for the large peak in TP loads shown in Figure 8G), the modeled flood peak occurred 5 days before the measured flood peak, which clearly affected the high underestimation of the TP load when water samples were measured.

3.2. Spatial Variability in NO₃-N and TP Emissions

In this section, we present an analysis of the calibrated model outputs for the Baseline scenario (2000–2011).

Figure 9 shows the mean emissions at the sub-basin level of NO_3-N (A) and TP (B) from land areas to the stream network. These emissions include all of the possible pathways of the studied constituents from the sub-basins to SWAT reaches via surface runoff, sub-surface runoff, tile drain outflow and base flow. The results are expressed per unit of catchment (not just agricultural) area. Therefore, the results indirectly incorporate the effects of different areas of agricultural land in different sub-basins. For nitrogen and phosphorus, the spatial variability of the calculated emissions is very high. The difference between the sub-basins with the highest and the lowest emissions is two orders of magnitude for NO_3-N and three orders of magnitude for TP.

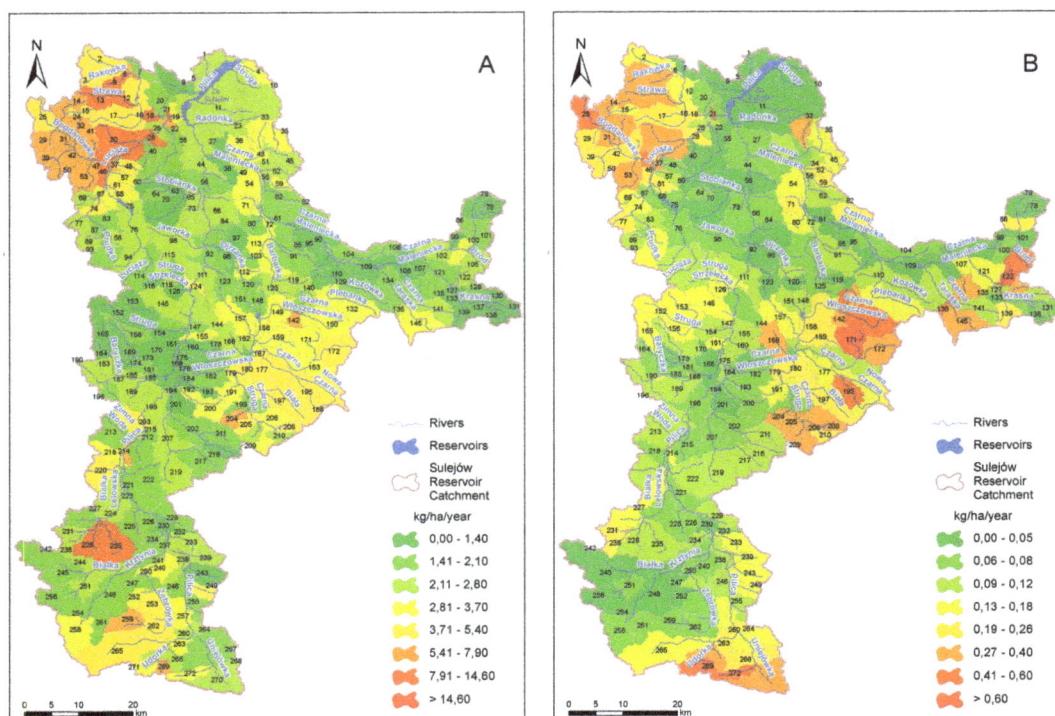

Figure 9. Mean emissions of NO_3-N (**A**) and TP (**B**) from land areas to stream networks for the baseline period 2000–2011 in the SRC. The units are in kilograms per hectare of sub-basin area per annum.

For NO_3-N, the highest emissions are concentrated in two regions of the SRC: (1) the Bogdanówka and Strawa sub-catchments in the northwest and (2) the Białka Lelowska catchment in the south. Both areas are characterized by relatively high proportions of agricultural land and high fertilizer rates (*cf.* Figure 4). The first area has the highest share of inhabitants not connected to sewage systems (*cf.* Figure 5A). The second area is covered by a large patch of loess soils that are less permeable than the neighboring sands and loamy sand. In addition, two large areas are present with moderately high emission rates: (1) the upper Pilica and Żebrówka sub-catchments in the south and (2) the Biała and

Nowa Czarna sub-catchments in the central portion of the SRC. As previously observed, it is clear that both agriculture and septic tank effluents play a critical role in the emission levels in these two areas.

The regions of high TP emissions only partly overlap with the regions of high NO_3-N emissions. The new regions with high TP emissions (in which NO_3-N emissions are not too high) include the headwater parts of the Czarna Włoszczowska and Czarna Maleniecka sub-catchments in the east and the Udorka and Uniejówka sub-catchments in the south. The latter area is also known for intensive head lettuce farming and for using large amounts of fertilizers. Moderately high TP emissions can be found in the Czarna Struga sub-catchment in the central region and in a few smaller isolated sub-catchments that are scattered around the SRC. Most of the mentioned regions overlap with areas receiving relatively high P fertilizer rates. However, this result does not only occur for the headwater area of the Czarna Maleniecka sub-catchment in the east. In this case, the emissions can be explained by the high septic tank effluent emissions from the households that are not connected to sewage systems (*cf.* Figure 5).

3.3. Buffer Zone Scenario Results

Figure 10 illustrates the locations of the agricultural HRUs with the highest NO_3-N and TP emission rates that were identified as the CSAs. Overall, 20% of the HRUs with the highest NO_3-N emissions are responsible for 36% of the total load, and the same amount of HRUs with the highest TP emissions is responsible for 51% of the total load. This finding shows that the magnitude of TP losses is more diversified than the magnitude of NO_3-N losses. The areas with the highest density of selected 606 HRUs largely correspond with the high emission regions described in Section 3.2. In addition, Figure 10 shows the mean difference in NO_3-N (A) and TP (B) emissions between the "Buffer zone" scenario and the Baseline scenario (negative values should be interpreted as the estimated reduction levels that are reached by applying buffer zones). The values are expressed in kg per unit of sub-basin area; thus, they are affected by the HRU-level efficiency of the buffer zone and the percentage of the selected HRUs in the sub-basins. The mean HRU-level reductions reached 0.82 kg NO_3-N $\cdot$ ha^{-1} and 0.18 kg TP $\cdot$ ha^{-1} (the values per hectare of HRU area), and the 90th percentiles reached 1.64 kg NO_3-N $\cdot$ ha^{-1} and 0.28 kg TP $\cdot$ ha^{-1}, respectively. However, at the sub-basin level, the efficiency is significantly reduced, with mean values of 0.16 kg NO_3-N $\cdot$ ha^{-1} and 0.03 kg TP $\cdot$ ha^{-1} (the values per hectare of sub-basin area), and 90th percentiles of 0.31 kg NO_3-N $\cdot$ ha^{-1} and 0.09 kg TP $\cdot$ ha^{-1}. When expressed as a percentage, the average reduction across all of the sub-basins where buffer zones were "implemented" is considerably higher for TP than for NO_3-N (19.4% compared to 5.9%).

A spatial analysis of Figure 10 results in the observation that the highest reductions of NO_3-N or TP generally correspond with the areas with the highest emissions (*cf.* Figure 9). However, this result does not occur in sub-basins where at least one of the two following circumstances occur: (1) the percentage of HRUs selected for this measure is low and (2) high emissions result from septic tank effluents rather than from agriculture. In several cases, the baseline emission level from some sub-basins was low. Thus, although the percent reduction reached 10% or 15%, it was too low in terms of the absolute value to appear on the map.

As mentioned in Section 2.7, the "Buffer zone" scenario implemented in SWAT in this study consisted of two items: (1) the application of the default SWAT VFS function dealing only with surface runoff and

(2) the incorporation of field monitoring-based reduction rates to the shallow groundwater component in SWAT. To verify how each item contributed to the final result, we created two additional scenarios, one that only incorporates feature No. 1 ("BZ-VFS"), and another that only incorporates feature No. 2 ("BZ-GW"). Next, we estimated the sub-basin level reduction rates for the "BZ-VFS" and "BZ-GW" scenarios and compared them with the results from the original "Buffer zone" scenario. Overall, the effect of effect of VFS (72% of the total load reduction) for NO_3-N was dominant over the effect of field monitoring-based parameters (28% of the total load reduction). By contrast, the contributions of each component to the total reduction of TP emissions in the SRC were similar: 46% and 54%, respectively. These modeling results showed that shallow groundwater reduction mechanisms are more effective for TP than for NO_3-N, which agrees with the calculated reduction rates from Section 2.6.

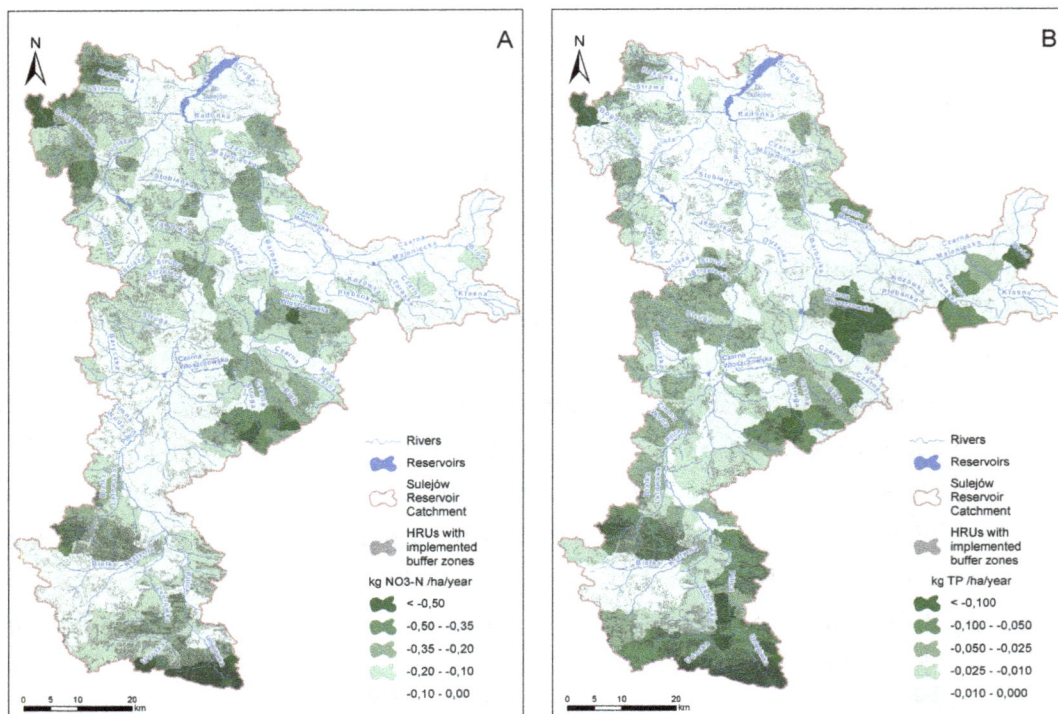

Figure 10. The modeled effects of buffer zones on agricultural pollution emissions from a land to stream network. The mean difference in NO_3-N (**A**) and TP (**B**) emission between the "Buffer zone" scenario and the Baseline scenario. Units are in kilograms per hectare of sub-basin area per annum.

3.4. Discussion

The performance of the SWAT model for simulating daily discharge in the SRC was spatially variable but generally good or satisfactory. The main downside was underestimation bias during the validation period, which occurred because of the significantly wetter hydrological conditions during this period compared with the calibration period. Because the calibrated parameter values are very sensitive to climatic conditions, the values calibrated for dry and short periods might not be suitable for simulating the opposite conditions [56,57], which results in lower performance statistics. Unfortunately, this bias in the validation period for discharge translates into an even greater bias for the NO_3-N and TP loads during this period. However, the reported values of PBIAS for most of the gauges are within the ranges of satisfactory performance for discharge (+/–25%) and the NO_3-N and TP loads (+/–70%) [49]. In summary, our results support the findings of Ekstrand *et al.* [58], who applied SWAT to model the TP losses in five catchments in central Sweden. Overall, Ekstrand *et al.* [58] observed that obtaining satisfactory results for a validation period often depends on whether the range of hydroclimatological conditions is similar (as in calibration).

An additional problem is that evaluating water quality simulations using a daily time step and only one measurement per month is not an optimal. Typically, model simulations are less accurate when shorter time steps are considered than when longer time steps are considered [59]. If sampling is performed during a flood event, which occurred several times (as shown in Figure 8), it is likely that (1) discharge estimations from SWAT for the sampling date are very far from the observations because of common underestimations and lag problems associated with flood peaks (*cf.* Section 3.1.1) We analyzed a few events with different magnitudes that occurred during different seasons and reaches and noted that SWAT was not capable of reproducing this kind of effect with reasonable accuracy. Regarding the problem of capturing peaks, we analyzed all NO_3-N and TP daily validation plots case by case. In six out of 11 plots (three per variable), we identified situations in which the observed peaks lagged behind or preceded the event by 2–15 days. Next, we matched the modeled and observed peaks (between two and three per plot) and recalculated the performance statistics. The model performance improved for each case and for each indicator (Electronic Supplement, Table S5). Increases in the NSE ranged between 0.15 and 0.57, increases in the R^2 value ranged between 0.1 and 0.54 and the positive values of PBIAS decreased by 2%–18%. This result demonstrates that the validation results were significantly impacted by a small number of missed peak events by the majority of gauges.

Furthermore, in five out of 11 cases, we identified another reason for poor validation results. We compared the mean observed discharges and loads between the calibration and validation periods (Electronic Supplement, Table S6). In all analyzed cases, (1) the PBIAS during the validation period was significantly higher than during the calibration period; (2) the PBIAS during the validation period was larger than or equal to 0.4; (3) the mean observed discharge during the validation period was significantly greater than that during the calibration period; and (4) the mean observed NO_3-N and TP loads were much greater during the validation period. These results clearly demonstrate that more than the hydrological conditions differed between these two periods. In addition, this analysis shows that the mean nutrient concentrations in some of the gauges were significantly greater in 2000–2005 than in 2006–2011. The first decade of the twentieth century in Poland has been marked by rapid development in the number of sewage treatment plants and by an increasing treatment level [60]. Because the majority of the input data used to build the model setup were valid for 2010 or later and may represent the period of 2000–2005, we hypothesized that this finding could partially explain the poor model performance during validation.

We applied SWAT in the SRC to spatially quantify NO_3-N and TP emission from various pollution sources. The purpose of this spatial quantification was to identify CSAs in which the buffer zones that mitigate pollutant emissions to the surface water could be implemented. In Table 2, we specified the mean annual TN and TP loads that originated from sewage treatment plants, septic tank effluents and fish pond releases. In Figure 9 we presented the spatially variable NO_3-N and TP loads that predominantly originated from cultivated land and, to a smaller extent, from septic tank effluent. Integration of the sub-basin data from Figure 9 to calculate the total catchment load and subtracting the load assigned to septic tank effluent provided a rough estimate of the mean annual diffuse pollution load in the SRC, which reached 1,240,000 kg NO_3-N and 60,700 kg TP. Although these values also include emissions from urban (very small percentage) and forest (very low emission) runoff, this estimate confirms the initial hypothesis that diffuse agricultural pollution is largely the dominant source of pollution in the SRC. Although the SRC, or more widely, the Pilica catchment, have recently attracted the attention of a number of researchers studying pollution emissions and transport [32,61,62], this finding is new and has certain implications regarding water management. Particularly, regarding the fact that Poland has been sent to court by the EU Commission for failing to guarantee that they are addressing water pollution by nitrates effectively [63]. However, no Nitrate Vulnerable Zone (NVZ) has been included in the SRC under current legislation. However, Figure 9 indicates that some portions of the catchment could easily be designed as NVZs.

Strong evidence for the contributions of agriculture to pollutant emissions to streams also strengthens the basis for scenarios that assume the application of buffer zones in the identified CSAs. Previous modeling attempts of buffer zones in Poland using SWAT [51] have shown its limitations (*i.e.*, the fact that the VFS sub-model only accounts for the trapping effect in surface runoff (*cf.* Section 2.2.2). Consequently, the efficiency of applying buffer zones described by Piniewski *et al.* [51] when measured at the catchment outlet was negligible for NO_3-N and small for PO_4-P. In this study, we used buffer zone field monitoring data from the studied catchment to improve the mechanisms by which SWAT reduces pollutant losses. The modeled reduction rates were spatially variable, but higher than those in the study of Piniewski *et al.* [51]. In addition, the results showed that, the average contributions of the "shallow groundwater" mechanism to total reduction reach 28% for NO_3-N and 54% for TP. This demonstrates that the "surface water" trapping mechanism by VFS in SWAT is not sufficient (*i.e.*, it overlooks an important pathway by which both NO_3-N and TP compounds can reach the stream network). The efficiencies of buffer zones critically depend on the mechanisms by which N and P are transported from the land to the stream [64]. Although we have only empirically tested SWAT in the SRC, it is likely that this limitation would affect other areas, particularly areas of the vast Polish Plain, which are characterized by physiographic conditions that are similar to those of the SRC. However, the approach we used to consider the field measurements of the buffer zones in SWAT was fairly simplistic and based purely on parameter modification. In the future, larger field monitoring samples (in space and time) should allow for the development of a new SWAT sub-routine that would better reflect the pollutant pathways from field areas through buffer zones to streams under variable hydrological conditions.

To assess the effects of buffer zones on the nutrient loads that enter the Sulejów Reservoir, we summed the mean annual loads from all eight reaches with their outlets at the reservoir shoreline (*cf.* Figure 2). The results showed that applying buffer zones in the selected CSAs (occupying 20% of the arable land area and culminating in 12.4% of the land in the catchment) would contribute to the reduction of the pollutant loads entering the reservoir by 7% for NO_3-N and 16% for TP. This outcome is particularly important for TP, which is mainly responsible for reservoir eutrophication and for intensity of toxic algal blooms [34,65].

The estimated buffer zone efficiency can be considered as substantial. However, it is clear that other conservation practices are important for obtaining more pronounced reductions in pollutant runoff. Particularly, the activities should focus on reducing the inputs of nutrients to the landscape in the form of mineral and organic fertilizers by convincing the farmers to use fertilization plans more widely. Examples of other measures include extension of the closed period for spreading organic fertilizers, elimination of soil cultivation during the autumn, the cultivation of catch crops, and the construction of wetlands. Spatially-explicit indications of CSAs provide an opportunity for selecting effective measures. In the second step, their precise and cost-efficient application substantially increases the chance of improving the water quality in the catchment.

4. Conclusions

This study demonstrated that diffuse agricultural pollution is the main contributor of elevated NO_3-N and TP concentrations in the surface waters of the SRC relative to point source pollution from sewage treatment plants, septic tank effluents and fish pond releases. The application of a semi-distributed water quality model and performing a comprehensive spatial calibration and validation allowed us to spatially quantify the emission rates at the HRU and sub-basin level, which helped identify Critical Source Areas. These CSAs were selected to test the efficiencies of riparian buffer zones. The default SWAT sub-model designed for simulating the effects of buffer zones only accounts for nutrient trapping in surface runoff and overlooks an important sub-surface pathway in which nutrients can be trapped. The monitoring data from the SRC showed that the mean field-level reductions in the concentrations in the shallow groundwater near the buffer zone average 56% for NO_3-N and 76% for TP. These empirical reduction rates were used to enhance the capability of SWAT for representing the effects of the buffer zone. The scenario results showed that the efficiency of the buffer zones at the catchment level is lower than that at the field level but still significantly contributes to reductions in pollutant emission to the nearest streams and to reductions of the total pollutant load entering the Sulejów Reservoir (by 7% for NO_3-N and 16% for TP). Only using the default SWAT function of the simulating buffer zones would lead to an underestimation of buffer efficiency, particularly for phosphorus (54%). Thus, we argue that empirical data are important for improving existing models that monitoring more samples in the future should allow us to develop new SWAT routines for simulating the sub-surface trapping effects of the buffer zones.

The poor model performance of the nutrient load simulation during the validation period indicates that the nutrient load estimates from the SWAT model of the SRC are highly uncertain. However, it can be argued that simulated percent reductions in pollutant emission due to the application of buffer zones are more reliable, because of known model bias.

The implications from this study are valuable for water managers and other decision-makers. The use of water quality mathematical models to address contemporary water management problems is still limited in many countries, including Poland. Our study shows how the SWAT model is useful for the (1) quantification of point and diffuse pollution sources; (2) identification of high emission areas (CSAs) where measure implementation should be prioritized; and (3) quantification of the efficiency of conservation practices. All these three aspects are vital for the development of medium and long-term water quality improvement strategies by river basin managers. Further progress can be achieved by including the economic functions representing implementation costs of different conservation practices.

Acknowledgments

We thank two anonymous reviewers for providing constructive comments on the manuscript. This paper is an outcome of the EKOROB project: Ecotones for reducing diffuse pollution (LIFE08 ENV/PL/000519). This project was supported by the LIFE+ Environment Policy and Governance Programme, National Fund for Environmental Protection and Water Management, and funding dedicated for science in the period 2012–2014, and granted for implementation by the co-financed international project No. 2539/LIFE+2007-2013/2012/2 (www.en.ekorob.pl). We would like to acknowledge the Institute of Meteorology and Water Management—National Research Institute (IMGW-PIB) for providing hydrometeorological data. The first author benefited from the START 2014 stipend from the Foundation for Polish Science and from the Humboldt Research Fellowship for Postdoctoral Researchers from the Alexander von Humboldt Foundation in the Potsdam Institute for Climate Impact Research during the manuscript preparation phase.

Author Contributions

Mikołaj Piniewski, Ignacy Kardel, Katarzyna Izydorczyk and Wojciech Frątczak developed the methodological framework. Ignacy Kardel and Wojciech Frątczak processed field monitoring data. Mikołaj Piniewski, Paweł Marcinkowski, Ignacy Kardel and Marek Giełczewski developed the model setup. Mikołaj Piniewski and Paweł Marcinkowski performed model calibration. Mikołaj Piniewski, Paweł Marcinkowski and Ignacy Kardel run the model scenarios. Mikołaj Piniewski wrote the manuscript with inputs from Paweł Marcinkowski, Katarzyna Izydorczyk and Wojciech Frątczak, Marek Giełczewski and Paweł Marcinkowski created the artwork.

Conflicts of Interest

The authors declare no conflicts of interest.

References

1. European Commission. Directive 2000/60/EC of 23 October 2000 of the European Parliament and of the Council establishing a framework for community action in the field of water policy. *Off. J. Eur. Communities* **2000**, *L327*, 1–72.
2. European Commission. Council Directive 91/676/EEC of 12 December 1991 concerning the protection of waters against pollution caused by nitrates from agricultural sources. *Off. J. Eur. Communities* **1991**, *L375*, 1–8.
3. Zalewski, M.; Janauer, G.A.; Jolankai, G. Conceptual background. In *Ecohydrology: A New Paradigm for the Sustainable Use of Aquatic Resources*; Zalewski, M., Janauer, G.A., Jolankai, G., Eds.; International Hydrobiological Programme UNESCO: Paris, France, 1997.

4. Zalewski, M. Ecohydrology for implementation of the EU water framework directive. *Proc. Instit. Civil Eng. Water Manag.* **2011**, *164*, 375–385.

5. Zalewski, M. Ecohydrology: Process-oriented thinking towards sustainable river basins. *Ecohydrol. Hydrobiol.* **2013**, *13*, 97–103.

6. Zalewski, M. Ecohydrology and hydrologic engineering: Regulation of hydrology-biota interactions for sustainability. *J. Hydrol. Eng.* **2015**, *20*, doi: 10.1061/(ASCE)HE.1943-5584.0000999.

7. Arnold, J.G.; Srinivasan, R.; Muttiah, R.S.; Williams, J.R. Large-area hydrologic modeling and assessment: Part I. Model development. *J. Am. Water Resour. Assoc.* **1998**, *34*, 73–89.

8. Gassman, P.W.; Sadeghi, A.M.; Srinivasan, R. Applications of the SWAT Model Special Section: Overview and Insights. *J. Environ. Qual.* **2014**, *43*, 1–8.

9. Niraula, R.; Kalin, L.; Srivastava, P.; Anderson, J.A. Identifying critical source areas of nonpoint source pollution with SWAT and GWLF. *Ecol. Model.* **2013**, *268*, 123–133.

10. Panagopoulos, Y.; Makropoulos, C.; Baltas, E.; Mimikou, M. SWAT parameterization for the identification of critical diffuse pollution source areas under data limitations. *Ecol. Model.* **2011**, *222*, 3500–3512.

11. Wu, Y.; Chen, J. Investigating the effects of point source and nonpoint source pollution on the water quality of the East River (Dongjiang) in South China. *Ecol. Indic.* **2013**, *32*, 294–304.

12. Wu, Y.; Liu, S. Modeling of land use and reservoir effects on nonpoint source pollution in a highly agricultural basin. *J. Environ. Monit.* **2012**, *14*, 2350–2361.

13. Lowrance, R.; Todd, R.L.; Fail, J., Jr.; Hendrickson, O., Jr.; Leonard, R.; Asmussen, L. Riparian forests as nutrient filters in agricultural watersheds. *BioScience* **1984**, *34*, 374–377.

14. Naiman, R.J.; Decamps, H.; Fournier, F. The role of land/water ecotones in landscape management and restoration: A proposal for collaborative research. In *MAB Digest4*; Naiman, R.J., Decamps, H., Fournier, F., Eds.; UNESCO: Paris, France, 1989.

15. Mander, U.; Hayakawa, Y.; Kuusemets, V. Purification processes, ecological functions, planning and design of riparian buffer zones in agricultural watersheds. *Ecol. Eng.* **2005**, *24*, 421–432.

16. Dosskey, M.G.; Vidon, P.; Gurwick, N.P.; Allan, C.J.; Duval, T.P.; Lowrance, R. The role of riparian vegetation in protecting and improving chemical water quality in streams. *J. Am. Water Resour. Assoc.* **2010**, *46*, 261–277.

17. Dosskey, M.G. Toward quantifying water pollution abatement in response to installing buffer on crop land. *J. Environ. Manag.* **2001**, *28*, 577–598.

18. Vought, L.B.M.; Pinay, G.; Fuglsang, A.; Ruffinoni, C. Structure and function of buffer strips from a water quality perspective in agricultural landscapes. *Lands. Urban Plan.* **1995**, *31*, 323–331.

19. Dillaha, T.A.; Sherrard, J.H.; Lee, D.; Mostaghimi, S.; Shanholtz, V.O. Evolution of vegetative filters strips as a best managements practice for feed lots. *J. Water Pollut. Control Fed.* **1988**, *60*, 1231–1238.

20. Dunn, A.M.; Julien, G.; Ernst, W.R.; Cook, A.; Doe, K.G.; Jackman, P.M. Evaluation of buffer zone effectiveness in mitigating the risks associated with agricultural runoff in Prince Edward Island. *Sci. Total Environ.* **2011**, *409*, 868–882.

21. Syversen, N. Effect and design of buffer zones in the Nordic climate: The influence of width, amount of surface runoff, seasonal variation and vegetation type on retention efficiency for nutrient and particle runoff. *Ecol. Eng.* **2005**, *24*, 483–490.

22. Mayer, P.M.; Reynolds, S.K., Jr.; McCutchen, M.D.; Canfield, T.J. Meta-analysis of nitrogen removal in riparian buffers. *J. Environ. Qual.* **2007**, *36*, 1172–1180.

23. Carlyle, G.C.; Hill, A.R. Groundwater phosphate dynamics in a river riparian zone: Effects of hydrologic flowpaths, lithology and redox chemistry. *J. Hydrol.* **2001**, *247*, 151–168.

24. Hoffmann, C.C.; Berg, P.; Dahl, M.; Lasen, S.E.; Andersen, H.E.; Andersen, B. Groundwater flow and transport of nutrients through a riparian meadow -field data and modelling. *J. Hydrol.* **2006**, *331*, 315–335.

25. Weller, D.E.; Jordan, T.E.; Correll, D.L. Heuristic models for material discharge from landscapes with riparian buffers. *Ecol. Appl.* **1998**, *8*, 1156–1169.

26. Larose, M.; Heathman, G.C.; Norton, D.; Smith, D. Impacts of conservation buffers and grasslands on total phosphorus loads using hydrological modeling and remote sensing techniques. *Catena* **2011**, *86*, 121–129.

27. Parajuli, P.B.; Mankin, K.R.; Barnes, P.L. Applicability of targeting vegetative filter strips to abate fecal bacteria and sediment yield using SWAT. *Agric. Water Manag.* **2008**, *95*, 1189–1200.

28. Park, Y.S.; Park, J.H.; Jang, W.S.; Ryu, J.C.; Kang, H.; Choi, J.; Lim, K.J. Hydrologic Response Unit Routing in SWAT to Simulate Effects of Vegetated Filter Strip for South-Korean Conditions Based on VFSMOD. *Water* **2011**, *3*, 819–842.

29. Shan, N.; Ruan, X.H.; Pan, Z.R. Estimating the optimal width of buffer strip for nonpoint source pollution control in the Three Gorges Reservoir Area, China. *Ecol. Model.* **2014**, *276*, 51–63.

30. White, M.J.; Storm, D.E.; Busteed, P.R.; Stoodley, S.H.; Phillips, S.J. Evaluating nonpoint source critical source area contributions at the watershed scale. *J. Environ. Qual.* **2009**, *38*, 1654–1663.

31. Bloschl, G. Scale and scaling in hydrology. In *Wiener Mitteilungen, Wasser-Abwasser-Gewasser*; Technical University of Vienna: Vienna, Austria, 1996.

32. Wagner, I.; Izydorczyk, K.; Kiedrzyńska, E.; Mankiewicz-Boczek, J.; Jurczak, T.; Bednarek, A.; Wojtal-Frankiewicz, A.; Frankiewicz, P.; Ratajski, S.; Kaczkowski, Z.; *et al.* Ecohydrological system solutions to enhance ecosystem services: The Pilica River Demonstration Project. *Ecohydrol. Hydrobiol.* **2009**, *9*, 13–39.

33. Tarczyńska, M.; Romanowska-Duda, Z.; Jurczak, T.; Zalewski, M. Toxic cyanobacterial blooms in drinking water reservoir-causes, consequences and management strategy. *Water Sci. Technol.* **2001**, *1*, 237–246.

34. Izydorczyk, K.; Jurczak, T.; Wojtal-Frankiewicz, A.; Skowron, A.; Mankiewicz-Boczek, J.; Tarczyńska, M. Influence of abiotic and biotic factors on microcystin content in Microcystis aeruginosa cells in a eutrophic temperate reservoir. *J. Plankton Res.* **2008**, *30*, 393–400.

35. Gągała, I.; Izydorczyk, K.; Jurczak, T.; Pawełczyk, J.; Dziadek, J.; Wojtal-Frankiewicz, A.; Jóźwik, A.; Jaskulska, A.; Mankiewicz-Boczek, J. Role of Environmental Factors and Toxic Genotypes in The Regulation of Microcystins-Producing Cyanobacterial Blooms. *Microb. Ecol.* **2014**, *67*, 465–479.

36. Williams, J.R. The Erosion-Productivity Impact Calculator (EPIC) Model: A Case History. *Philos. Trans. B* **1990**, *329*, 421–428.

37. Brown, L.C.; Barnwell, J.T.O. *The Enhanced Stream Water Quality Models QUAL2E and QUAL2E-UNCAS: Documentation and User Manual*; EPA-600/3-87/007; U.S. Environmental Protetction Agency: Athens, GA, USA, 1987.

38. White, M.J.; Arnold, J.G. Development of a simplistic vegetative filter strip model for sediment and nutrient retention at the field scale. *Hydrol. Proc.* **2009**, *23*, 1602–1616.

39. Munoz-Carpena, R.; Parsons, J.E.; Gilliam, J.W. Modeling hydrology and sediment transport in vegetative filter strips. *J. Hydrol.* **1999**, *214*, 111–129.

40. Williams, J.R. Sediment-yield prediction with universal equation using runoff energy factor. In *Proceedings of the Sediment-Yield Workshop*; USDA Sedimentation Laboratory: Oxford, MS, USA, 1975.

41. Szcześniak, M.; Piniewski, M. Improvement of hydrological simulations through applying daily precipitation interpolation schemes in meso-scale catchments. *Water* **2015**, *7*, 747–779.

42. Kowalczak, P.; Kundzewicz, Z.W. Water-related conflilicts in urban areas in Poland. *Hydrol. Sci. J. Spec. Issues* **2011**, *56*, 588–596.

43. Siegriest, R.L.; McCray, J.; Weintraub, L.; Chen, C.; Bagdol, J.; Lemonds, P.; van Cuyk, S.; Lowe, K.; Goldstein, R.; Rada, J. *Quantifying Site-Scale Processes and Watershed-Scale Cumulative Effects of Decentralized Wastewater Systems*; Colorado School of Mines: Golden, CO, USA, 2005.

44. Arnold, J.G.; Kiniry, J.R.; Srinivasan, R.; Williams, J.R.; Haney, E.B.; Neitsch, S.L. Soil and Water Assessment Tool Input Output Documentation, Version 2012; Texas Water Resources Institute TR-439; Texas Water Resources Institute: College Station, TX, USA, 2012.

45. Barszczewski, J.; Kaca, E. Water retention in ponds and the improvement of its quality during carp production. *J. Water Land Dev.* **2012**, *17*, 31–38.

46. Vsetickova, L.; Adamek, Z.; Rozkosny, M.; Sedlacek, P. Effects of semi-intensive carp pond farming on discharged water quality. *Acta Ichthyo. Piscat.* **2012**, *42*, 223–231.

47. Abbaspour, K. *SWAT-CUP2: SWAT Calibration and Uncertainty Programs—A User Manual*; Swiss Federal Institute of Aquatic Science and Technology (Eawag): Duebendorf, Switzerland, 2008; p. 95.

48. Abbaspour, K.C.; Johnson, C.A.; van Genuchten, M.T. Estimating uncertain flow and transport parameters using a sequential uncertainty fitting procedure. *Vadose Zone J.* **2004**, *3*, 1340–1352.

49. Moriasi, D.N.; Arnold, J.G.; van Liew, M.W.; Bingner, R.L.; Harmel, R.D.; Veith, T.L. Model evaluation guidelines for systematic quantification of accuracy in watershed simulations. *Trans. ASABE* **2007**, *50*, 885–900.

50. Marcinkowski, P.; Piniewski, M.; Kardel, I.; Giełczewski, M.; Okruszko, T. Modelling of discharge, nitrate and phosphate loads from the Reda catchment to the Puck Lagoon using SWAT. *Ann. Warsaw Univ. Life Sci. SGGW Land Reclam.* **2013**, *45*, 125–141.

51. Piniewski, M.; Kardel, I.; Giełczewski, M.; Marcinkowski, P.; Okruszko, T. Climate change and agricultural development: Adapting Polish agriculture to reduce future nutrient loads in a coastal watershed. *Ambio* **2014**, *43*, 644–660.

52. Santhi, C.; Kannan, N.; Arnold, J.G.; di Luzio, M. Spatial calibration and temporal validation of flow for regional scale hydrologic modelling. *J. Am. Water Resour. Assoc.* **2008**, *44*, 829–846.

53. Qi, C.; Grunwald, S. GIS-based hydrologic modelling in the Sandusky watershed using SWAT. *Trans. ASAE* **2005**, *48*, 169–180.

54. Piniewski, M.; Okruszko, T. Multi-site calibration and validation of the hydrological component of SWAT in a large lowland catchment. In *Modelling of Hydrological Processes in the Narew Catchment, Geoplanet: Earth and Planetary Sciences*; Świątek, D., Okruszko, T., Eds.; Springer: Berlin, Germany, 2011; pp. 15–41.

55. Van Liew, M.W.; Garbrecht, J. Hydrologic simulation of the Little Washita river experimental watershed using SWAT. *J. Am. Water Resour. Assoc.* **2003**, *39*, 413–426.

56. Vaze, J.; Post, D.A.; Chiew, F.H.S.; Perraud, J.M.; Viney, N.R.; Teng, J. Climate non-stationarity—Validity of calibrated rainfall-runoff models for use in climate change studies. *J. Hydrol.* **2010**, *394*, 447–457.

57. Zhu, Q.; Zhang, X.; Ma, C.; Gao, X.; Xu, Y. Investigating the uncertainty and transferability of parameters in SWAT model under climate change. *Hydrol. Sci. J.* **2015**, doi: 10.1080/02626667.2014.1000915.

58. Ekstrand, S.; Wallenberg, P.; Djodjic, F. Process based modelling of phosphorus losses from arable land. *Ambio* **2010**, *39*, 100–115.

59. Engel, B.; Storm, D.; White, M.; Arnold, J.G. A hydrologic/water quality model application protocol. *J. Am. Water Resour. Assoc.* **2007**, *43*, 1223–1236.

60. Kowalkowski, T.; Pastuszak, M.; Igras, J.; Buszewski, B. Differences in emission of nitrogen and phosphorus into the Vistula and Oder basins in 1995–2008—Natural and anthropogenic causes (MONERIS model). *J. Mar. Syst.* **2012**, *89*, 48–60.

61. Wagner, I.; Zalewski, M. Effect of hydrological patterns of tributaries on biotic processes in a lowland reservoir - consequences for restoration. *Ecol. Eng.* **2000**, *16*, 79–90.

62. Kiedrzyńska, E.; Kiedrzyński, M.; Urbaniak, M.; Magnuszewski, A.; Skłodowski, M.; Wyrwicka, A.; Zalewski, M. Point sources of nutrient pollution in the lowland river catchment in the context of the Baltic Sea eutrophication. *Ecol. Eng.* **2014**, *70*, 337–348.

63. Environment: Commission Takes Poland to Court over Nitrates and Water Pollution. Available online: http://europa.eu/rapid/press-release_IP-13-48_en.htm (accessed on 30 January 2015)

64. Heathwaite, A.L.; Johnes, P.J. Contribution of nitrogen species and phosphorus fractions to stream water quality in agricultural catchments. *Hydrol. Proc.* **1996**, *10*, 971–983.

65. Izydorczyk, K.; Skowron, A.; Wojtal, A.; Jurczak, T. The stream inlet to a shallow bay of a drinking water reservoir a "Hot-Spot" for *Microcystis* blooms initiation. *Int. Rev. Hydrobiol.* **2008**, *93*, 257–268.

2

Upflow Evapotranspiration System for the Treatment of On-Site Wastewater Effluent

Sean Curneen [†] **and Laurence Gill** [†,*]

Department of Civil, Structural and Environmental Engineering, Trinity College Dublin, Dublin 2, Ireland; E-Mail: curneens@tcd.ie

[†] These authors contributed equally to this work.

[*] Author to whom correspondence should be addressed; E-Mail: laurence.gill@tcd.ie;

Academic Editor: Michael O'Driscoll

Abstract: Full-scale willow evapotranspiration systems fed from the base with septic tank or secondary treated domestic effluent from single houses have been constructed and instrumented in Ireland in order to investigate whether the technology could provide a solution to the problem of on-site effluent disposal in areas with low permeability subsoils. Continuous monitoring of rainfall, reference evapotranspiration, effluent flows and water level in the sealed systems revealed varying evapotranspiration rates across the different seasons. No system managed to achieve zero discharge in any year remaining at maximum levels for much of the winter months, indicating some loss of water by lateral exfiltration at the surface. Water sampling and analysis however, showed that the quality of any surface overflow from the systems was similar to rainfall runoff. The performance results have then been used to formulate design guidelines for such systems in Ireland's temperate maritime climate. The effect of varying different combinations of design parameters (plan area, soil depth, *etc.*) has been evaluated with respect to the simulated number of overflow days over a five-year period using a water balance model. Design guidelines have then been based upon minimising the amount of runoff, in conjunction with other practical and financial considerations.

Keywords: on-site wastewater; low permeability; evapotranspiration; willow; passive treatment

1. Introduction

The domestic wastewater of approximately one third of the population in Ireland (~500,000 dwellings) is treated on-site by domestic wastewater treatment systems (DWWTS) of which more than 87% are septic tanks [1]. If situated and constructed incorrectly, the potential impacts of such on-site effluent include the pollution of either groundwater and/or surface water. In particular, areas with inadequate percolation due to low-permeability subsoils and/or insufficient attenuation due to high water tables and shallow subsoils present the greatest challenge in Ireland for dealing with effluent from DWWTS. If there is insufficient permeability in the subsoil to take the effluent load, ponding and breakout of untreated or partially treated effluent at the surface may occur with associated serious health risks. There will also be a risk of effluent discharge/runoff of pollutants to surface water and to wells which lack proper headwork or sanitary grout seals [2]. It is estimated that the overall proportion of the country with inadequate conditions for DWWTS either all year round or that can arise intermittently during wet weather conditions is 39% [3].

The specification [4] of a lower limit to subsoil permeability (defined according to an on-site falling head percolation test) for effluent discharge to ground, in conjunction with the fact that surface water discharges are generally not being licensed for one-off housing, means that many areas will be deemed unsuitable for single house development. To address these problematic areas and allow development while protecting water resources from the risk of pollution by on-site effluent, alternative wastewater treatment and disposal options are needed. One option that has been considered is the concept of discharging the on-site wastewater effluent into a sealed basin and relying on the net evapotranspiration (ET) from willow trees to exceed the rainfall and effluent hydraulic loads, thereby creating a zero-discharge treatment solution. This technology has been introduced into Denmark for on-site wastewater treatment with some success [5,6] with national guidelines produced [7]. Willows are highly suitable for such an application in such temperate climates due to their high transpiration rates throughout the growing season [8,9], efficient uptake of nutrients [10,11], tolerance of flooded soils and oxygen shortage in the root zone [12] and resilience to pollutants [13]. In more tropical climates other species have been proven to perform such a role; for example, trials in self-recirculating systems in Australia have demonstrated zero discharge using mixed species of bamboo and citrus fruits [14].

Hence, a series of full-scale trials was established in Ireland to evaluate the use of sealed basin ET systems to treat on-site wastewater effluent in Ireland. These systems have been monitored over a five-year period and the results obtained from these trials have been used to draft national design criteria and to determine the operating limits for these systems in an Irish context.

2. Materials and Methods

2.1. Construction of Full-Scale ET Systems

Thirteen full scale willow systems on 10 sites were constructed to treat domestic wastewater effluent from single house dwellings across Ireland: 10 in County Wexford, 1 in County Limerick and 2 in County Leitrim, as shown on Figure 1. The sites were located in areas with very low permeability (*i.e.*, clayey) subsoils, in which effluent percolation would not be a feasible disposal method for the on-site effluent. The systems were designed to specifications that would allow for the long-term study and comparison of key parameters (effluent type, willow species, plan area, aspect ratio and effluent distribution) as detailed in Table 1. The systems were designed (area and depth) based upon a modeled water balance between the typical wastewater effluent rates expected from the dwelling, the local rainfall and the estimated evapotranspiration from the basin over a four year period. Realistic time-varying fluctuations in flows (from basis of previous research carried out by Gill *et al*. [15]), local meteorological conditions and estimated crop coefficients were input into the water balance model which then simulated willow evapotranspiration across a four year period. The crop coefficients were assumed to be 1 throughout the winter months, increasing up to maximum values of 2.65 across the summer months. At each time step, the calculated stored volume was then divided by the porosity of the soil expected within each willow system (estimated at 30%) to determine the depth fluctuations. Different combinations of area and depth were compared in order to determine the design dimensions of each system which should (theoretically) ensure zero discharge across the 4 year simulation period. The choice of depth was also guided by previous Scandinavian experience with such ET systems using the same willow varieties [7].

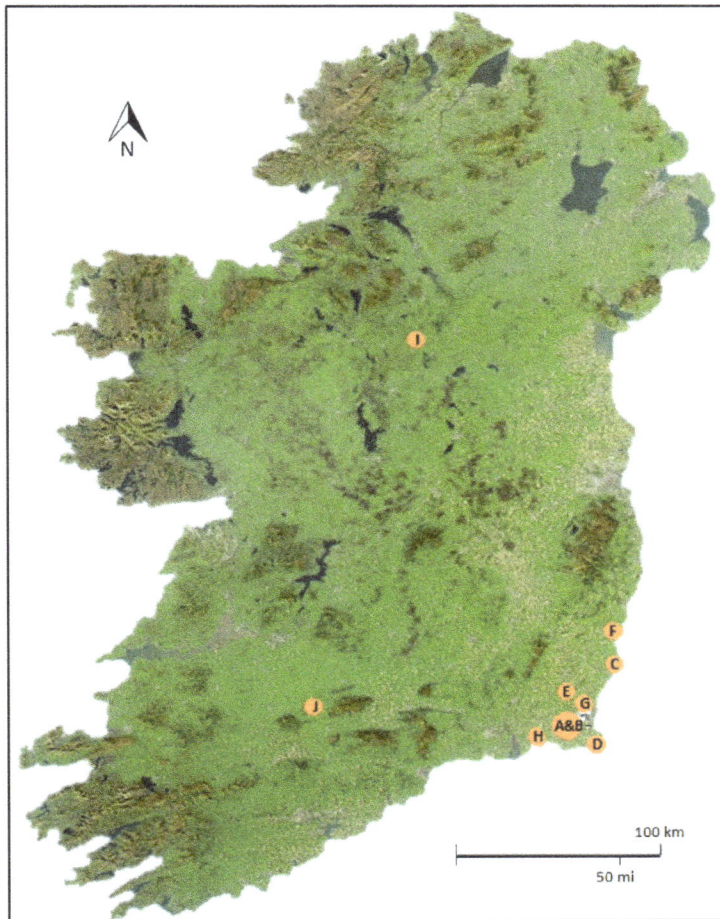

Figure 1. The location of the 13 full-scale ET systems researched in Ireland.

The basins were excavated to a depth of 1.8 m with a geosynthetic barrier laid along the bottom and sides of the excavated basin. An impermeable membrane (minimum 0.5 mm thickness butyl rubber or sealed low-density polyethylene (LDPE) was then laid on top of the geosynthetic barrier followed by a second geosynthetic barrier laid on top of the impermeable membrane. For most systems, the effluent was distributed either by pumping or gravity flow into 110 m rigid diameter pipes at 4 m spacing along the base of the basins bedded in 300 mm depth of 8–32 mm diameter gravel. The excavated soil was then back-filled into the basins up to ground level. On Sites A and B however, the effluent was pumped on a volume-dose basis into a semi-rigid 40 mm diameter distribution pipe at 1 m depth. It should be noted that the excavated soil was used to re-fill the systems due to economic considerations. Importing such a quantity of material would also require more than 50 truckloads of imported earth for a single house which was deemed to be unacceptable for a single house, both logistically and environmentally.

Table 1. Summary of design parameters compared between 13 full-scale systems.

Site	Constructed	Effluent [a]	Area (m^2)	Aspect Ratio (length:width)	Distribution	Willow Variety
A	May 2009	STE	296	4.6:1	pumped	*Bjorn, Tora, Jorr*
B	May 2009	STE	570	1.6:1	pumped	*Bjorn, Tora, Jorr*
C	April 2010	STE	420	2.9:1	pumped	*Tora, Torhild, Tordis*
D	May 2010	STE	464	1.8:1	gravity	*Tora, Torhild, Olof*
E	April 2010	STE	24	2.7:1	gravity	native Irish species
F(1)	July 2010	SE	900	1.4:1	gravity	*Tora, Tordis, Olof*
F(2)	July 2010	SE	900	9.0:1	gravity	*Tora, Tordis, Olof*
G	September 2010	SE	340	3.4:1	gravity	*Tora, Tordis, Olof*
H(1)	April 2011	STE	260	4.1:1	pumped	*Tora, Tordis, Inger*
H(2)	April 2011	STE	260	4.1:1	pumped	*Tora, Tordis, Inger*
I(1)	May 2011	SE	216	3.4:1	gravity	*Tora, Torhild, Olof*
I(2)	May 2011	SE	216	3.4:1	gravity	*Tora, Tordis, Torhild*
J	May 2012	STE	170	6.8:1	pumped	*Tora, Torhild, Tordis*

Notes: [a] STE—septic tank effluent; SE—secondary treated effluent.

Willow cuttings (see Table 1 for varieties) were planted at a density of 3 per m^2. Coppicing of all the willows was then carried out, in accordance with standard short rotation coppice guidelines [16,17] whereby all shoots were cut back as close to the ground as practically possible in the first dormant season to encourage vigorous growth and multiple shoot development in the following season. Cutback was carried out in a three year cycle thereafter, with alternating thirds of the willow system cut every year.

2.2. Instrumentation

Each ET system was designed in order to quantify the hydraulic loading via rainfall and effluent from which a water balance calculation was used to determine the removal of water from the system via evapotranspiration. A tipping bucket device was used to measure flow continuously on the sites that had a sufficient slope for a gravity feed from the proprietary system to the willow system (as well as splitting the flows evenly between systems on sites with two willow systems). A reed switch on the tipping bucket was connected to a Solinist® Rainlogger Model 3002 (Georgetown, ON, Canada) datalogger. On sites where pumping of effluent to the willow system was required, OTT Thalimedes (Kempten, Germany) water level meters (accuracy: ±0.002 m) were installed in the sump to measure continually the water level. The water level within each ET system was measured on an hourly basis using an OTT Orpheus Mini groundwater datalogger (accuracy: ±0.05%). Campbell Scientific (Loughborough, UK) weather stations were installed to record temperature, relative humidity, wind speed, solar radiation and net radiation. Campbell Scientific ARG100 tipping bucket rain gauges were installed on all sites and either connected to the Campbell Scientific dataloggers or to Solinist® Rainlogger (Model 3002) dataloggers.

2.3. Water Quality Analysis

All septic tank effluent (STE), secondary treated effluent (SE), samples of stored water within each system (taken from the inspection wells) and ponded surface water were periodically analysed for chemical oxygen demand (COD), total nitrogen (TN), ammonium (NH_4^+–N), nitrite (NO_2^-–N), nitrate (NO_3^-–N), orthophosphate (PO_4^{3-}–P), chloride (Cl^-) and sulphate (SO_4^{2-}) using a Merck Spectroquant Nova 60® spectrophotometer (Darmstadt, Germany). In addition, indicator bacteria of faecal contamination, Total Coliforms (TC) and *E. coli* were analysed for using the IDEXX Colilert®-18 test (Westbrook, MA, USA) with enumeration carried out using IDEXX Quanti-Tray®/2000, a semi-automated quantification method based on the Standard Methods Most Probable Number (MPN) model.

2.4. Soil Porosity

In order to calculate the water balance accurately, the available pore space in the soil within the willow ET system needed to be calculated from which the depth of effluent in the system could be converted into a stored volume. Replicate 200 mm deep soil samples were collected in 100 mm diameter steel cores at a variety of depths from within the ET systems on each site at the end of the growing season when the water levels within the willow systems were low. The samples were transported back to the laboratory and extruded and then weighed immediately to represent the weight of a fully drained (*in situ*) sample. The sample length was then submerged in water until saturation point was reached. Upon removal, the soil sample was then reweighed to determine the water content at the saturation point. In addition, the resultant change in water level within each ET system in response to single peak, short duration, large rainfall events were analysed from the field data. Such short rainfall events were used to minimise the effect that any ongoing evapotranspiration would have on the water level. Knowing the net rainfall falling onto the system and respective rise in water level, the effective porosity (usable storage volume) could be calculated and then averaged out for a series of events. It should be noted that the effective porosity of any gravel in the system was also included in the depth-volume calculations when appropriate, using a measured gravel porosity of 32%.

3. Results from the Field Trials

3.1. Loadings

A summary of the data collected at each site is shown in Table 2 with examples of rainfall and potential ET (ET_o) from 2013. A more detailed discussion of the field results is discussed in Curneen and Gill [18].

Table 2. Summary hydraulic loading data from ET system trials.

Site	Effluent Production Started	Population Equivalent (PE)	Mean Daily Flow (L/d)	Effective Soil Porosity (%)	Annual Rainfall (2013) (mm)	Annual ET_o (2013) (mm)
A	October 2010	4	628.4 [a]	15.1	981	585
B	January 2012	1	443.1 [a]	15.6	981	585
C	December 2010	2	356.7	11.0	933	584
D	May 2010	4	483.3	11.0	847	592
E	April 2011	n/a [b]	3.0	21.0	833	571
F(1)	January 2013	3	306.6	18.0	811	587
F(2)	January 2013	3	306.6	19.0	811	587
G	n/a [b]	0	0	9.6	928	610
H(1)	March 2012	2	181.3	9.5	930	602
H(2)	March 2012	2	181.3	9.5	930	602
I(1)	November 2012	2	163.4	16.0	1158	465
I(2)	November 2012	2	163.4	20.0	1158	465
J	November 2012	3	330.3	19.0	1039	526

Notes: [a] suspected rainwater infiltration into effluent; [b] an ET system was constructed for a new house but no-one moved into the house during the monitoring period.

3.2. Water Level Fluctuations

Although all the willow ET systems had been designed to operate as zero discharge systems on the basis of a theoretical water balance, the water level data (as well as observations at the sites) revealed that no system managed to achieve zero discharge in any year, remaining full for much of the winter months. This indicated ongoing loss of water by lateral exfiltration at the surface during these winter periods. Overflows did also occur to a certain extent during the summer of 2012 when precipitation levels were exceptionally high.

An example of annual water level profiles within two willow systems is shown in Figure 2. The equilibrium level (the level above which overflow from the system occurs) along with the soil surface level are illustrated, both of which are measured relative to the base of the system. For example, at Site A (Figure 2a) overflow was determined to be occurring at a depth of 1.02 m even though the soil surface level was 1.35 m, suggesting that either the impermeable liner did not reach up to the top of the system or it had been breached at this depth. It can also be observed that the system was empty for most of the summer months, meaning that optimal evapotranspiration performance was probably not achieved during that period. This did not occur at Site C (Figure 2b), primarily due to the system being much deeper with an overflow level at 1.73 m from the base. It is also interesting to note the effect of large October rainfall events on the water level in the systems, in response to which both systems refilled to capacity in a very short space of time, which is indicative of the low effective soil porosities.

(a)

(b)

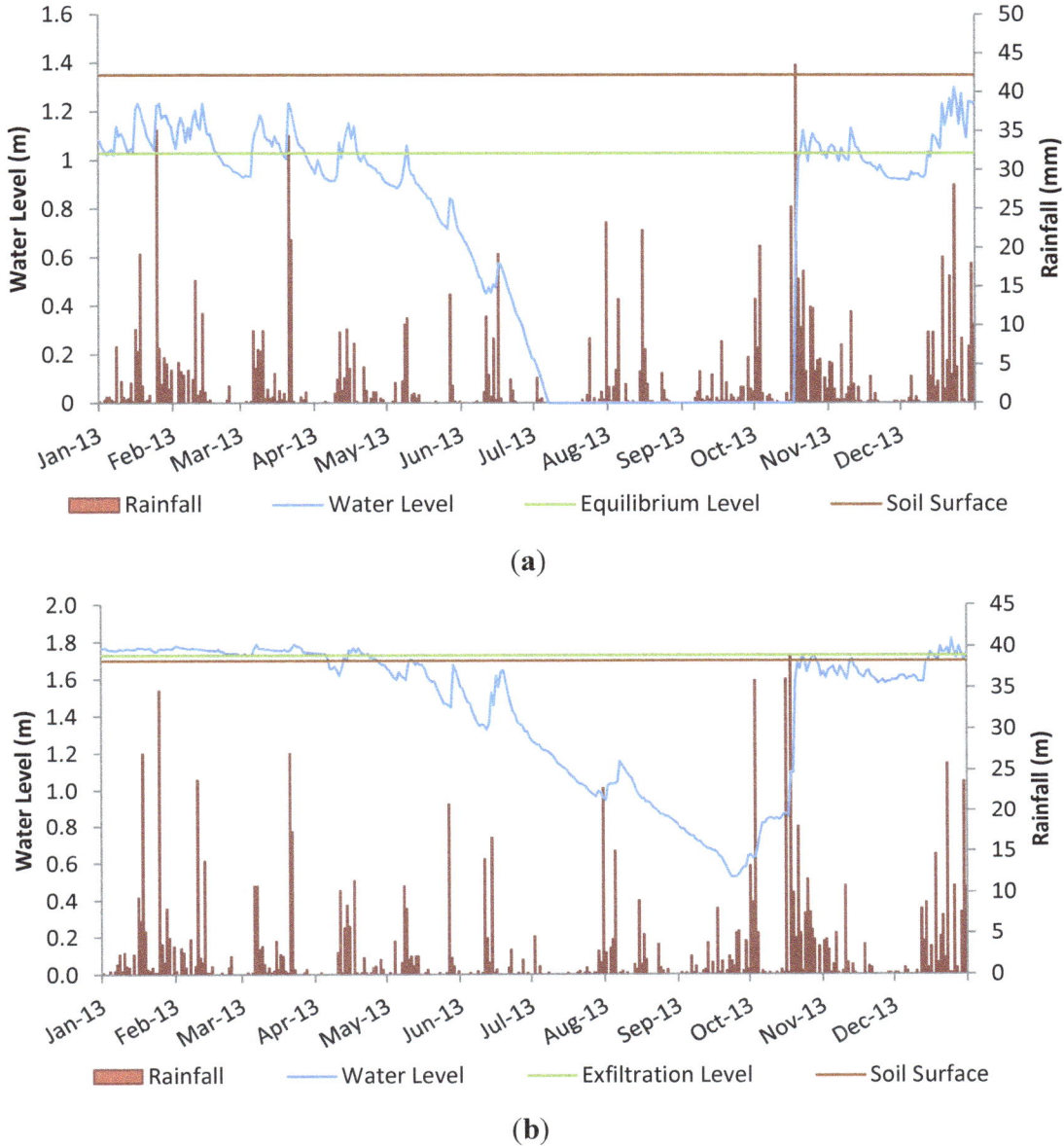

Figure 2. Water level and daily rainfall (2013) for (**a**) Site A; (**b**) Site C.

3.3. Water Balance (ET and Crop Factors)

The actual evapotranspiration from each system throughout the monitoring period was determined on the basis of a water balance taking into account the measured hydraulic loadings from rainfall and wastewater, as shown in Equation (1):

$$\text{ET}_{\text{wilow}} = R + I + (\Delta H \times V_u \times A) \qquad (1)$$

where, $\text{ET}_{\text{willow}}$ = evapotranspiration from willow system; R = rainfall; I = influent (wastewater); ΔH = change in water level within system; V_u = usable water volume storage within soil/gravel (*i.e.*, effective porosity); A = surface area.

Evapotranspiration rates attained by the willow systems were calculated on a daily basis with the results then averaged and presented on a monthly basis. At such times when the water level was above the overflow point, $\text{ET}_{\text{willow}}$ could not be accurately calculated using water balance Equation (1), as the volume of water leaving the system due to any exfiltration/overflow could not be quantified. Hence, for such flooded periods during the growing season, mean daily ET rates for those days were assumed to be the same as the closest days during which ET_{willow} rates could be rigorously calculated using Equation (1). In the winter months when the trees were dormant, it was assumed that ET_{willow} was equal to ET_o on days where the water level was above the overflow level: on all other days, the actual ET_{willow} was calculated according to Equation (1). Monthly crop coefficients (K_c) were calculated by dividing the assumed monthly ET_{willow} by the ET_o for the month.

The total assumed ET_{willow} for the monitored period in each calendar year for each willow system shown in Table 3 reveals significant variation between the willow systems, even for those systems in which the willow cuttings had been planted in the same year. Several reasons are thought to attribute to this including different effluent loadings, different soil porosities and different rates of establishment/development of the willow trees. The highest ET_{willow} rates were calculated for Sites A, B, C, and F(2), which was expected as these sites undoubtedly had better willow tree development (also aided by effluent loading from the outset for Sites A and C).

Table 3. Annual calculated ET_{willow} for each willow system.

System	2010		2011		2012		2013		2014	
	ET_{willow}	months	ET_{willow}	months	ET_{willow}	months	ET_{willow}	months	ET_{willow}	months
A	755	10	929	12	725	12	928	12	-	-
B	554	7	638	12	765	12	723	12	-	-
C	358	4	717	12	670	12	752	12	845	12
D	-	-	558	12	495	12	487	12	-	-
E	282	7	573	12	669	12	90	4	-	-
F(1)	-	-	511	10	595	12	577	12	598	12
F(2)	-	-	641	10	827	12	817	12	930	12
G	-	-	341	7	574	12	445	8	-	-
H(1)	-	-	-	-	864	9	788	12	-	-
H(2)	-	-	-	-	771	9	576	12	-	-
I(1)	-	-	-	-	424	9	613	12	-	-
I(2)	-	-	-	-	402	9	540	12	-	-
J	-	-	-	-	-	-	680	12	-	-

Typical ET_{willow} results and crop coefficients from two of the longest operating systems (Site A and Site C) are shown in Figure 3. As expected, the highest ET_{willow} rates were attained in the summer months during the willow trees' growing season. It might have been expected that the ET_{willow} rates would increase year on year, as the willow trees developed and became larger. However, the mean daily ET_{willow} for each month on Site A shows that this was not always the case. While the mean daily ET_{willow} did increase year on year from 2010 to 2012 for the months May, June, July and September, it decreased for all these months (except June) in 2013. This is probably due to the low water availability within the system at these periods, thus preventing the willows from being able to attain their optimum potential evapotranspiration, and possibly an indication that the roots had not developed down to the base of the system.

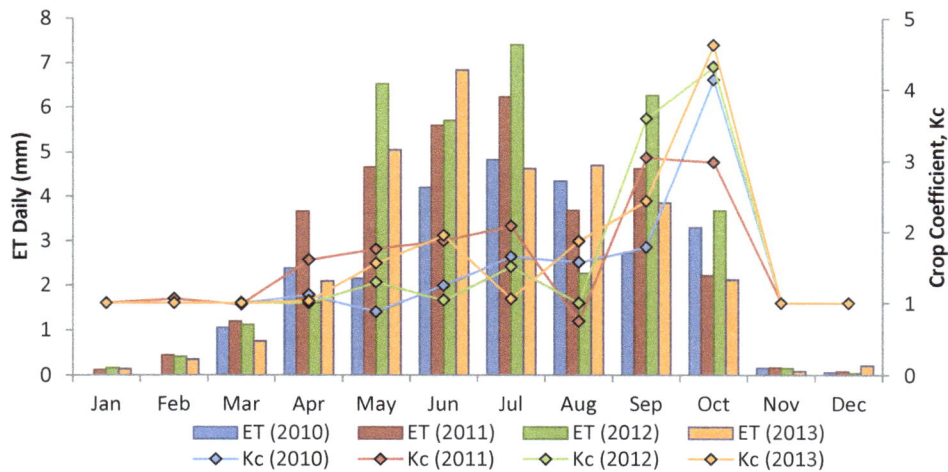

(a)

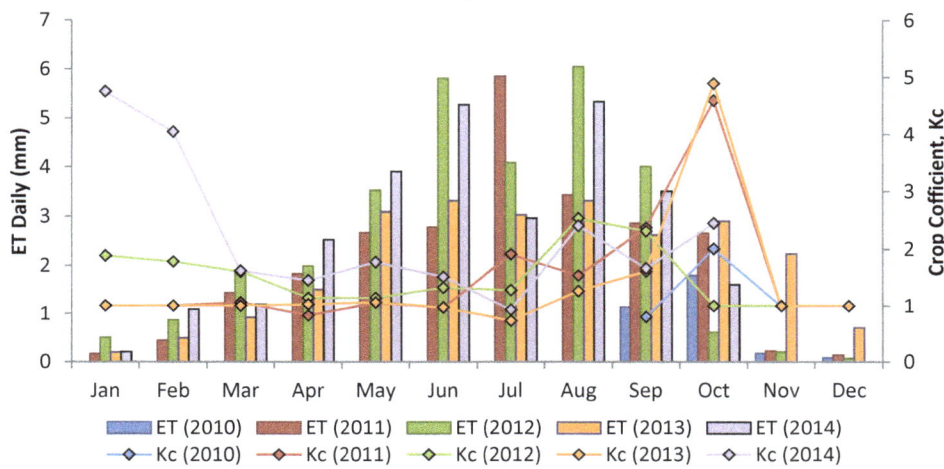

(b)

Figure 3. Monthly average ET_{willow} rates for (**a**) Site A and (**b**) Site C.

A comparison of the crop coefficients also shows a similar result. Most of the monthly ET_{willow} highs (as well as associated crop coefficients) throughout the growing season occurred in 2012 when water availability was at its highest due to the high rainfall levels, not in 2013 when the willow trees were more developed and reference evapotranspiration rates were higher. It is also interesting to note the relatively high crop coefficients in September and October, which shows that there was still a significant amount of water being evapotranspired even though the ET_o values were relatively low.

3.4. Water Quality

The quality of the effluent discharged into the systems, the water in each system's monitoring well and ponded water on the surface during winter is shown in Table 4. An interesting comparison can be made between the water quality in the sumps from those systems that were receiving effluent and those monitored before effluent was introduced. As would be expected, the nutrient concentrations were marginally higher in the samples from the systems receiving effluent compared to the samples from the systems with no effluent. However, the results show that the willow systems acted as an excellent pollutant attenuation process on the influent, promoting greatly reduced organic, nutrient and indicator bacteria concentrations. Equally, the quality of standing water during winter periods on top of the systems receiving effluent proved similar to the sump samples from the systems that had not received effluent, with the exception of very low levels of indicator bacteria.

Table 4. Mean wastewater influent, internal sump water quality and ponded surface water.

water quality parameter	STE	ET Systems Receiving Effluent		ET Systems with No Effluent
		ET System Sump	Ponded Water on Surface	ET System Sump
COD (mg/L)	377 ± 96	42.6 ± 8.9	29.8 ± 6.7	32.9 ± 7.1
Total-N (mg/L)	50.3 ± 8.7	9.8 ± 2	12.4 ± 4.1	8.1 ± 1.4
Ortho-P (mg/L)	6.3 ± 1.2	0.55 ± 0.4	0.24 ± 0.14	0.30 ± 0.08
E. coli no./100 mL *	9.2×10^5	22.1	8.3	<1

Note: * geometric mean.

4. Evaluation of Design Parameters on Performance

4.1. Model to Simulate ET System Performance

In order to use the results of the field trials to develop appropriate national design guidelines, modelling was first carried out to predict the effect of varying the different design parameters using realistic crop factors and meteorological data as determined on site from the field trials. The design model for sizing of a willow system was based on a daily water balance (*i.e.*, Equation (1)) which incorporated hydraulic loading into the system via rainfall and effluent and removal of water from the system via evapotranspiration. Daily rainfall and reference evapotranspiration measured on Sites A and B over the duration of the monitoring period was used as inputs into the model to produce the resultant daily water level profiles over a five-year monitored period (January 2010–December 2014).

The usable effective porosities used in the model varied according to the medium at and below any particular water level, as shown schematically on Figure 4: changes in water level occurring within the soil medium assumed a V_u value of 15%; changes within the gravel layer used a V_u of 33%; and changes in water level above the surface of the soil assumed V_u to be 100%.

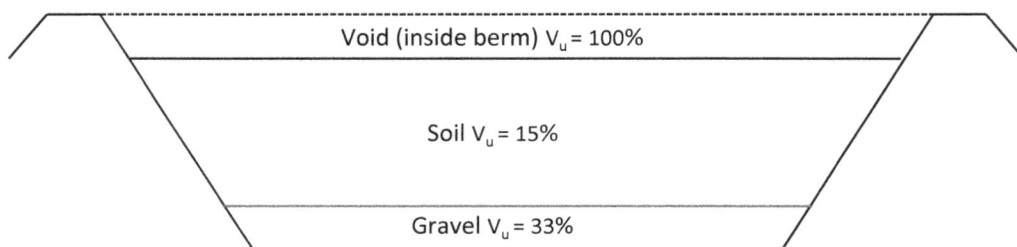

Figure 4. Sample cross-section of ET system with relevant effective porosities (V_u).

The ET rates and crop coefficients (K_c) used were based on the results from the systems at Sites A, B, C and F(2) which were deemed to have performed to the expected standard over the duration of the research project as they had not been compromised by poor growth or low effluent loading as many of the other sites had. For Sites A and B, it was considered that conservatively low crop coefficients were calculated across some of the summer months due to the fact that the systems had emptied out quite early in two of the monitored growing seasons, leaving little water readily available for evapotranspiration. This is illustrated in Figure 5 which shows a comparison between the crop coefficients for the two systems with those determined in irrigated lysimeter experiments carried out by Guidi *et al.* [19] for willows in the first and second growing seasons: the crop coefficients on Sites A and B were equal to or higher than the lysimeter experiments for May and June but then dropped back for the rest of the summer months when the systems were empty. As such, it was deemed that these crop coefficients were lower than values that could be expected had water been more readily available if the systems had been deeper. Hence, the representative K_c values used for the model shown in Table 5 were chosen based the crop coefficients determined in the field trials but also influenced by the higher values determined by Guidi *et al.* [19] in August. The dormant season crop coefficients were determined by taking the average values from all systems as the results showed that willow development appeared to have a minimal bearing on these values. The annual crop coefficient used to model the ET systems works out at 2.2 which, for a mean annual ET_o on Sites A and B of 517 mm across the monitoring period, would give an expected annual total of 1135 mm of evapotranspiration from the modelled system.

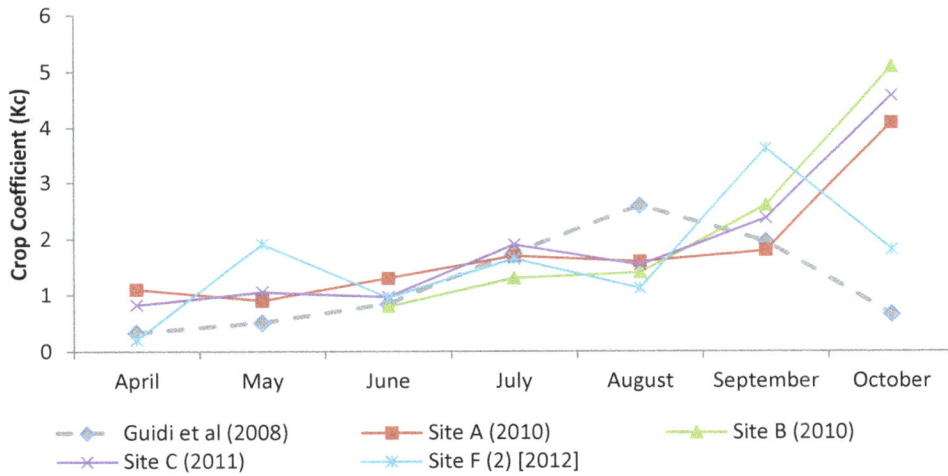

(a)

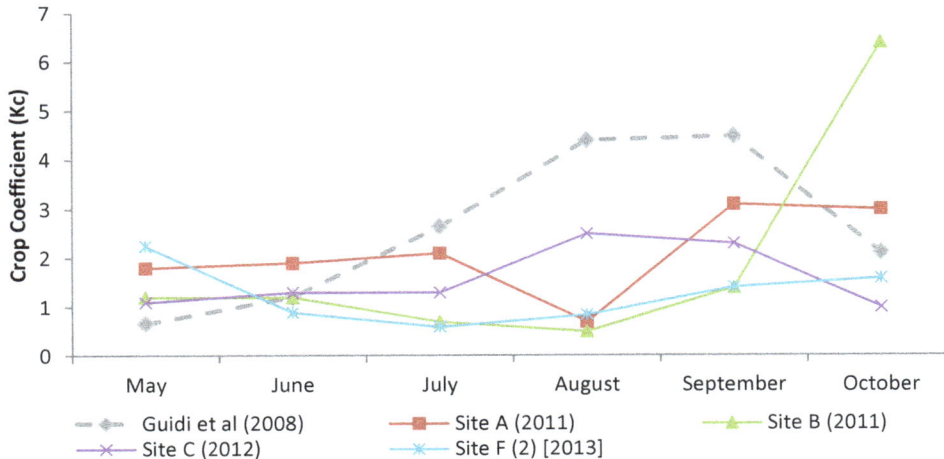

(b)

Figure 5. *Cont.*

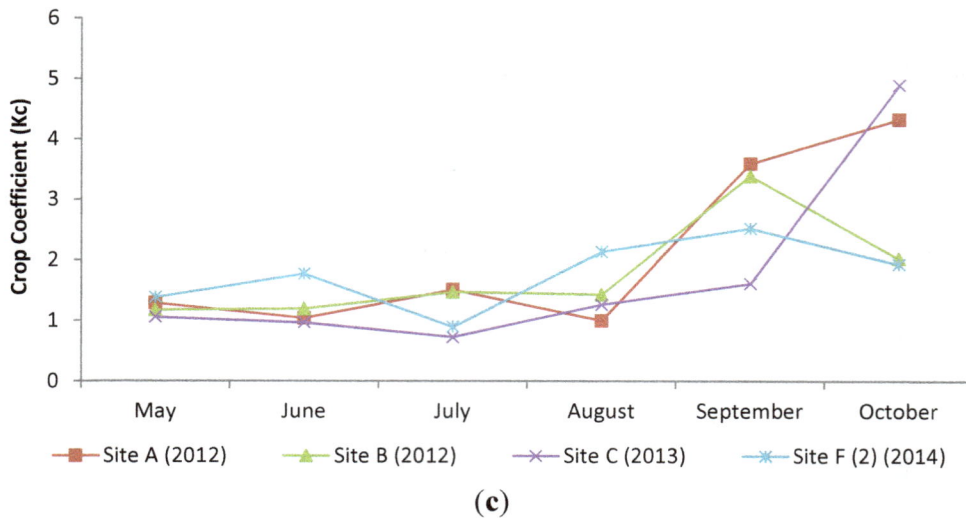

(c)

Figure 5. Crop coefficients for Sites A, B, C and F(2) compared to those determined by Guidi *et al.* [18] for (**a**) first growing season; (**b**) second growing season and (**c**) third growing season.

Table 5. Representative K_c values and resulting system ET rates in the model.

Month	Assumed K_c	Mean ET_o (mm)	ET_{willow} (mm)
January	1.5	4.4	9.1
February	1.4	11.0	20.8
March	1.4	32.1	48.0
April	1.3	59.2	77.0
May	1.5	82.2	123.4
June	1.9	92.1	175.0
July	2.1	84.4	177.2
August	3.4	76.7	260.9
September	3.6	46.7	168.1
October	3.2	20.9	66.8
November	1.4	4.9	6.8
December	1.2	1.9	2.3
Annual Mean	2.2	516.6	1135.4

4.2. Comparison of Design Parameters

The ET_{willow} water balance model was then used to predict the water level profile response of the system to various different physical design factors: the plan area of system, the effective porosity of the storage medium (soil and gravel), the depth of soil within the willow system basin, the depth of gravel layer within the system and the depth of free space available above soil surface (maintained by a raised boundary (bund)). For each simulation, the five years of rainfall and reference evapotranspiration data from the field trials (2010 to 2015) were used as inputs. The model then simulated the expected water level profile and any predicted overflow for any defined wastewater loading. Various combinations (within practical constraints) of the above factors were used in order to determine the minimum area required to try to reduce any runoff from the systems as much as possible throughout the year. The model showed that the evapotranspiration that could be expected from a willow ET system in a given year in Ireland's climate would generally be lower than the corresponding hydraulic load (rainfall and effluent), resulting in overflow from the system occurring, as was the case in 2012. This mirrored the findings on most of the systems monitored. Hence, simulations were carried out to seek to optimise the system in terms of lowering the amount of overflow days that may occur, while also taking practical area and financial constraints into account.

4.2.1. Plan Area and Overall Depth

The effect of plan area and overall system depth on water level depth was examined using simulations based on a dwelling with a population equivalent (PE) of 4 with the other design parameters listed in Table 6, fixed. The effect of increasing the area on the number of overflow days is illustrated in Figure 6 which shows that significant gains with respect to decreasing the amount of overflow days can be made by increasing the plan area of the ET system from 250 up to 500 m^2. However, any further gains made by increasing the surface area beyond 500 m^2 appear to be marginal and would probably not justify the significant additional construction costs required.

Table 6. Fixed parameters used for comparative model analyses of system design.

Area (m^2)	Effluent (L/c.day)*	Soil Porosity (%)	Gravel Porosity (%)	Depth of Gravel Layer (m)	Depth to Soil Surface (m)	Depth of Free Space (m)
500	100	15	35	0.3	1.8	0.3

Note: * litres per capita per day

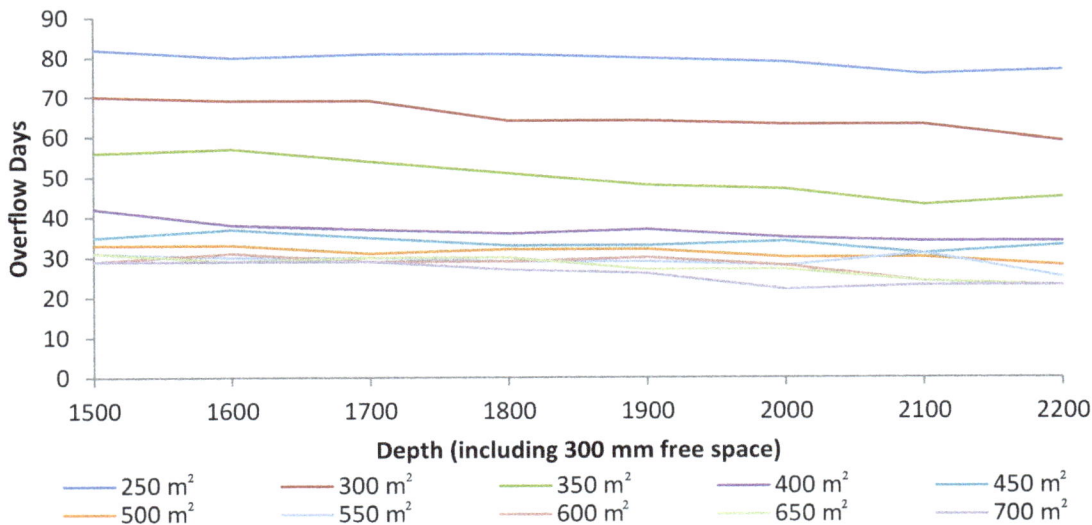

Figure 6. Effect of plan area on the mean number of overflow days simulated per year.

An example of a water level profile produced by the model for a 500 m^2 plan area system for a 4 PE residence is shown in Figure 7. The system was determined to have overflow for 141 days across the five-year period corresponding to the monitoring of the full-scale systems. It is worth noting that water level would be predicted to reach the bottom of the system by the end of August in four out of five of the growing seasons monitored.

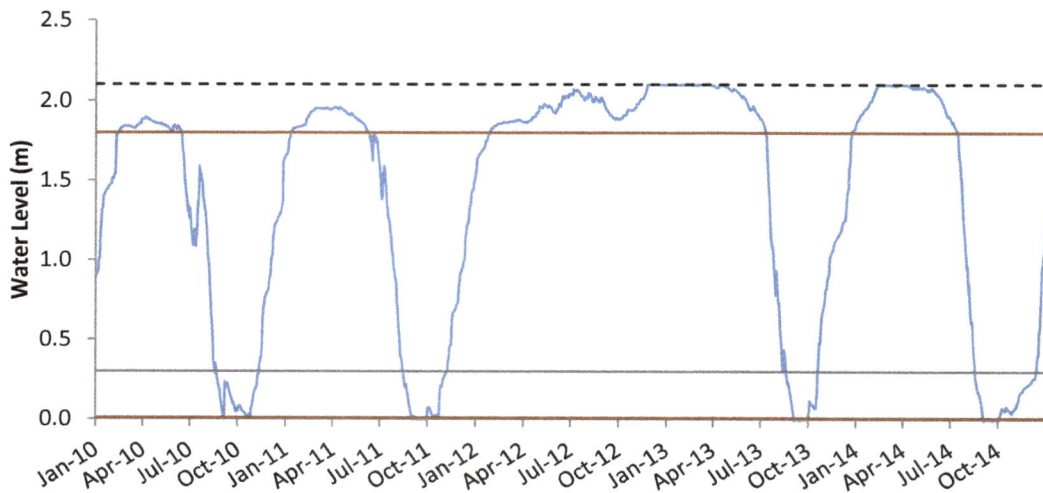

Figure 7. Water level profile simulated by the water balance model for a 500 m^2 ET system receiving effluent from a 4 PE house.

The effect of increasing the overall depth on the amount of overflow from the willow system in Figure 8 shows that it would be beneficial for smaller areas (<350 m^2) and to a lesser extent for larger areas (>550 m^2). For the range of areas in between, however, it appears that not much benefit can be gained by increasing the depth of the system.

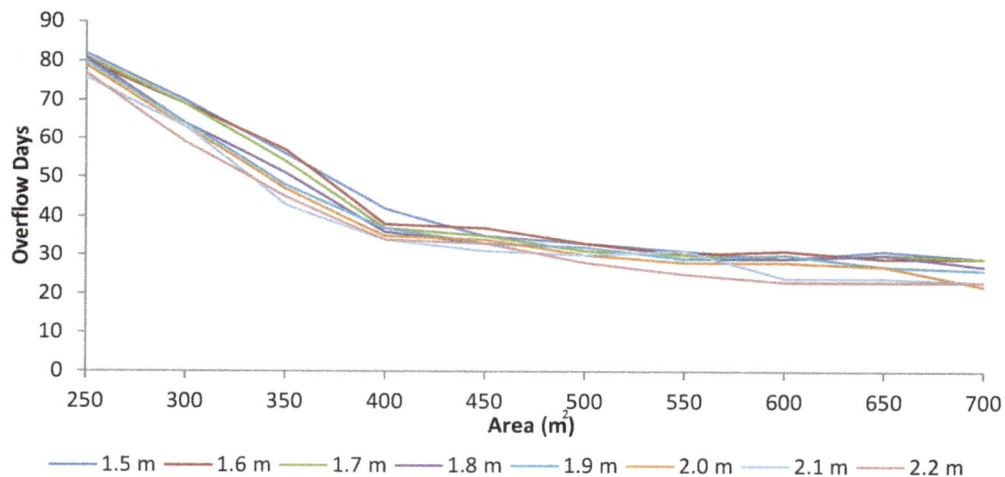

Figure 8. Effect of increasing the depth on the mean number of overflow days simulated per year.

A matrix highlighting the predicted overflows for different combinations of area and depth is illustrated in Table 7. As would be expected, the combinations of higher area and depth result in lower predicted days with overflow. Several combinations result in approximately 30 to 35 days of overflow per year, which is probably the realistic optimal minimum, taking practical and financial constraints into account. The effect on the water level profile of increasing the depth of a 500 m^2 willow system from 1.5 to 1.8 m, for example, has a limited effect and the number of overflow days, decreasing from 35 to 33. A further increase in depth of the 500 m^2 area system to 2.1 m again has limited effect, with a predicted decrease of just 1 overflow day per year over the duration of the modelled period. The overflow days can be reduced to as low as 22 days per year for a 4 PE dwelling, but this would require a significantly larger system of area 700 m^2 and overall depth of 2.0 m, at which point such ET systems for single houses would not be deemed to be financially viable in an Irish context.

Table 7. Matrix showing the mean number of overflow days simulated per year from the willow system for different area and depth combinations.

Area	Depth of ET System							
	1.5 m	1.6 m	1.7 m	1.8 m	1.9 m	2.0 m	2.1 m	2.2 m
250 m^2	82	80	81	81	80	79	76	77
300 m^2	70	69	69	64	64	63	63	59
350 m^2	56	57	54	51	48	47	43	45
400 m^2	42	38	37	36	37	35	34	34
450 m^2	35	37	35	33	33	34	31	33
500 m^2	33	33	31	32	32	30	30	28
550 m^2	31	30	30	29	29	28	31	25
600 m^2	29	31	29	29	30	28	24	23
650 m^2	31	29	30	30	27	27	24	23
700 m^2	29	29	29	27	26	22	22	22

The result of running the model again for four different effluent loadings is shown in Figure 9 and Table 8. The loading rates were 5, 6, 8, and 10 PE, while assuming a daily effluent production of 100 L/c.day. While this effluent loading rate is lower than that assumed in the Irish Code of Practice [4], it is based on the assumption that any dwelling using a willow ET treatment system should have up-to-date water saving appliances (e.g., dual flush toilet, low-flow shower head, tap aerators) whereby such a wastewater production would be realistic.

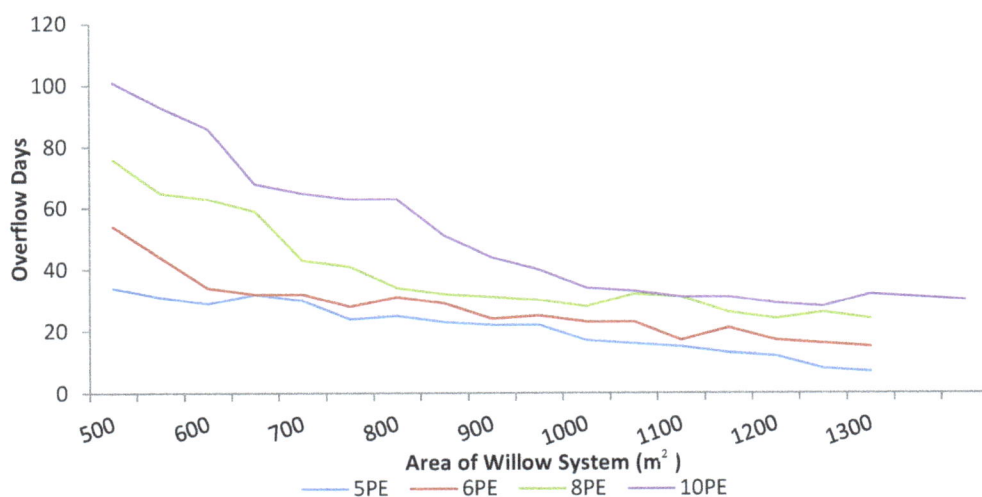

Figure 9. Effect of increasing the plan area of the ET system on the mean number of overflow days simulated per year for different hydraulic loading scenarios.

Table 8. Matrix showing the effect of varying the area on the mean number of overflow days simulated per year generated by four different dwelling sizes.

Area (m^2)	5PE	6PE	8PE	10PE
500	34	54	76	101
550	31	44	65	93
600	29	34	63	86
650	32	32	59	68
700	30	32	43	65
750	24	28	41	63
800	25	31	34	63
850	23	29	32	51
900	22	24	31	44
950	22	25	30	40
1000	17	23	28	34
1050	16	23	32	33
1100	15	17	31	31
1150	13	21	26	31
1200	12	17	24	29
1250	8	16	26	28
1300	7	15	24	32
1350	-	-	-	31
1400	-	-	-	30

This analysis shows that reductions in the number of overflow days plateau off once a certain area is reached. Lowering the number of overflow days below the 30–35 range requires a significant increase in area, which again would not be considered feasible from a financial point of view. Hence, the practical considered optimum areas for different loadings are shown in Table 9. It should be noted that these areas would give an effluent hydraulic loading of 292 mm per year, based on an effluent production of 100 L/c.day.

Table 9. Practical optimum ET system area for different hydraulic loadings from a dwelling.

	Hydraulic loading					
	2PE	4PE	5PE	6PE	8PE	10PE
Optimum Area (m^2)	250	500	650	750	1000	1200

4.2.2. Free Surface Depth to Overflow

The optimum depth of free space above the surface of the soil maintained by a raised bund provides a significant amount of potential hydraulic storage although, ideally, long-term ponding of water on top of the willow system is not desirable. However, the organic and nutrient analysis carried out (see Section 3.4) showed that pollutant concentrations in any surface water were relatively low. Furthermore, no foul odours have been detected on any visits to the sites. The effect of varying the depth of free space on a 500 m^2 plan area willow system was conducted by entering a range of depths between 0 and 350 mm (50 mm increments) for three different overall willow system depths (1.5, 1.8, and 2.1 m). Overall depths greater than 2.1 m were not considered because of practical constraints during the excavation phase of construction. The change in the number of overflow days by varying the depth of free space is shown in Table 10.

Table 10. Effect of varying the depth of free space above the surface of the willow system on the mean number of overflow days simulated per year for three different system depths.

Depth of Free Space (mm)	Depth of ET System		
	1.5 m	1.8 m	2.1 m
0	76	65	59
50	72	61	55
100	62	47	40
150	58	44	38
200	42	35	34
250	37	33	34
300	33	32	28
350	30	29	27

Similar to the effect of increasing the area to lower the number of days with overflow, the gains made by increasing the free space appear to be less significant beyond a certain point. The optimum depth of free space in systems with a total depth of 2.1 and 1.8 m is approximately 200 mm (with 34 and 35 overflow days per year respectively), while the optimum depth for a system of depth 1.5 m is 300 mm (33 overflow days).

4.2.3. Hydraulic Load

The benefit of decreasing the hydraulic loading on the system by using an impermeable membrane on the surface of the willow system to direct a proportion of the rainfall away from the basin was also assessed. This analysis assumed that under such a scenario there would be no free storage space available above the soil. As the proportion of rainfall that would be diverted away from the willow system by using an impermeable membrane at the surface is unknown, a hypothetical analysis has been carried out where the rainfall is assumed to be reduced by varying degrees. The model again was based upon a willow system area of 500 m^2 receiving effluent from a dwelling with 4 PE with the results of the analysis shown in Table 11. This shows that the ET system could be modified to act as a fully zero-discharge treatment system if 60% or more of the annual rainfall falling onto the system is diverted. However, it should be recognised that the impact of covering the surface of the soil with an impermeable membrane may produce deleterious effects with respect to oxygen transfer down into the soil. This would also reduce soil evaporation.

Table 11. Matrix showing the mean number of overflow days simulated per year from the willow ET system for different reduced rainfall and depth combinations over a five-year period.

Rainfall	Depth of ET System						
	1.5 m	1.6 m	1.7 m	1.8 m	1.9 m	2.0 m	2.1 m
100%	76	73	68	65	64	62	59
90%	68	64	63	61	58	51	46
80%	58	55	51	43	39	36	34
70%	44	37	33	30	28	25	21
60%	28	24	19	16	13	12	9
50%	16	13	9	8	6	5	4
40%	6	4	3	1	0	0	0
30%	0	0	0	0	0	0	0
20%	0	0	0	0	0	0	0

4.2.4. Optimum Design (All Parameters)

In order to further refine the design parameters for an ET willow system, different combinations of plan area, overall depth, depth of gravel layer and depth of free space to overflow were varied to minimise the number of overflow days using the Excel Solver optimisation routine whereby the target parameter to be minimised was the number of overflow days. The constraints for the design parameters were as follows:

- Area (up to a maximum of the value detailed for the corresponding PE in Table 8).
- Depth (up to a maximum overall depth of 2.1 m due to practical constraints)
- Depth of gravel layer (up to a maximum of 350 mm due to cost constraints)
- Depth of free space at the surface (up to a maximum of 300 mm)

The optimum target parameter values (*i.e.*, the smallest values) for different PEs are shown in Table 12. As would be expected, the area required to give the lowest number of overflow days is approximately equal to the maximum area constraint for all cases. The overall optimum depth is also equal to the maximum that the constraint in the model would allow (2.1 m) for all loading rates. There is some variance in the depths of the gravel layer determined for each the different PE, ranging from 279 mm to the maximum of 300 mm. The optimal free space depth above the surface is close to the maximum of 300 mm in all cases.

Table 12. Design parameters for willow system determined from optimisation.

PE	Area (m²)	Overall Depth (mm)	Depth of Gravel Layer (mm)	Depth of Free Space (mm)	Overflow Days/yr
2	250	2095	300	295	30
4	498	2100	285	297	30
5	646	2083	279	298	30
6	750	2100	300	300	30
8	991	2100	295	295	30
10	1200	2100	300	300	31

4.2.5. Aspect Ratio

Regression analysis on the field trial results found no conclusive evidence that wind speed had a significant bearing on the ET rate from any willow system (ET_{willow}) with the exception of the 2011 ET_{willow} rate on Site B (p-value < 0.05) and the 2013 ET_{willow} rates on Site A being just outside the 95% confidence interval (p-value $= 0.092$). Equally, sensitivity analysis, in which the relative effect of the different meteorological parameters on ET_o was assessed, revealed that wind speed had a low Spearman Rank-order correlation coefficient with ET_o. Analysis also showed that wind speed had a relatively low influence on the ET_o: for example, a one standard deviation increase in wind speed would result in an increase of just 0.15 mm ET_o, compared to a one standard deviation increase in net radiation which would result in an increase of 1.38 mm ET_o. The effect of wind speed has been reported to have a greater effect on ET_o for taller crops although such research [20] has shown that the wind speed needs to be relatively high (>6 m/s) before any significant increase on expected crop evapotranspiration is gained. The average wind speed at the sites monitored as part of this research project was between 2 and 3 m/s. Hence, the field study results indicate that the impact of wind speed on ET_{willow} is not likely to be a critical factor in such a temperate, high humidity climate (such as Ireland's) and so any benefit derived from tailoring the willow system design to take advantage of wind speed via a large aspect ratio would be modest. However, if space is available on the site, then it is recommended to aim for a high aspect ratio design, as some benefit will be gained from the exposure to higher wind speeds throughout the willow standing.

5. Discussion and Design Criteria

The optimal design dimensions for a willow ET system under different PE loadings (Table 11) indicate that some hydraulic overflow from the sealed systems would be expected during most years due to lower ET levels than the net hydraulic loading from effluent and rainfall. If the wastewater is introduced into the system via distribution pipes in the gravel layer at the base then the effluent will accumulate from the bottom up, compared to the rainwater on the surface which will accumulate from the top down. These hydraulic levels of effluent and rainfall in such an ET system have been simulated on a monthly basis across a hydrological year (*i.e.*, the beginning of October to the end of September) to assess when an overflow would be expected and whether it would consist of primarily rainwater runoff or wastewater, as illustrated in Figure 10. This somewhat simple, yet instructive, analysis assumed that no mixing of effluent and rainwater occurred in the soil within the willow system but also that any water uptake by the willows is preferentially taken from the flooded level downwards (*i.e.*, any ET will preferentially take stored monthly rainfall first before effluent). The depth of the ET system was set at 1.8 m with an overflow at 1.9 m in order to minimise the depth of any standing water and also encourage faster runoff of any rainfall events. The system was designed for a 4 PE residence producing effluent at a rate of 100 L/c.day. The analysis applied a water balance equation using mean annual monthly rainfall and other monitored meteorological data from the weather stations.

The results show that some overflow would be expected from the system during four months of the year (January–April). However, the respective levels of stored effluent from the base and rainfall from the surface indicate that any overflow should be rainwater runoff, which would thus present very little pollution threat to the environment. Indeed, the analysis shows that the majority of the water stored throughout the winter period would be rainwater. From May onwards, the wastewater would begin to make up the majority of stored water, but this would all be contained within the willow system with the maximum depth of wastewater only predicted to reach just over 1 m. It should be noted again that this analysis is based on the conservative assumption that the willow trees would preferentially evapotranspire the rainfall at the top before any deeper wastewater which is unlikely to be the case in reality, as the willow trees would have a preference for the nutrient rich wastewater.

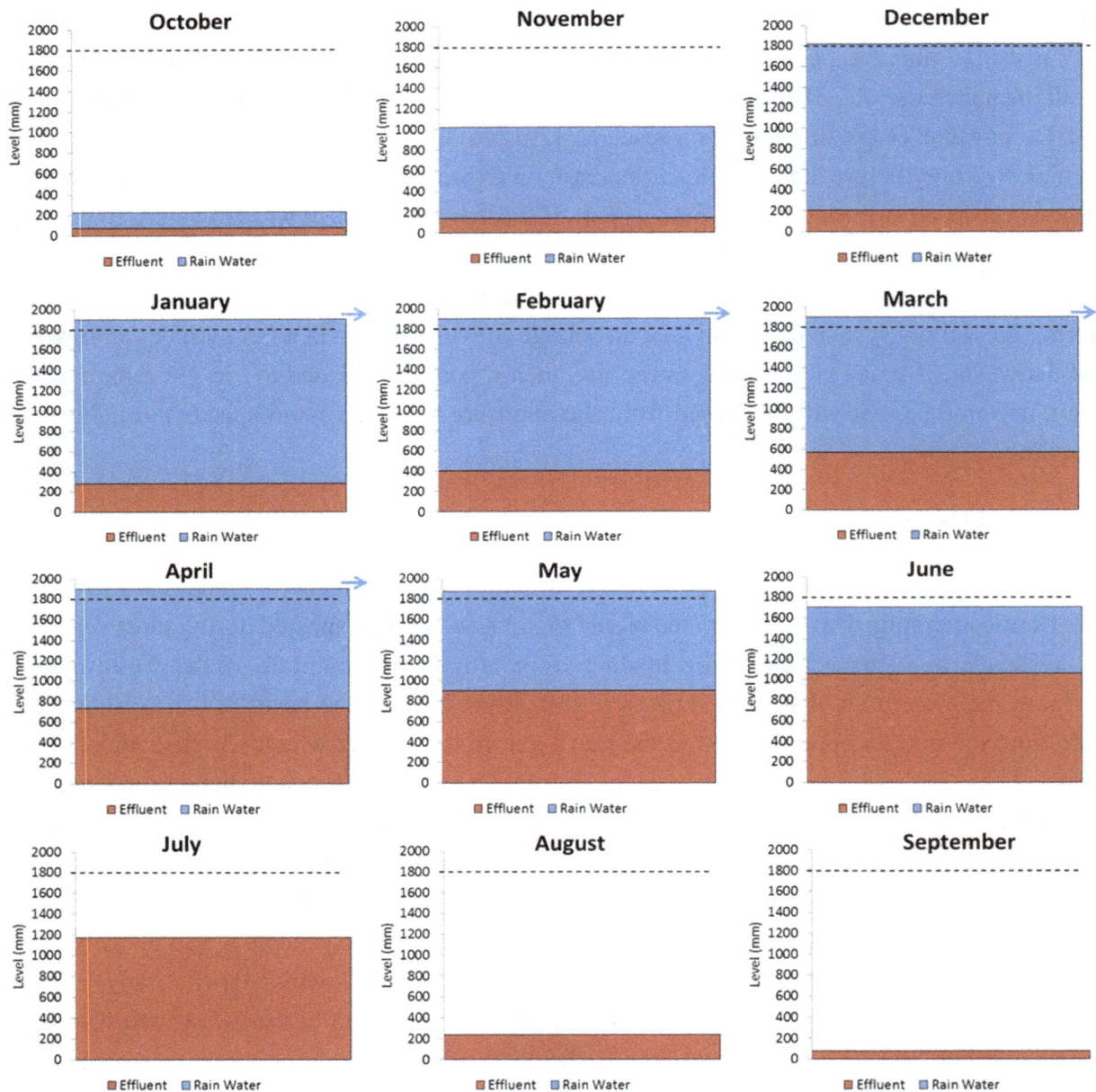

Figure 10. Expected water levels in willow ET evapotranspiration system across an annual cycle fed with effluent from the base.

Hence, this analysis provides a useful insight into the worst-case scenario that could occur and also demonstrates that the system can potentially operate effectively as a zero discharge systems with regards to the effluent loading. It is also important to recognise that this analysis was based upon an effluent production rate of 100 L/c.day (which equates to an aerial hydraulic loading of just less than 300 mm/year). This is similar to the effluent production measured on many of the sites throughout the monitoring period as well as in other research projects in Ireland [15,21] but significantly lower than the 150 L/c.day assumed in the national on-site wastewater Code of Practice [4] for a typical single dwelling. The effect of increasing the effluent loading rate to 150 L/c.day in the simulation is contained in Table 13 where it can be seen that the mean number of overflow days would increase by 60% to 80% for the same area of ET system for any given PE. If such a difference was to be negated by an increase in area, it would require an increase of between 40% and 50%, resulting in a very significant increase in cost.

Table 13. Design parameters for willow systems determined from optimisation.

| PE | Area (m²) | Overflow Days per Year | | Area | |
		100 l/c.day	150 l/c.day	Increased Area (m²)	% increase
2	250	55	85	350	40%
4	500	55	85	675	35%
5	600	51	92	890	48%
6	750	55	85	1020	36%
8	1000	55	85	1350	35%
10	1200	51	92	1770	48%

Hence, it is imperative that, in such a climate where the expected annual ET_{willow} is going to be marginally higher than the expected annual rainfall at best, that wastewater production is kept as low as possible. Any design guidelines should include the requirement for modern water-efficient plumbing and appliances to be installed in the dwelling at the time of construction. It is also imperative that the rainfall runoff from the roof of the house is not plumbed into the wastewater system, as this would cause a significant increase in the hydraulic loading of the ET system. This practice appeared to be happening at Sites A and B with the corresponding enhanced hydraulic loads listed in Table 2.

6. Conclusions

A series of full-scale field trials on willow-based ET systems sited in areas of low permeability subsoil has produced data from which realistic ET values from such systems in the temperate, maritime, Irish climate have been gained. The field trials showed that none of the systems were able to evapotranspire all the effluent and rainfall hydraulic loading which were therefore deemed to be losing water by overflow or exfiltration at some times of the year. The effective porosities from the backfilled low porosity soils (*i.e.*, storage space for effluent and rainfall) were also measured on site and found to be much lower than might typically be assumed for soil void ratios, ranging from just 9% to 20%. However, the quality of water stored in the systems, according to samples taken from the inspection wells, indicate that the systems act as very effective passive treatment systems. Furthermore, samples of ponded water on the surface of systems during winter periods have shown that the quality of any potential runoff would be similar to that of water in systems that had never received any effluent.

These field data have then been used to compare the influence of many design parameters on overall ET performance and so develop advice towards the formulation of generic national design guidelines.

Realistic dimensions of ET systems for the treatment of wastewater from single houses have been established in order to minimize the numbers of days of any potential overflow. The plan areas of these systems are based on an annual effluent loading of just less than 300 mm per year. Further analysis has shown that if the effluent is fed from the base of the ET system designed to such a specification, this should ensure any overflow during winter periods will be predominantly rainfall runoff and that the net annual wastewater production is effectively fully evapotranspired.

Acknowledgments

The research project was mainly financed by Wexford County Council, Ireland with some additional funding from the Environmental Protection Agency, Ireland. The authors also wish to acknowledge the contribution of Brendan Cooney (Wexford County Council) and Arne Backlund (Backlund ApS) to the research.

Author Contributions

The two co-authors contributed equally to this work. Both authors read and approved the final manuscript.

Conflicts of Interest

The authors declare no conflict of interest.

References

1. Central Statistics Office (CSO). *Census 2011, Principal Demographic Results*; Central Statistics Office, Government of Ireland, Stationery Office: Dublin, Ireland, 2012.

2. Hynds, P.D.; Misstear, B.D.R.; Gill, L.W. Development of a microbial contamination susceptibility model for private domestic groundwater sources. *Water Resour. Res.* **2012**, *48*, doi:10.1029/2012WR012492.

3. Environmental Protection Agency (EPA). *A Risk-Based Methodology to Assist in the Regulation of Domestic Waste Water Treatment Systems*; EPA: Wexford, Ireland, 2013.

4. Environmental Protection Agency (EPA). *Code of Practice: Wastewater Treatment and Disposal Systems Serving Single Houses*; EPA: Wexford, Ireland, 2009.

5. Gregersen, P.; Brix, H. Zero-discharge of nutrients and water in a willow dominated constructed wetland. *Water Sci. Technol.* **2001**, *44*, 407–412.

6. Arias, C.A. Current state of decentralized waste water treatment technology in Denmark. In Proceedings of the EPA International Symposium on Domestic Wastewater Treatment and Disposal Systems, Dublin, Ireland, 10–11 September 2012.

7. Miljøstyrelsen. Retningslinier for Etablering af Pileanlæg op til 30 PE; In *Økologisk byfornyelse og spildevandsrensning, Nr. 25*; Miljøstyrelsen: Copenhagen, Denmark, 2003. (In Danish)

8. Rosenqvist, H.; Aronsson, P.; Hasselgren, K.; Perttu, K. Economics of using municipal wastewater irrigation of willow coppice crops. *Biomass Bioenergy* **1997**, *12*, 1–8.

9. Pauliukonis, N.; Shneider, R. Temporal patterns in evapotranspiration from lysimeter with three common wetland plant species in the eastern United States. *Aquat. Bot.* **2001**, *71*, 35–46.

10. Elowson, S. Willow as a vegetation filter for cleaning of polluted drainage water from agricultural land. *Biomass Bioenergy* **1999**, *16*, 281–290.

11. Dimitriou, I.; Aronsson, P. Wastewater and sewage sludge application to willows and poplars grown in lysimeters–Plant response and treatment efficiency. *Biomass Bioenergy* **2011**, *35*, 161–170.

12. Kuzovkina, Y.; Knee, M.; Quigley, M. Soil compaction and flooding effects on the growth of twelve Salix L. species. *J. Environ. Hort.* **2004**, *22*, 155–160.

13. Bialowiec, A.; Wojnowska-Baryla, I.; Agopsowicz, M. The efficiency of evapotranspiration of landfill leachate in the soil-plant system with willow Salix amygadlina. *Ecol. Eng.* **2007**, *30*, 356–361.

14. Kele, B.; Midmore, D.J.; Harrower, K.; McKenniary, B.J.; Hood, B. An overview of the Central Queensland University self-contained evapotranspiration beds. *Water Sci. Technol.* **2005**, *51*, 273–281.

15. Gill, L.W.; O'Luanaigh, N.; Johnston, P.M.; Misstear, B.D.R.; O'Suilleabhain, C. Nutrient loading on subsoils from on-site wastewater effluent, comparing septic tank and secondary treatment systems. *Water Res.* **2009**, *43*, 2739–2749.

16. Department for Environment, Food and Rural Affairs (DEFRA). *Growing Short Rotation Coppice, Best Practice Guidelines*; Defra Publications: London, UK, 2002.

17. Caslin, B.; Finnan, J.; McCracken, A. *Short Rotation Coppice Willow Best Practice Guidelines*; Teagasc, Crops Research Centre: Carlow, Ireland, 2010.

18. Curneen, S.; Gill, L.W. Evapotranspiration Systems using willows for the treatment of on-site wastewater effluent in areas of low permeability subsoils. *Ecol. Eng.* **2015**, in press.

19. Guidi, W.; Piccioni, E.; Bonari, E. Evapotranspiration and crop coefficient of poplar and willow short-rotation coppice used as vegetation filter. *Bioresour. Technol.* **2008**, *99*, 4832–4840.

20. Allen, R.G.; Pereira, L.S.; Raes, D.; Smith, M. Crop Evapotranspiration—Guidelines for computing crop water requirements. *FAO Irrig. Drain. Paper* **1998**, *55*, 227.

21. Dubber, D.; Gill, L.W. Application of on-site wastewater treatment in Ireland and perspectives on its sustainability. *Sustainability* **2014**, *6*, 1623–1642.

3

Synchronous Oscillations Intrinsic to Water: Applications to Cellular Time Keeping and Water Treatment

D. James Morré * and Dorothy M. Morré

MorNuCo, Inc., 1201 Cumberland Avenue, Suite B, Purdue Research Park, West Lafayette, IN 47906, USA; E-Mail: dj_morre@yahoo.com

* Author to whom correspondence should be addressed; E-Mail: dj_morre@yahoo.com;

Academic Editor: Marc Henry

Abstract: A homodimeric, growth-related and time-keeping hydroquinone oxidase (ENOX1) of the eukaryotic cell surface capable of oxidizing intracellular NADH exhibits properties of the ultradian driver of the biological 24 h circadian clock by exhibiting a complex 2 + 3 set of oscillations of copper salts and appear to derive from periodic variations in the ratio of ortho and para nuclear spins of the paired hydrogen atoms of the elongated octahedral structure of the ENOX1 protein bound copper II hexahydrates. A corollary of these observations is that the ortho/para oscillations must occur in a highly synchronized matter. Our findings suggest that water molecules communicate with each other via very low frequency electromagnetic fields and that these fields also appear to be generated by the energetics of the synchronous ortho to para interconversions of the nuclear spin pairs of the water hydrogens. Further evidence for energy absorbed and emitted by water and correlated with ortho/para oscillations of ortho/para spin pairs of water hydrogens is indicated from the auto-oscillations in water luminescence. The emissions oscillate with period lengths of 18.8 min that agree with our previously found period of oscillation of about 18 min for pure water, reflective of ortho to para spin isomers based on measurements of redox potential. The period length of pure water (increased by about 25% in D_2O) and varies depending on the dominant cation present (copper salts in solution are unique in that the period length is exactly 24 min). Synchrony is maintained through generation of and response to LFEMF generated by the ortho-para spin pairs. Changes in redox potential sufficient to catalyze NADH oxidation were used to monitor synchronous water oscillations that appear to extend indefinitely over great distances in contiguous bodies of either still or flowing water. Adjacent out-of-phase water samples contained in thin plastic cuvettes auto-synchronize in a matter of seconds when placed side by side. Potential applications from water treatment along with opportunity related to human health are anticipated to derive from a better understanding of how water synchrony is generated and maintained, and to be aided by methodological advances in measurement and analysis.

Keywords: oscillations intrinsic to water; oscillations in ortho-para nuclear spins of paired hydrogens of water; time-keeping; water coherence; water treatment; electrodermal sensing

1. Introduction

Over 70% of the human body is water. Water participates in virtually every function of life. It is essential to most biochemical reactions, a predominant constituent of bodily fluids—blood, lymph, cerebrospinal fluid, saliva and other digestive secretions, and is essential for joint lubrication, for detoxifications, and for blood pressure maintenance. More than just a solvent, water has a complex and dynamic structure that is sensitive and responsive to solutes and to its surroundings. Both water and living systems appear to be equally sensitive even to a single quantum of magnetic flux [1]. In addition, water exhibits long-range ordering features hitherto suspected but largely undocumented.

In this report, we provide evidence for the concept that physical properties of water are the responsible time keepers for the cells' biological clock. Other external clock interacting influences, *i.e.*, day-night cycles, geomagnetic fields, melatonin, *etc.* phase, but do not necessarily drive, clock related rhythms. If physical properties of water serve as the time keeper of cellular clocks, then what are the fundamental properties of water that might be responsible? Water would be an ideal medium to serve as a universal time keeper. It permeates every region of the cell and is capable of interaction with both small and large molecules.

This universal property of water in biological time-keeping may also contribute to the maintenance of synchrony of most, if not all, physical, chemical and biological processes involving water. An application to water treatment along with implications related to human health are discussed along with evidence for synchrony with systems of contiguous water molecules as well as the physical basis for how the fundamental water energy oscillations are generated and synchronized. We will summarize evidence that the underlying phenomena for water oscillations important to cellular time keeping are the result of a largely unrecognized property of water, that of oscillations in the ratios of ortho to para spin pairs of water hydrogens.

2. Time-Keeping ENOX Proteins

A link to tie water to cellular time-keeping was the result of the discovery in our laboratory of a family of growth-related cell surface ENOX proteins with alternating NADH oxidase and protein disulfide-thiol interchange that oscillated with a period length of 24 min [2–4]. All are located on the external surface of the plasma membrane and are shed into the circulation [5]. Site directed mutagenesis and overexpression studies correlated changes in period lengths of the activity oscillations of the ENOX proteins directly with corresponding changes in cellular circadian rhythms [6,7]. More importantly, the period lengths of the activity oscillations of the ENOX proteins where water was replaced by D_2O were increased from 24 min to 30 min [5]. Organisms grown in D_2O consistently exhibit circadian periods of about 30 h compared to the normal period length of 24 h [8]. Further studies revealed a requirement for bound copper[II] to sustain the periodicity of the ENOX proteins [9,10]. When examined in the absence of protein, copper[II] hexahydrate in aqueous solution also revealed oscillatory changes in redox potential sufficient to catalyze the oxidation of NADH with a period length of 24 min.

A characteristic of the oscillatory pattern of the time-keeping, clock-related ENOX proteins that distinguishes it from all other known oscillatory phenomena, has been its characteristic asymmetry (Figure 1). Each period is defined by a series of five maxima, two of which, designated ① and ② are strictly separated by 6 min and three of which, designated by ③, ④ and ⑤ are separated from each other and from the adjacent maxima ① and ② by 4.5 min [6 min + (4 × 4.5 min) = 24 min].

The ENOX proteins are further characterized by having two distinct biochemical activities both of which are essential to their functioning in the growth process—that of NADH or hydroquinone oxidation (Figure 2A) and that of protein disulfide-thiol interchange (Figure 2B). With NADH or hydroquinone oxidation, maxima ① and ② are most evident (Figure 2A) whereas with protein disulfide-thiol interchange, assayed from the cleavage of an artificial dithiodipyridine (DTDP) substrate, maxima ③, ④ and ⑤ dominate. Figures 2–4 are averages of 3 to 6 decomposition fits [11] ± standard deviations that trace the specific 2 + 3 oscillations from ENOX proteins to water as the ultimate source. For consistency, the data are generated in the same manner for each figure following the methods for measurements of NADH oxidation (Figure 2A) where the decrease in A_{340} from spectrophotometer traces averaged over exactly 1 min are taken at intervals of 1.5 min (for details see [5]).

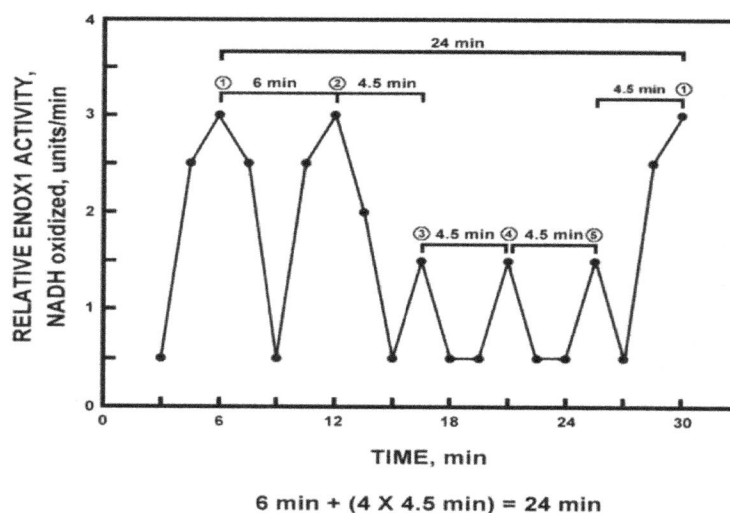

Figure 1. Diagrammatic representation of the signature ENOX cycle with measurements averaged over 1 min at intervals of 1.5 min. Typically there are five maxima, two of which ① and ② are separated by 6 min and three of which ③, ④ and ⑤ are separated from each other and from ① and ② by 4.5 min. The asymmetry defines a 24 min period (6 min + (4 × 4.5 min)). Reproduced from [4] with permission from Springer Science + Business Media.

All ENOX proteins require bound Cu^{II} for activity [9,10]. Activity is lost when the ENOX proteins are unfolded in the presence of the copper chelator bathocuproine and then refolded. Unfolding and refolding in the presence of Cu^{II} and absence of chelator were without effect on activity. In control experiments, the oscillatory activity was recovered quantitatively with the removed Cu^{II} in preparations devoid of protein, as measured either as a change in redox potential sufficient to drive NADH oxidation (Figure 3A) or by NADH oxidation per se (Figure 3B). Both showed a similar oscillatory pattern with a precise 24 min period recapitulating the periodicity of NADH (hydroquinone) oxidation/protein disulfide-thiol exchange of the Cu^{II} bearing ENOX proteins (Figure 2). The principle difference was that Cu^{II} alone did not catalyze protein disulfide-thiol interchange such that each of the five maxima for NADH oxidation were of nearly equal magnitude (Figure 3B).

Thus, the copper bound to the ENOX protein, as a Cu^{II} hexahydrate, emerged as the driver of the protein oscillations, a conclusion supported by Extended X-ray Absorption Fine Structure Spectroscopy (EXAFS) measurements using the Advanced Photon Source of the Argonne National Laboratories [12].

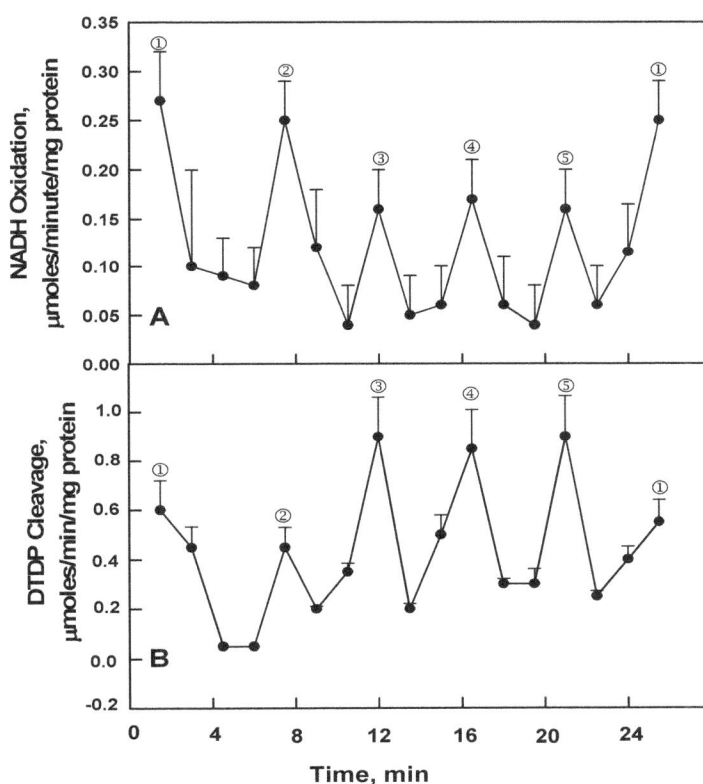

Figure 2. Averages of 3 (B) or 4 (A) decomposition fits [11] ± standard deviations of oscillatory patterns of human ENOX1 activity with a period length of 24 min. (**A**) NADH oxidation measured as the decrease in A_{340} averaged over 1 min every 1.5 min; (**B**) Protein disulfide-thiol interchange activity measured from the cleavage of the artificial substrate dithiodipyridine (DTDP). Note that the two activities tend to alternate with maxima ① and ② separated by 6 min enhanced for NADH oxidation whereas maxima ③, ④ and ⑤ separated from each other and maxima ① and ② by 4.5 min are more conspicuous for DTDP cleavage.

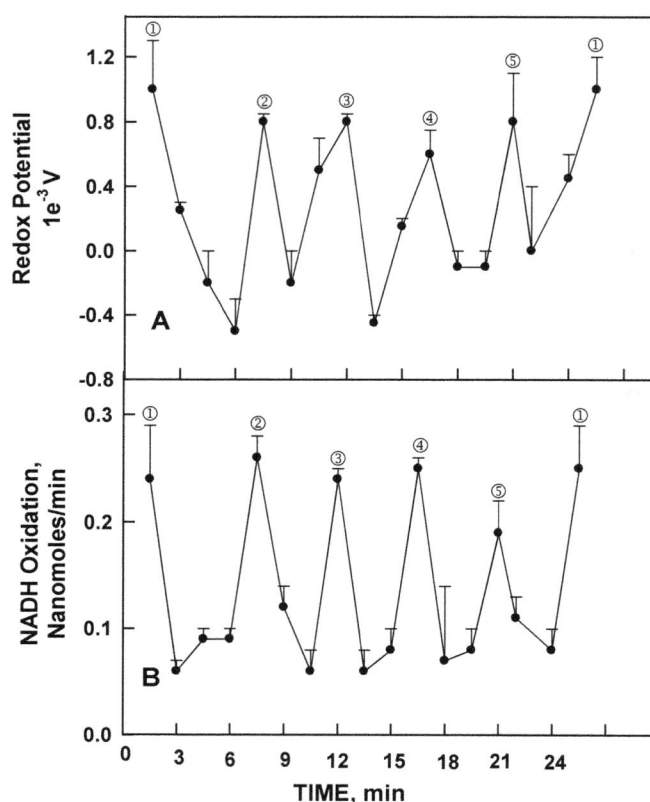

Figure 3. Averages of 3 (A) or 4 (B) decomposition fits [11] ± standard deviations for changes in redox potential (**A**) and rate of NADH oxidation measured as for Figure 2A (**B**) for 10 (A) or 25 (B) mM solutions of $Cu^{II}Cl_2$ each with a period length of 24 min. The asymmetric oscillatory pattern of Figures 1 and 2 are recapitulated. The major difference being that $CuCl_2$ solutions are unable to catalyze protein disulfide-thiol interchange so that all maxima are of nearly equal magnitude.

Ultimately, similar oscillations in both redox potential and in NADH oxidation were observed in pure water with a period length of about 18 min (24 min for D_2O) and also confirmed by EXAFS [13]. Pure water (HPLC grade) essentially devoid of metal cations, exhibited a pattern of alternation in redox potential with the first two maxima separated by six min (Figure 4A), sufficient to drive NADH oxidation (Figure 4B). The principle feature is that the period length was about 18 to 20 min as compared to 24 min for ENOX proteins and aqueous solutions of Cu^{II}, with the method of rate determinations being based on averages over 1 min at intervals of 1.5 min. With this method, however, maxima ④ and ⑤are often incompletely resolved. However, continuous measurements or sequential measurements at 1 min intervals did resolve maxima ④ and ⑤ for water although not as widely separated as for the 24 min oscillations. Maxima labeled ① and ② were phased by exposure to low frequency electromagnetic fields (LFEMF) [14] with synchrony maintained over extended periods of time. The length of the period was independent of temperature suggesting a physical rather than chemical basis as the source of the oscillations [3].

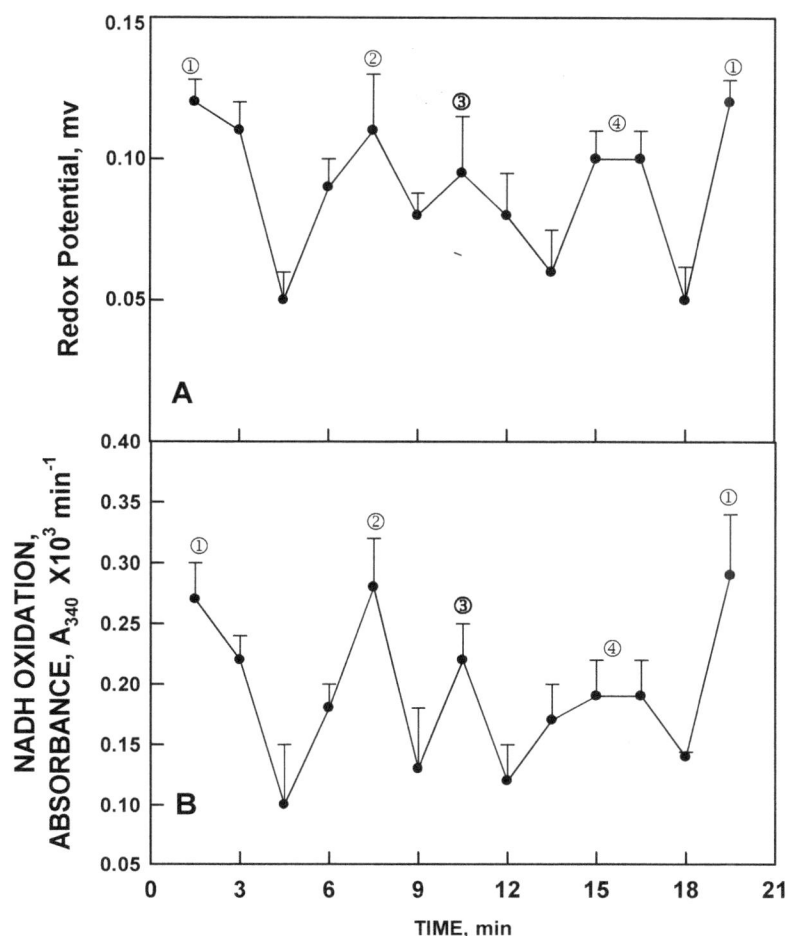

Figure 4. Decomposition fits ($n = 3$) as in Figure 3 except for pure water. The oscillatory pattern is similar except that in the absence of Cu^{II}, the period length is reduced from 24 min to about 18 min and maxima ④ and ⑤ are incompletely resolved to yield a 4 maxima pattern with (**A**) maxima ① and ② separated by 6 min and (**B**) maxima ③ and ④ separated from each other and from maxima ① and ② by 4 min.

3. Disequilibrium of Ortho: Para Spin States in Liquid Water that Oscillate

That the underlying mechanism of the biological clock was physical rather than chemical, provided an explanation for two characteristics universally associated with biological timing mechanisms that have been increasingly overlooked in recent years in the molecular analyses of circadian clock mechanisms. One of these, never adequately explained, is the temperature independence of the period lengths of circadian rhythms [15]. Temperature independence implies an underlying physical rather than a chemical or a metabolic basis. The second was that heavy water, deuterium oxide, was the only substance known that consistently changed the length of the circadian day in a wide range of organisms from a 24 h day to a *ca.* 30 h day [8]. While the D_2O observations clearly implicated a role for water in biological time-keeping, at the time, there was nothing associated with the atomic structure of water sufficiently slow or well characterized to link a time-keeping function to water.

A clue emerged with the report of Tikhonov and Volkov [16] where equilibration of pure ortho water vapor required 4 h to approach a 3:1 ratio of ortho to para water and appeared to oscillate with a period length of about 18 min in doing so.

Evidence that water vapor is a mixture to two independent fractions of ortho and para spin isomers is consistent with the existence of two independent states of liquid water with long life times and two hydrogen bond types differing in energy as reported by Pershin [17]. Specifically, the energy of the hydrogen bond between ortho isomers of molecules which always rotate was seen to be lower than between para molecules, a part of which cannot rotate at room temperature. The conclusion was reached that this overheating—overcooling process observed by Pershin [17] was a fundamental property of water approximated by a harmonic function without noticeable damping.

Subsequent studies have contributed evidence of ortho-para spin conversions [13,18–21] evidence that was missing from our information prior to the report of Tikhonov and Volkov [14].

3.1. Spectral Evidence

In our own work [13], FTIR spectroscopic measurements of ortho-H_2O/para-H_2O interconversions were carried out in the middle infra-red spectra region above a water sample surface determined at 3801 and 2779 cm^{-1} respectively in a manner similar to those for NAD(P)H oxidation and redox potential described above (Figure 5).

The ratios of the two wave lengths exhibited a repeating pattern of oscillations which when analyzed by decomposition fits using an imposed period length of 18 min revealed the typical oscillatory pattern of the biological clock with recurrent maxima, two of which labeled ① and ②, were separated by about 6 min and either 2 or 3 additional maxima separated in time by 3–4 min. Maxima labeled ④ and ⑤ are incompletely resolved and ③ and ④ + ⑤ are separated from each other and from maxima ② by 4 min (6 min + (3 × 4 min) = 18 min) which may represent the more typical water pattern. For water, maxima are either very close together or unresolved.

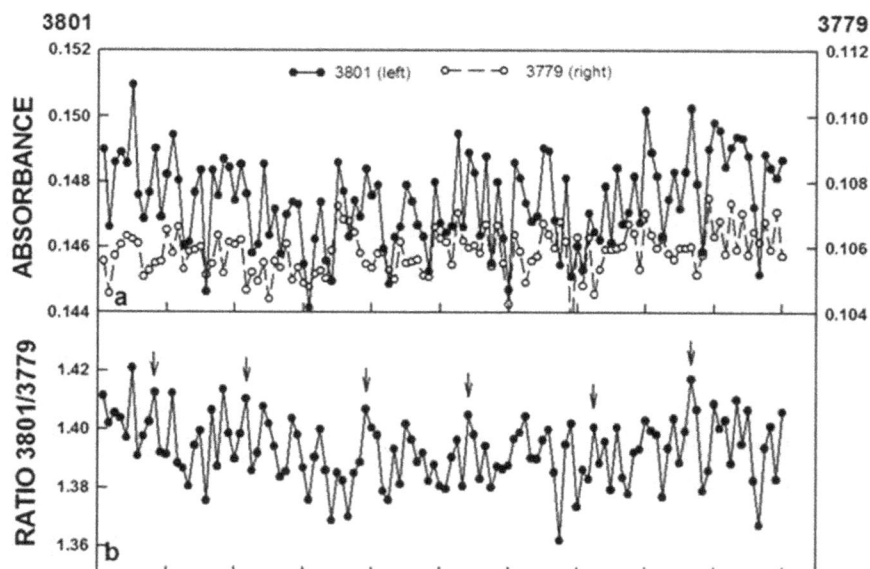

Figure 5. *Cont.*

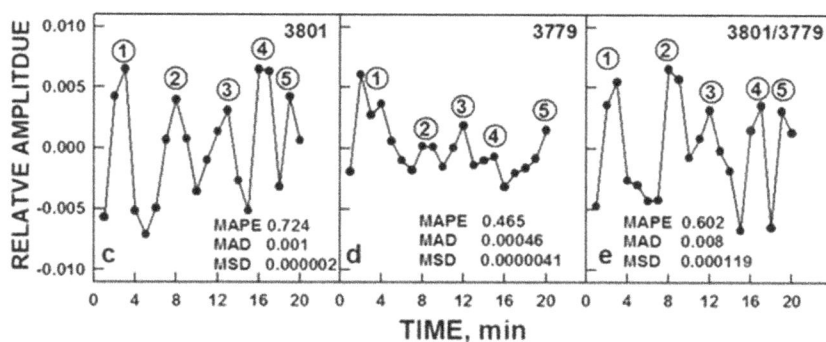

Figure 5. FTIR spectroscopic measurement of the ratio of para-H_2O/ortho-H_2O above a water sample surface determined at 3801 and 3779 cm^{-1}, respectively. (**a**) The ratio of the two wave lengths exhibited a repeating pattern of oscillations of five maxima at intervals of about 18 min (arrows); (**b**) Decomposition fits using an imposed period length of 18 min of data collected at 3801 cm^{-1}; (**c**) and at 3779 cm^{-1} (**d**), as well as the ratio of the two, revealed the oscillatory pattern typical of water with five recurrent maxima, two of which, labeled ① and ②, were separated in time by about 4.5 min and three additional maxima separated in time by 3.4 min, labeled ③, ④ and ⑤. The accuracy measures, MAPE, MAD, and MSD are indicative of a close fit between the original and the fitted data. Reproduced from [13] with permission from Elsevier.

3.2. Water Luminescence a Result of Ortho-Para Oscillations

In the study of Gudkov *et al.* [22], air-saturated double distilled water was exposed for 5 min to low-intensity laser infrared radiation at the wavelength of the electronic transition of dissolved oxygen to the singlet state (1264 nm). After a latent period of more than 2 h, auto-oscillations of water luminescence in the blue-green region were observed over several h without indications of damping. The laser-enhanced auto-oscillations were not observed if water was irradiated beyond the oxygen absorption band or if oxygen was removed from the water. Wavelet transform analysis of the luminescence records indicated two characteristic periods of pulsations of about 300 and 1150 s. These times correspond to the oscillation periods of the concentration ratios of the ortho and para spin isomers of water molecules reported by us [3].

The assumption that follows is that the oscillations in luminescence are the result of an alternation of the two energetic water states (ortho and para) with luminescence augmented auto-catalytically on the laser-induced steady state formation of singlet oxygen. The oscillations in the luminescence data may result from energy discharged as the synchronized paired hydrogens of the water transition from their high energy to their low energy states. Two such different energetic states would be consistent with alternation of the ratios of ortho and para nuclear spin isomers. Thus, water luminescence may represent the release of energy in the form of light as the highly synchronized ortho-para disequilibration returns from the high energy to the low energy state.

The latter supposition is supported by observations that oscillatory changes in redox potential and NADH oxidation in pure water occur in solutions purged of dissolved oxygen and occur in a manner where both amplitude and period length are independent of the concentration of dissolved oxygen and are unresponsive to infra-red irradiation..

We observed a similar phenomenon in pure water in the absence of excitation by laser radiation although at an intensity *ca.* 25% of that following laser excitation and presumably not requiring involvement of singlet oxygen in the excitation process (Figure 6). Decomposition fits of those data show a clearly defined pattern of oscillations with the two of the maxima labeled ① and ② separated by an interval of 6 min. With the moving-averaged data collected at 1 s intervals of Figure 7, the decomposition fit clearly resolved the primary data into four maxima where two of the maxima were separated by 6 min following the same pattern as described for Figure 4.

3.3. Limit Oscillator Model

An appropriate model to provide sustained ortho-para oscillations comes from studies of limit oscillators [5] of which control of heart beat rate is the classical example. That certain materials may preferentially adsorb para water due to non-rotational ground state [22] implies that populations of water molecules might spontaneously form some sort of collective order whose structure subtly favor formation of para water. In the manner of limit oscillators, as para water forms, it might locally add to the field generated by the collective order which would further favor the accumulation of para water for some period of time. However, also, in keeping with the limit oscillator model, the initial conditions favoring para water formation would produce an increased potential that would be released or discharged only after some minimum threshold of activation energy was achieved. Once the minimum threshold was achieved, a discharge of potential and conditions favorable to ortho water formation would result and the process would now run for a time in the opposite direction until some threshold level of ortho water was restored and the cycle would reverse.

The conclusion was reached that this overheating-overcooling process may be a fundamental property of water, which manifests itself at any temperature and is not the result of perturbation of overcooled water by an optical pulse. Pershin's [17] findings show "*that such an evolution of the band center is steadily observed in water and at room temperature. Moreover, this overheating-overcooling process is approximated by a harmonic function without noticeable damping*".

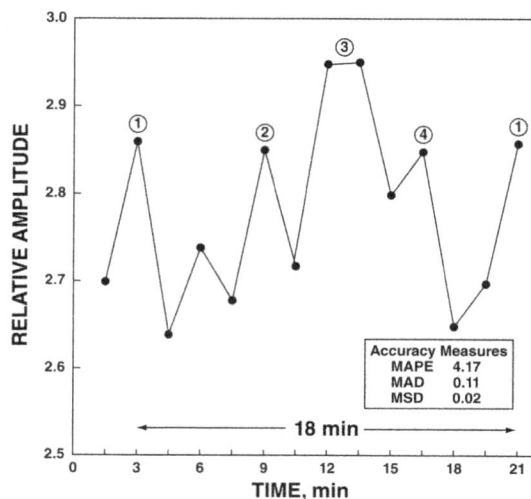

Figure 6. Decomposition fit [11] of luminesce data of Gudkov *et al.* [23] collected at intervals of 1 sec averaged over 1 min every 1.5 min prior to infrared laser activation. Maxima labeled ① and ② are separated by 6 min as for Figure 4.

The differences in energies of the different spin isomers of water are small (less than 10^{-24} erg) [24] and are much lower than the energy of thermal motion. Therefore, a spin-only interaction would not be expected to affect intermolecular interactions. On the other hand, the absorption rates of ortho and para water from water vapor to various organic and inorganic sorbents have been observed to differ markedly with the binding of the para isomer. Estimates by these authors of the energy barriers that determine rates of absorption suggest that the difference in free energy barriers may exceed the energy of spin-spin and spin-orbit interaction by many orders and raises the possibility that the spin state of water may substantially influence physical, chemical and biological phenomena in parallel to changes in redox potential.

The method of complex wavelet transforms may be the most adequate method for analysis of problems of this kind [25]. Periods of approximately 18 and 5 min have been observed previously by the wavelet transform analysis method [26].

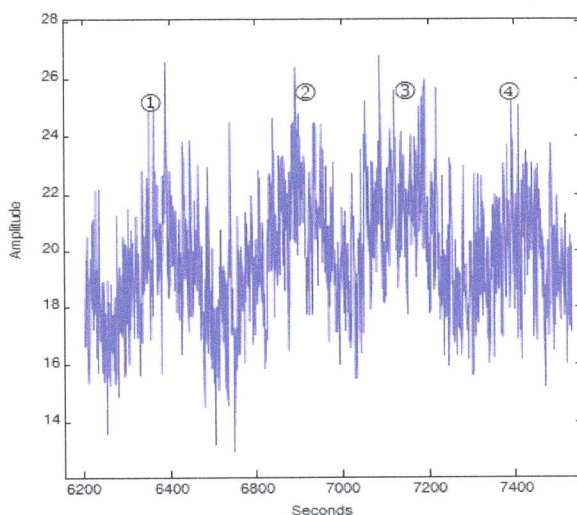

Figure 7. Decomposition fit [11] of moving averages of luminescence data of Gudkov *et al.* [18] collected at 1 sec corresponding to a Fourier signal at 2874 s = *ca.* 4.8 min revealed within each 18.8 min period consisting of four unevenly spaced maxima, two of which, labeled ① and ② were separated by six min and two of which ③ and ④ were sparated from each other and from maxima ① and ② by about 4 min (6 min + 3 (4 min) = 18 min) recapitulating the overall pattern of Figure 4.

3.4. Oscillations in NADH Oxidation are Driven by Oscillations in Redox Potential

Rates of NADH oxidation correlated with fluctuations in continuous traces of redox potential, the latter being sufficient to drive the oxidation of NADH (Figure 8). Figure 8 illustrates simultaneous measurements of redox potential and NADH oxidation for a solution of copper chloride. For both, the pattern consists of five maxima two of which are separated by 6 min. The redox measurements are from a continuous trace (Figure 8A) whereas the rates of NADH oxidation were measured in parallel over 1 min every 1.5 min to improve resolution (Figure 8B). Two of the maxima, ① and ② in the figures are separated by 6 min. The remaining three maxima are separated from each other and from maxima ① and ② by 4.5 min. These intervals confer the characteristic asymmetry to the pattern of oscillations, and for solutions of Cu^{II}, repeat every 24 min.

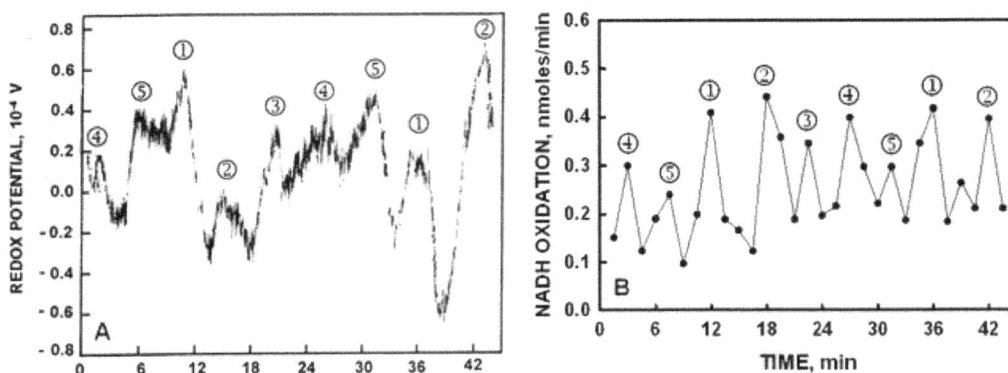

Figure 8. The redox potential (**A**) of an aqueous solution of copper chloride measured continuously showing a 2 + 3 pattern of oscillation with NADH oxidation measured over 1 min at intervals of 1.5 min (**B**) in parallel. The period length of both is 24 min. Results with pure water are similar except that the period length is now about 18 min.

3.5. Oscillations not Unique to Copper II

Chlorides of cations other than Cu^{II} in solution also exhibited oscillations but the period lengths were longer or shorter than those of Cu^{II} [3]. The property of these different cations that correlated with the period length of the oscillations was ionic radii (Figure 9). Period lengths of the oscillations of an aqueous solution were in direct proportion to the ionic radius of the cation and independent of the concentration.

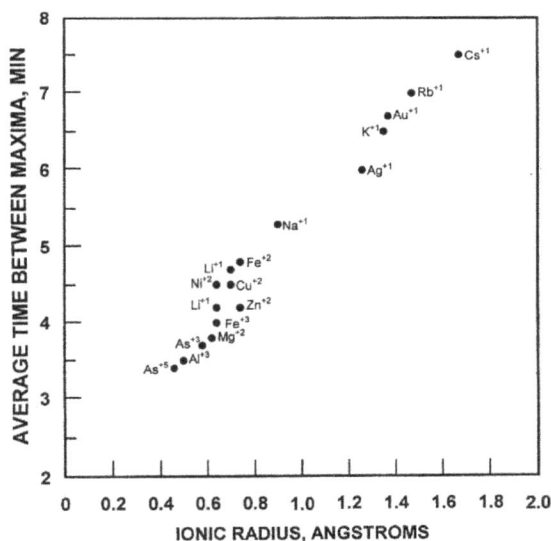

Figure 9. Period length of the oscillations of an aqueus solution is directly proportional to the ionic radius of the cation present and independent of cation concentration. All solutions were tested as the chloride at a final concentration of 10 μM. Only with Cu^{2+} (replaceable by Ni^{2+}) was the asymmetric period length of 24 min observed. Reproduced from Morré and Morré [3] with permission from Elsevier.

3.6. The Redox Potential and Rate of NADH Oxidation Still Oscillate even with Pure Water in the Absence of Cations

Pure water exhibits periodic oscillations in redox potential and NADH oxidation even in the absence of cations. However, the overall period length is about 18 min. As with Cu^{II} solutions, maxima ① and ② are separated by 6 min and the remaining maxima are truncated as discussed for Figure 4.

3.7. Water Oscillations are Highly Synchronous

As reported by Morré and Morré [27], synchronization of asynchronous water samples was achieved through communication through a thin plastic barrier. When two samples of HPLC grade water contained in plastic spectrophotometer cuvettes, with one sample phased using LFEMF (30 s 40 µT) in a manner to be out of phase with the original water sample, were placed adjacent to each other for only a few min, both samples began to oscillate in parallel. Thus, synchronization of asynchronous water samples was achieved by communication through the thin plastic barrier. When the two samples placed side by side were separated by a thin barrier of sheet copper sufficient to block LFEMF, the phasing of the two water samples was prevented. To test the prediction that contiguous water samples oscillate in phase through LFEMF communication, water was sampled in a shallow pond at a distance of 100 ft apart. Also, water was sampled at opposite ends of a 20 acre lake. As reported previously [26], in both instances, the samples were synchronous in their oscillatory pattern. Two samples of West Lafayette, Indiana tap water collected simultaneously at two different locations separated by a linear distance of 5 miles (8.3 km) also were synchronous.

Water samples from a flowing stream at $t = 0$ and then 15 min later were synchronous. It was estimated that at least 500,000 gallons of water had passed the test site between sample collections. The largest bodies of fresh water samples have been from Lake Ontario and from the Niagra River with sampling points separated by approximately 20 miles for each. Samples from both were synchronous. Simultaneous sampling of water from the Pacific Ocean at San Diego, California and of the Atlantic Ocean at Ocean City, Maryland, also revealed the water at the two sampling points to be in phase with each other (Figure 10). In these measurements, the period length was approximately 36 min reflective of sodium being the dominant cation (see Figure 9). These findings, which were confirmed in two separate experiments several months apart, support the concept of water coherence advanced by others [28,29] but to an extent greater than that which may have been previously anticipated.

Frequency of the EMF associated with coherent water is estimated to be 1.2 µm in the infrared [30]. One coherent domain "*speaks*" to another domain to align the contained water molecules. The frequency of transmission might be a few kHz. However, the waves are trapped within the water (generated and absorbed by contiguous water molecules) and, for the most part, do not escape [30]. A seeming exception is access across a plastic or glass surface where transmission apparently does take place to effect synchrony of water in an adjacent plastic or glass container.

A more puzzling exception was subsequently encountered in our attempts to understand how a unique water purification system developed by Grander Wasserbelebung of Jochburg, Austria might function.

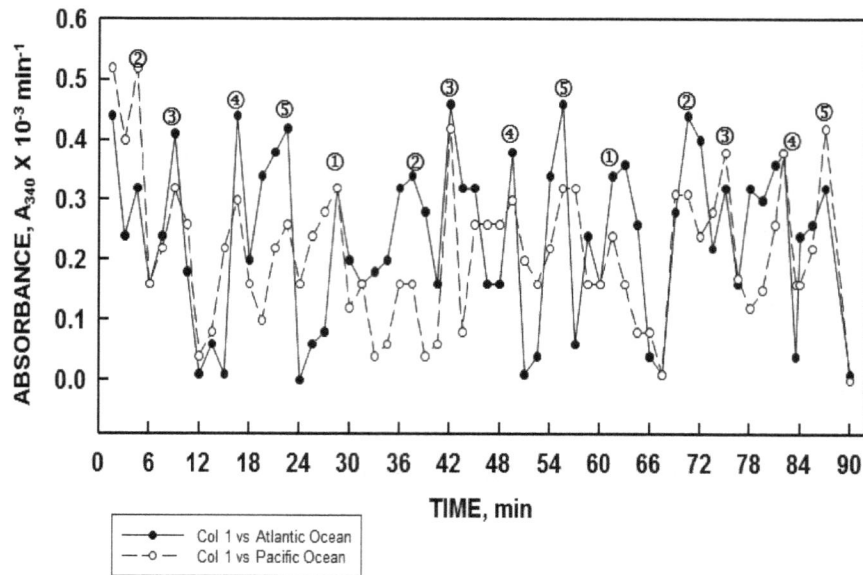

Figure 10. Water sampled simultaneous from the Atlantic Ocean at Ocean City, Maryland (solid line, closed symbols) and from the Pacific ocean at San Diego, California (dashed line, open symbols) exhibited similar synchronous oscillatory patterns when analyzed in parallel.

4. Trans-Metal Driven Direct Contact Water Communication

The Grander water purification device is based on the principle that water inside a metal chamber causes a fundamental alteration in the properties of the water passing though the device (Figure 11). Industrial water passing through the device containing the pure (Grander) water on the inside of the stainless steel encased chamber exits with properties of pure water after only a few seconds of transit through the device. An alternative device is a copper coil where the pure (Grander) water is contained inside the coil and the water to be purified is placed on top of the coil in a glass or plastic container.

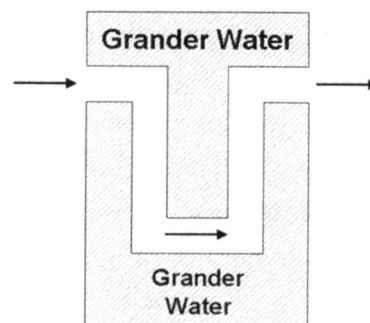

Figure 11. Diagrammatic representation of the Grander water purification device constructed of stainless steel. A transit time through the device of less than 1 min is sufficient to effect the transition.

Some cited benefits of Grander-treated water include: More palatable to drink, bacteria will not grow, reduction in scale buildup, increased service life of equipment, reduced allergies and skin problems, use of less soap or detergents when washing clothes and elimination of the need for a water softener.

In our initial experiments, the device consisting of the copper coil was used. The test sample in a plastic container containing 120 µM NADH as the redox indicator to measure the length and phase of the period was placed over the copper coil with no liquid inside (empty) or with water or copper chloride (100 µM) inside. With a coil without liquid inside, there was no effect on the length or phase of the water in the test solution placed over the coil (Figures 12A and 13A). When the coil contained Grander or HPLC water and the solution in the plastic container was 100 µM CuCl$_2$, the result was that the water in the plastic container exhibited the 4-maxima pattern with an 18 min period characteristic of pure water in the coil (Figure 12B) and not the 5 maxima pattern with a 24 min period characteristic of CuCl$_2$ actually present in the container. Alternatively with a coil containing 120 µM CuCl$_2$ and with Grander or HPLC water placed over the coil, the Grander or HPLC water in the plastic container assumed the 5-maxima pattern and 24 min period characteristic of the CuCl$_2$ solutions (Figure 13B). In all of these experiments Grander water or HPLC grade water produced indistinguishable responses. Whatever water, whether Grander or HPLC, or CuCl$_2$ prepared in either Grander or HPLC water, was placed in the container on top of the coil, it assumed the same frequency and phase of oscillations characteristic of the water inside the coil with no change in ionic or solute composition. Impure water, exposed to pure water inside the coil, behaved as pure water. Less than 1 min exposure was required to affect the transformation.

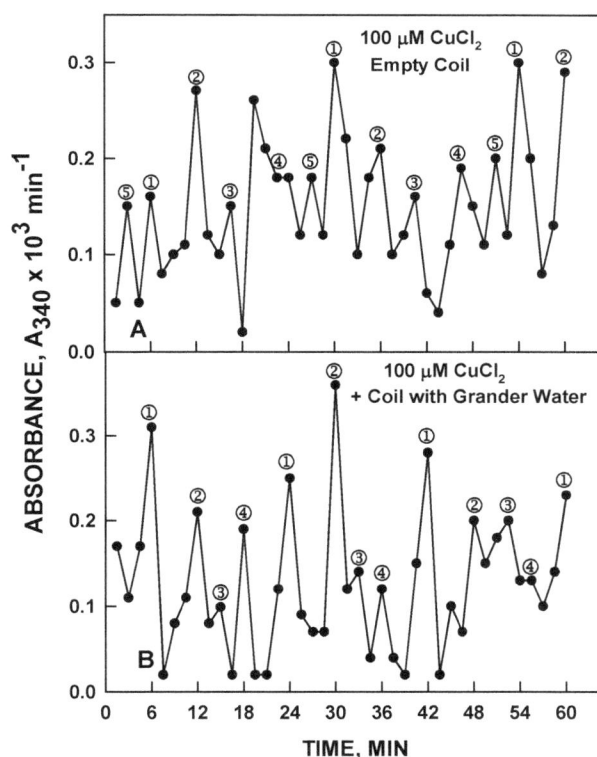

Figure 12. NADH, 120 µM, was prepared in water or 100 µM CuCl$_2$ with a standard copper coil filled with Grander water. The pattern shown in (**A**) was observed. The empty coil had no effect. With the coil containing Grander water, the result was the 4-peak pattern with the 18 min period of water shown in (**B**) and not the 5 peak pattern with a 24 in period characteristic of 100 µM CuCl$_2$ shown in A. Results with HPLC water in the coil were equivalent.

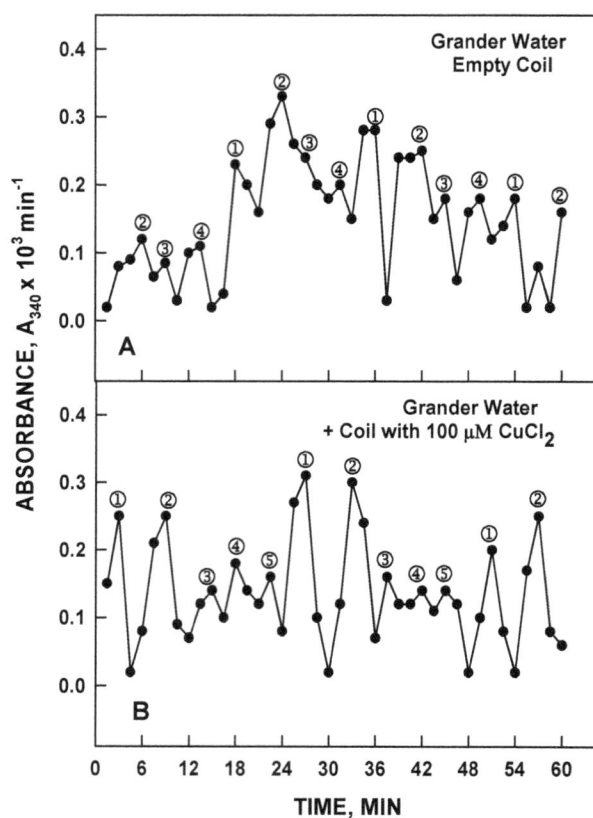

Figure 13. The reverse experiment gave the reverse result. With HPLC water (4 maxima, 18 min period) (**A**) placed over the coil now filled with 100 μM CuCl₂, the result was that the HPLC water now oscillated with the 5 peak 24 min period characteristic of the 100 μM CuCl₂ (**B**).

The metal of the chamber needed not be copper nor was it necessary for the device to be in the form of a coil. Identical results were obtained with a fluid-filled chamber of stainless steel similar to the Grander device (Figure 14). With the CuCl₂ solution placed over the stainless steel chamber containing Grander or HPLC water, the CuCl₂ solution now oscillated with the 4 maxima pattern and 18 min period typical of HPLC or Grander water instead of the maxima pattern and 24 min period characteristic of the CuCl₂ solution (Figure 14B). The inverse experiment with CuCl₂ inside the chamber and HPLC or Grander water placed on top resulted in the water exhibiting the oscillatory pattern of CuCl₂.

As measured with a shielded DC magnetometer, water inside the coil or chamber generated a *ca.* 7 milligaus LFEMF at the surface of the coil or chamber. A thin sheet of copper foil interposed between the coil or chamber and the external sample blocked the communication consistent with an LFEMF signal. It would appear that the LFEMF at the coil or chamber surface is of the same frequency and phase as the aqueous liquid inside the coil or chamber and is capable of transmitting that frequency to aqueous liquids external to the metal coil or chamber.

How does water at the surface of a metal barrier communicate with the metal of the barrier to generate the measured LFEMF at the outer barrier surface? A possible answer is that the electromagnetic carrier wave from the water in direct contact with a conducting metal surface conducts that signal via the conductor to a suitable detector in exactly the same way as do electrodermal sensors [31].

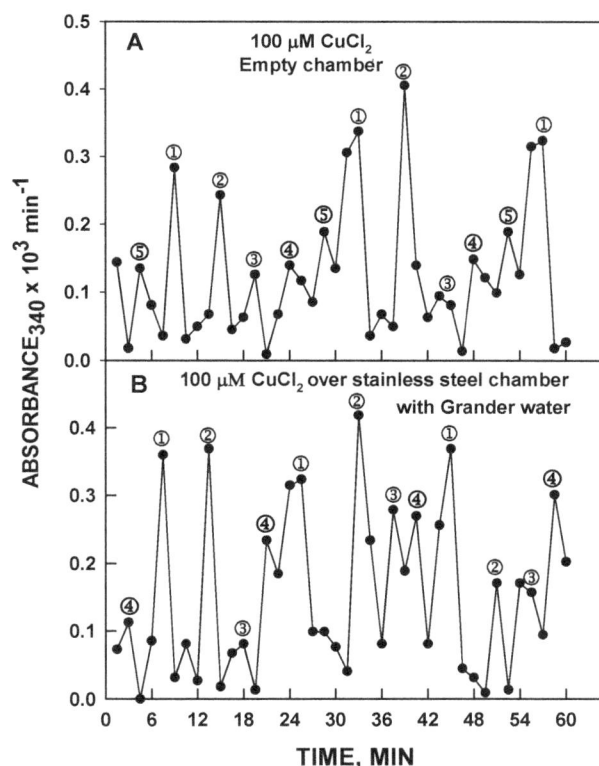

Figure 14. The metal need not be copper, and neither does the container need to be in the form of a coil. Identical results were obtained with a water filled chamber of stainless steel. (**A**) 120 μM NADH prepared in 100 μM $CuCl_2$ yielded the 5 peak 24 min period characteristic of $CuCl_2$; (**B**) With the $CuCl_2$ solution placed over the stainless steel chamber containing Grander or HPLC water, the $CuCl_2$ solution now oscillated with the 4 peak pattern and 18 min period typical of water.

Why is the frequency and phase of the oscillation of the aqueous solution inside the device not influenced by the frequency and phase of the aqueous solution at the outer surface of device? Perhaps the LFEMF signal at the outer barrier surface is always that of the liquid inside is equivalent to inside solution on the outside such that there is no opportunity for reverse communication of anything other than a pure water signal.

So what certain experts have said was a preposterous supposition that could not be tested scientifically has been tested scientifically. Moreover, the findings support the contentions of Benvenist and coworkers [32] and, more recently, those of others [33], that *"signaling of EM waves are potentially transmissible to cells and water by electromagnetic means."*

An analogous situation may account for information generated by electrodermal sensing where electrical resistance at the skin's surface is measured. The expressed purpose of electrodermal sensing is to detect *"energy"* imbalance along invisible lines of the body described by acupuncturists as meridians. The device sends an electrical current, too small to be detected by the subjects, through a probe. When a second probe is touched to another part of the body, a low voltage electrical circuit is completed and a computer screen or needle on a guage reads out a number between 0 and 100 as an indication of whether the patient's *"energy"* is in or out of balance.

Samples of various remedies such as homeopathic liquids and dietary or vitamin supplements may be tried as the probe is touched to the problem area. These different substances are tested until one is found that "*balances*" the energy disturbance. However, what do electrodermal diagnostic instruments actually measure? Might such measurements provide an early warning system of body pathology as claimed? Until now, the connection between those skin measurements and functioning of internal organs has been obscure. The answer may lie in water generated electromagnetic fields influenced by the environment of the water following the different meridians. From a theoretical standpoint, William Tiller [31] wrote "*one expects that cooperative cellular oscillations channel in the body, the electromagnetic radiation waves of appropriate wavelength from this generated organ spectrum will be guided away from the organ environs and out to the skin through specific acupuncture points.*"

If water moving through the body provides, as it seems, a long range communication channel, the opportunities that arise are truly of potential medical importance. Not only might it be possible to obtain wellness information electrodermally but also transmit information back to the body to help correct the reported imbalance. As we have concluded earlier [27]: "*Opportunities afforded may be not only to listen in but to talk back as well.*"

Acknowledgments

We thank Marc Henry and Stanislav Zakarov for valuable discussions and Peggy Runck for preparing the manuscript.

Author Contributions

D. James Morré and Dorothy M. Morré contributed equally to the development of the study including conceptualization, data acquisition, experimental design and manuscript preparation. D. James Morré and Dorothy M. Morré read and approved the final manuscript.

Conflicts of Interest

The authors declare no conflict of interest.

Since this manuscript was submitted for publication, a paper appeared in which dynamics of physical characteristics of water, studied using various methods of molecular structure analysis: IR spectroscopy, Raman spectroscopy, microwave radiometry and nuclear magnetic resonance, revealed similar reproducible harmonic components based on wavelet analysis that included periods of *ca* 18 min which were proposed to result from intramolecular interactions due to spin isomerism of the water molecules.

References

1.   Smith, C.W. Quanta and coherence effects in water and living systems. *J. Altern. Complement. Med.* **2004**, *10*, 69–79.

2.   Morré, D.J.; Chueh, P.-J.; Pletcher, J.; Tang, X.; Wu, L.-Y.; Morré, D.M. Biochemical basis for the biological clock. *Biochemistry* **2002**, *41*, 11941–11945.

3. Morré, D.J.; Morré, D.M. ENOX proteins, copper hexahydrate-based ultradian oscillations of the cells' biological clock. In *Ultradian Rhythms from Molecules to Mind: A New Vision of Life*; Lloyd, D., Rossi, E., Eds.; Springer: New York, NY, USA, 2008; pp. 43–84.

4. Morré, D.J.; Morré, D.M. Water in biological time keeping. In *Water and Society*; Pepper, D.W., Brebbia, C.A., Eds.; WIT Press: Southampton, Boston, MA, USA, 2012; pp. 13–23.

5. Morré, D.J.; Morré, D.M. *ECTO-NOX Proteins*; Springer: New York, NY, USA, 2013.

6. Chueh, P.-J.; Kim, C.; Cho, N.; Morré, D.M.; Morré, D.J. Molecular cloning and characterization of a tumor-associated, growth-related and time-keeping hydoquinone (NADH) oxidase (NOX) of the HeLa cell surface. *Biochemistry* **2002**, *41*, 3732–3741.

7. Jiang, Z.; Gorenstein, N.M.; Morré, D.M.; Morré, D.J. Molecular cloning and characterization of a candidate human growth-related and time-keeping constitutive cell surface hydroquinone (NADH) oxidase. *Biochemistry* **2008**, *47*, 14028–14038.

8. Pittendrigh, C.S.; Caldarola, P.C.; Cosbey, E.S. A differential effect of heavy water on temperature-dependent and temperatue-compensated aspects of circadian system of *Drosophila pseudoobscura*. *Proc. Natl. Acad. Sci. USA* **1973**, *70*, 2037–2041.

9. Jiang, Z.; Morré, D.M.; Morré, D.J. A role for copper in biological time-keeping. *J. Inorg. Biochem.* **2006**, *100*, 2140–2149.

10. Tang, X.; Chueh, P.-J.; Jiang, Z.; Layman, S.; Martin, B.; Kim, C.; Morré, D.M.; Morré, D.J. Essential role of copper in the activity and regular periodicity of a recombinant, tumor associated, cell surface, growth-related and time-keeping hydroquinone (NADH) oxidases with protein disulfide-thiol interchange activity (ENOX2). *J. Bioenerg. Biomembr.* **2010**, *42*, 355–360.

11. Foster, K.; Anwar, N.; Pogue, R.; Morré, D.M.; Morré, D.J. Decomposition analyses applied to a complex ultradian biorhythm: The oscillating NADH oxidase activity of plasma membranes having a potential time-keeping (clock) function. *Nonlinearity Biol. Toxicol. Med.* **2003**, *1*, 51–70.

12. Morré, D.J.; Heald, S.; Coleman, J.; Orczyk, J.; Jiang, Z.; Morré, D.M. Structural observations of time dependent oscillatory behavior of $Cu^{II}Cl_2$ solutions measured via extended X-ray absorption fine structure. *J. Inorg. Biochem.* **2007**, *100*, 715–726.

13. Morré, D.J.; Orczyk, J.; Hignite, H.; Kim, C. Regular oscillatory behavior of aqueous solutions of Cu^{II} salts related to effects on equilibrium dynamics of ortho/para hydrogen spin isomers of water. *J. Inorg. Biochem.* **2008**, *102*, 260–267.

14. Morré, D.J.; Jiang, Z.; Marianovic, M.; Orczyk, J.; Morré, D.M. Response of the regulatory behavior of copper^{II} containing ECTO-NOX proteins and $Cu^{II}Cl_2$ in solution to electromagnetic fields. *J. Inorg. Biochem.* **2008**, *102*, 1812–1818.

15. Edmunds, L.N., Jr. *Cellular and Molecular Basis of Biological Clocks*; Springer: New York, NY, USA; Berlin, Germany, 1988; p. 497.

16. Tikhonov, V.I.; Volkov, A.A. Separation of water into its ortho and para isomers. *Science* **2002**, *296*, 2363.

17. Pershin, S.M. Two-liquid water. *Phys. Wave Phenom.* **2005**, *13*, 192–208.

18. Bunkin, A.F.; Pershin, S.M.; Nurmatov, A.A. Four-photon spectroscopy of ortho/para spin-isomer H_2O molecule in liquid water in sub-millimeter range. *Laser Phys. Lett.* **2006**, *3*, 275–277.

19. Bunkin, A.F.; Lebedenko, S.L.; Nurmatov, A.A.; Pershin, S.M. Four-photon raleigh-wing spectroscopy of the aqueous solution of α-chymotrysin protein. *Quantum Electr.* **2006**, *36*, 612–615.

20. Sliter, R.; Gish, M.; Vilesov, A.F. Fast nuclear spin conversion in water clusters and ices: A maxtrix isolation study. *J. Phys. Chem.* **2011**, *1115*, 9682–9688.

21. Veber, S.L.; Bagryanskaya, E.G.; Chapovsky, P.L. On the possibility of enrichment of H$_2$O nuclear spin isomers by adsorption. *J. Exp. Theor. Phys.* **2006**, *102*, 76–83.

22. Potekhin, S.A.; Khusainova, R.S. Spin-dependent absorption of water molecules. *Biophys. Chem.* **2005**, *118*, 84–87.

23. Gudkov, S.V.; Bruskov, V.I.; Astashey, M.D.; Chernikov, A.V.; Yaguzhinsky, L.S.; Zakharov, S.D. Oxygen-dependent auto-oscillations of water luminescence triggered by the 1264 nm radiation. *J. Phys. Chem B* **2011**, *115*, 7693–7698.

24. Emsley, J.M.; Feeney, J.; Sutcliffe, L.H. *High Resolution Nuclear Magnetic Resonance Spectroscopy*; Pergamon Press; Oxford, UK, 1965; Volume 1.

25. Peng, C.K.; Buldyrev, S.V.; Havlin, S.; Simons, M.; Stanley, H.E.; Goldberger, A.L. Mosaic organization of DNA nucleotides. *Phys. Rev. E* **1994**, *49*, 1685–1689.

26. Astafieva, N.M. Wavelet analysis: Basic theory and some applications. *UFN* **1996**, *166*, 1145–1170; English Version: *Phys. Uspikhi* **1996**, *39*, 1085–1108.

27. Morré, D.J.; Morré, D.M. *Water Talks to Water. Might We Listen In*; Chapter, XI, Lo Nostro, P., Ninham, B.W., Eds.; Aqua Incognita, Conner Court, Vallarat: Victoria, Australia, 2014.

28. Del Giudice, E.; Spinetti, P.R.; Tedeschi, A.L. Water dynamics at the root of metaorphorsis in living organisms. *Water* **2010**, *2*, 566–586.

29. Zhen, J.-M.; Pollack, G.H. Solute exclusion and potential distribution near hydrophilic surfaces. In *Water and the Cell*; Pollack, G.H., Cameron, L., Wheatly, D.N., Eds.; Springer: Dordrecht, The Netherlands, 2006; pp. 165–174.

30. Bono, I.; Del Giudice, E.; Gamberale, L.; Henry, M. Emergence of the coherent structure of liquid water. *Water* **2012**, *4*, 510–532.

31. Tiller, W.A. What do electrodermal diagnostic acupuncture instruments really measure. *Am. J. Acupunct.* **1987**, *15*, 15–23.

32. Thomas, Y.; Schiff, M.; Belkade, L.; Jungens, P.; Kahhak, L.; Benveniste, J. Activation of human neutrophils by electronically transmited phorbol myristate. *Med. Hypothesis* **2000**, *54*, 33–39.

33. Montagnier, L.; Lavallee, C.; Aissa, J. General Procedure for the Identification of DNA Sequences Generating Electromagnetic Signals in Biological Fluids and Tissues. U.S. Patent No. 2012/24701, 2 February 2012.

4

Preliminary Toxicological Evaluation of the River Danube Using *in Vitro* Bioassays

Clemens Kittinger [1,*], **Rita Baumert** [1], **Bettina Folli** [1], **Michaela Lipp** [1], **Astrid Liebmann** [1], **Alexander Kirschner** [2,3], **Andreas H. Farnleitner** [3,4], **Andrea J. Grisold** [1] and **Gernot E. Zarfel** [1]

[1] Institute of Hygiene, Microbiology and Environmental Medicine, Medical University Graz, Graz 8010, Austria; E-Mails: rita.baumert@medunigraz.at (R.B.); bettina.folli@medunigraz.at (B.F.); michaela.lipp@medunigraz.at (M.L.); astrid.liebmann@gmx.at (A.L.); andrea.grisold@medunigraz.at (A.J.G.); gernot.zarfel@medunigraz.at (G.E.Z.)

[2] Division Water Hygiene, Institute for Hygiene and Applied Immunology, Medical University of Vienna, Vienna 1090, Austria; E-Mail: alexander.kirschner@meduniwien.ac.at

[3] Interuniversity Cooperation Centre for Water and Health, Vienna, Austria; E-Mail: andreas.farnleitner@wavenet.at

[4] Institute of Chemical Engineering, Research Group Environmental Microbiology and Molecular Ecology, Vienna University of Technology, Vienna 1090, Austria

* Author to whom correspondence should be addressed; E-Mail: clemens.kittinger@medunigraz.at;

Academic Editor: Say-Leong Ong

Abstract: The Joint Danube Survey 3, carried out in 2013 was the world's biggest river research expedition of its kind. The course of the second largest river of Europe passes large cities like Vienna, Budapest and Belgrade and is fed from many tributaries like Inn, Thisza, Drava, Prut, Siret and Argeş. During the 6 weeks of shipping the 2375 km downstream the River Danube from Germany to the Black Sea an enormous number of water samples were analyzed and collected. A wide spectrum of scientific disciplines cooperated in analyzing the River Danube waters. For toxicological analysis, water samples were collected on the left, in the middle, and on the right side of the river at 68 JDS3 sampling points and frozen until the end of the Danube survey. All samples were analyzed with two *in vitro* bioassays tests (umuC and MTS). Testing umuC without S9 activation and MTS test did not show positive signals. But umuC investigations of the water samples came up with toxic signals on two stretches, when activated with S9 enzymes. The override of the limiting value of the umuC investigation with prior S9 activation started downstream Vienna (Austria) and was prolonged until Dunaföldvar (Hungary). This stretch of the River Danube passes a region that is highly industrialized, intensively used for agricultural purposes and also highly populated (Vienna, Bratislava and Budapest). The elevated values may indicate these influences.

Keywords: Joint Danube Survey; Joint Danube Survey 3 (JDS3); UV mutagenesis gene C (umuC); 3-(4,5-dimethylthiazol-2-yl)-5-(3-carboxymethoxyphenyl)-2-(4-sulfophenyl)-2H-tetrazolium (MTS); toxicity; river; surface water

1. Introduction

The 2872 km long River Danube, the second longest river in Europa, passes ten countries until it flows into the Black Sea forming a large river delta. The drainage basin is around 817,000 km^2 large, including the waste waters of this mostly densely populated area. The course of the river passes large cities like Vienna, Budapest and Belgrade and is fed from many tributaries like Inn, Tisza, Drava, Sava, Pruth, Siret and Argeş.

The Joint Danube Survey 3 (JDS3) 2013 was the world's biggest river research expedition of its kind [1]. Until now the JDS has been carried out three times every six years. Between 13 August and 26 September, samples were taken along a 2563 km stretch of the River Danube starting in Böfinger Halde (Germany, river 2581 km) to the Danube Delta (river 18 km). Besides collecting water samples and directly surveying the microbiological status, many other river relevant parameters from water, sediment and suspended solids were evaluated by laboratories all across Europe: e.g., hydromorphology, basic chemistry, biological key elements like fish, macrozoobenthos, phytobentos, phytoplankton, macrophytes, *etc.*

One aspect of the investigation was the primary evaluation of the toxicological burden over the whole river course. In order to provide a first toxicological investigation and status assessment of the River Danube, two widely used, easily applicable toxicological tests were applied for all JDS3 samples (umuC and MTS). These tests have been used for the investigation of surface waters by other groups [2–6] and have been additionally established and used for the investigation of the River Mur in Styria, Austria [7]. The investigation of the water samples with these protocols is very reliable in terms of unspecific screening for toxic signals in surface or waste water samples [4–6]. These tests need only a small amount of the test liquid and can react on high numbers of mutagenic and cytotoxic substances and are therefore suitable for looking for unknown hazardous substances originating from all sources [3,8,9]. Compared with other investigations carried out during the JDS3, the results of the toxicity survey may lead to a new discussion on the methodology in the search for toxic substances and to new insights into the toxicological burden of the Danube.

2. Materials and Methods

2.1. Water Samples

Samples were taken all over the River Danube course at 68 positions (Figure 1). At a sampling point (SP) samples were always taken from the left side (L), in the middle (M) and from the right side (R) (resulting in 171 samples) of the River Danube, with the exception of the tributary samples that were mostly collected only once in the middle (11) (Table 1).

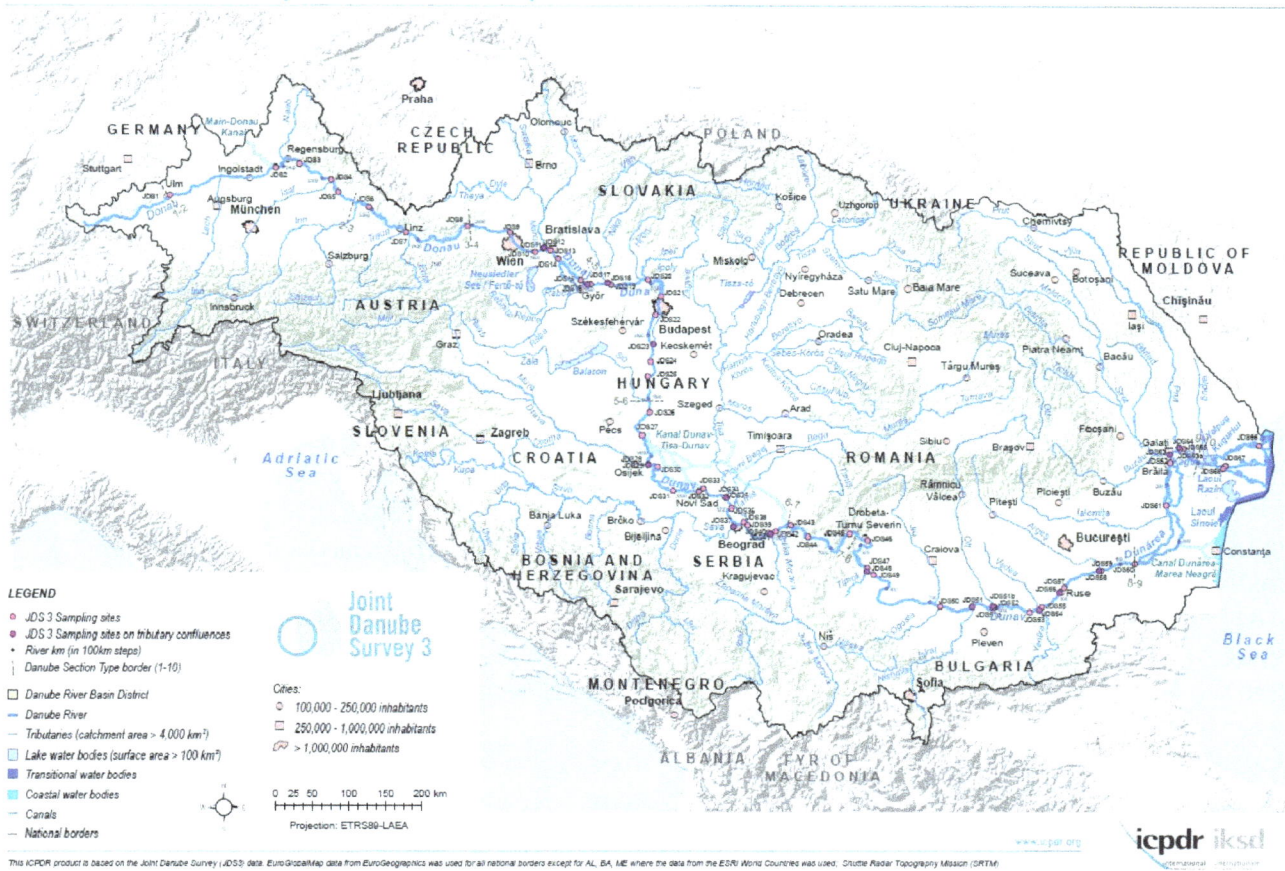

Figure 1. Overview of the Joint Danube Survey 3 (JDS3) sampling points along the river Danube. The map was taken with kind permission of the ICPDR.

Subsamples of 50 mL from the sample bottle taken for the microbiological investigations (surface water collected 0.3 m under the river surface) were filled into sterile non-toxic 50-mL plastic vials and immediately stored at −20 °C until analysis in the home laboratory. Before being used in the experiments, the samples, were thawed on ice, vortexed and filtrated to eliminate bacteria via 0.45 μm syringe filter (TPP, Techno Plastic Products, Switzerland). Freezing of the samples might alter the composition and amount of toxic compounds in the sample. Although studies of Armishaw *et al.* showed for pesticide spiked material no alteration over 168 days of freezer storage, this cannot be predicted for hundreds of toxic substances in surface water [10]. The stability of the JDS3 water samples stored at 4 °C was also investigated on three exemplary samples during the study and showed that most substances were relatively stable over a period of 173 days [1]. The small sample volume, the storage at −20 °C and the possibility to test a large sample number was a requirement for the screening investigation.

Table 1. List of the JDS3 sampling points (SP), the orange highlighted sampling points were only collected midstream.

SP	Name of SP	River km	SP	Name of SP	River km
JDS1	Böfinger Halde	2581	JDS35	Tisa	1215
JDS2	Kelheim, gauging station	2415	JDS36	DS Tisa/US Sava (Belegis)	1200
JDS3	Geisling power plant	2354	JDS37	Sava	1170
JDS4	Deggendorf	2285	JDS38	Upstream Pancevo	1159
JDS5	Mühlau	2258	JDS39	DownstreamPancevo	1151
JDS6	Jochenstein	2204	JDS40	Upstream Vel. Morava	1107
JDS7	US dam Abwinden-Asten	2120	JDS41	Velika Morava	1103
JDS8	Oberloiben	2008	JDS42	DS Velika Morava	1097
JDS9	Klosterneuburg	1942	JDS43	Banatska Palanka	1071
JDS10	Wildungsmauer	1895	JDS44	IGR Golubac/Koronin	1040
JDS11	US Morava (Hainburg)	1881	JDS45	IGR Tekija/Orsova	954
JDS12	Morava	1880	JDS46	Vrbica/Simijan	926
JDS13	Bratislava	1869	JDS47	Upstream Timok	849
JDS14	Gabcikovo reservoir	1852	JDS48	Timok	845
JDS15	Medvedov/Medve	1806	JDS49	Pristol/Novo Salo	834
JDS16	Moson Danube	1794	JDS50	Downstream Kozloduy	685
JDS17	Klizska Nema	1790	JDS51	Iskar	637
JDS18	Vah	1766	JDS52	Downstream Olt	602
JDS19	Iza/Szony	1761	JDS53	Downstream Zimnicea/Svistov	550
JDS20	Szob	1707	JDS54	Jantra	537
JDS21	US Budapest - Megyeri Bridge	1660	JDS55	Downstream Jantra	532
JDS22	DS Budapest—M0	1632	JDS56	Russenski Lom	498
JDS23	Rackeve-Soroksar arm-end	1586	JDS57	Downstream Ruse	488
JDS24	Dunaföldvar	1560	JDS58	Arges	432
JDS25	Paks	1533	JDS59	Downstream Arges	429
JDS26	Baja	1481	JDS60	Chiciu/Silistra	378
JDS27	Hercegszanto	1434	JDS61	Giurgeni	235
JDS28	US Drava	1384	JDS62	Braila	167
JDS29	Drava	1379	JDS63	Siret	154
JDS30	DS Drava (Erdut/Bogojevo)	1367	JDS64	Prut	135
JDS31	Ilok/Backa Palanka	1300	JDS65	Reni	130
JDS32	US Novi Sad	1262	JDS66	Vilova/Kilia Arm	18
JDS33	DS Novi Sad	1252	JDS67	Sulina Arm	26
JDS34	US Tisa (Stari Slankamen)	1216	JDS68	St.Gheorge Arm	104

2.2. Toxicity Assay: umuC

An SOS/umuC assay was carried out to search for mutagenicity. The assay was carried out according to Reifferschied *et al.*, following the modifications of the ISO 13829 standard [11]. The umuC assay was conducted with or without S9 enzymatic activation (Trinova Biochem, Gießen, Germany). Filtrated water samples as described above were applied to the test without pH correction as the pH values were between 8.0 and 8.5 over the whole stretch of the Danube River [1]. Tests were carried out in 96 well plates (TPP, Techno Plastic Products, Trasadingen, Switzerland). The absorbance at 600 nm and 420 nm was measured with a Zenyth 3100 Multimode Detector (Beckman Coulter, Austria). All experiments were carried out in triplicates and mean and standard error of the mean (SEM) were calculated. According to the ISO 13829 the growth rate (G) was calculated with Equation (1).

$$G = \frac{OD600 sample - OD600 blank}{OD600 control - OD600 blank} \tag{1}$$

A growth reduction of 25% compared to the growth control was considered to be a cytotoxic water sample. The induction rate (IR) was calculated with Equation (2):

$$IR = \frac{1}{G} \times \frac{A420 sample - A420 blank}{A420 control - A420 blank} \tag{2}$$

According to ISO 13829 an induction rate of ≥ 1.5 was taken as a signal for mutagenic potency in the water samples.

2.3. Cytotoxicity Assay: MTS

For determination of cytotoxic potential of the water samples a MTS test (Promega, Mannheim, Germany) was carried out. The test is based on the yellow salt [3-(4,5-dimethylthiazol-2-yl)-5-(3-carboxymethoxyphenyl)-2-(4-sulfophenyl)-2H-tetrazolium, inner salt; MTS] which is converted into the blue/violet water insoluble salt formazan. The conversion into formazan is mediated by dehydrogenases of intact mitochondria and therefore provides insight into cell viability. HepG2 (DSMZ ACC 180) cells were used for cytotoxicity assays. HepG2 cells are capable of phase one liver enzymatic reaction and are highly sensitive against polycyclic aromatic hydrocarbons and genotoxic effects can be seen after challenging with carcinogenic mycotoxins. These cells also react positively to Arsenic and carcinogenic metals like Cadmium [12]. Cells were cultivated in Dulbecco's Modified Eagle Medium (DMEM, Promega, Vienna, Austria) with 10% fetal bovine sera (FBS, Promega, Austria) and 100 U/mL penicillin/streptomycin (Sigma Aldrich, Vienna, Austria) at 37 °C and 5% CO_2. Passages 3 to 6 were taken for the experiments. Cell number was titrated to find out the best ratio between cell number and maximum signal response. A cell number of 1×10^4 cells/well was found to be ideal. For the cytotoxic analysis, cells were freshly seeded into 96 well plates (Thermo Scientific, Vienna, Austria) and allowed to attach for 4 h. After that, 40% of the medium was replaced by filtrated water samples and incubated for 20 h at 37 °C and 5% CO_2. After the incubation, 20 µL of the Dye Solution was added. The plates were incubated for up to 4 h at 37 °C in a humidified, 5% CO_2 atmosphere. The absorbance at 492 nm was measured with a Zenyth 3100 Multimode Detector (Beckman Coulter, Vienna, Austria). Deionized water served as control. Experiments were carried out in triplicates. Viability (VC) of the cells incubated with deionized water was taken 100% and the viability of the river samples was put into relation to them and calculated with Equation (3).

$$V_C = \frac{100 \times Abs492 \; Water \; Sample}{Abs492 \; Control \; Sample} \tag{3}$$

A reduction of the viability to 70% compared to the test sample was taken as a cytotoxic response [3].

3. Results and Discussion

All samples of the JDS3 sampling points were investigated for a toxic signal with the umuC test and the MTS. Experiments were carried out in triplicates and means and standard deviations are given as line and error bars in the figures.

3.1. UmuC Results without Enzymatic S9 Activation

The umuC investigation of the River Danube Samples without S9 activation did not show any raised values (Figure 2). The only exception was one value of the triplicates at sample position JDS31 M that was elevated to 1.79. But because the two other midstream values were 0.95 and 0.92, this high single value of 1.79 has to be interpreted as an outlier. In addition, the mean value was below the limit value of 1.5. The results go also well with previous river studies, were the samples without S9 activation did not came up with a toxic signal [7]. Evaluation of growth of the umuC *Salmonella* as requested in ISO 13829 did also not show any inhibition.

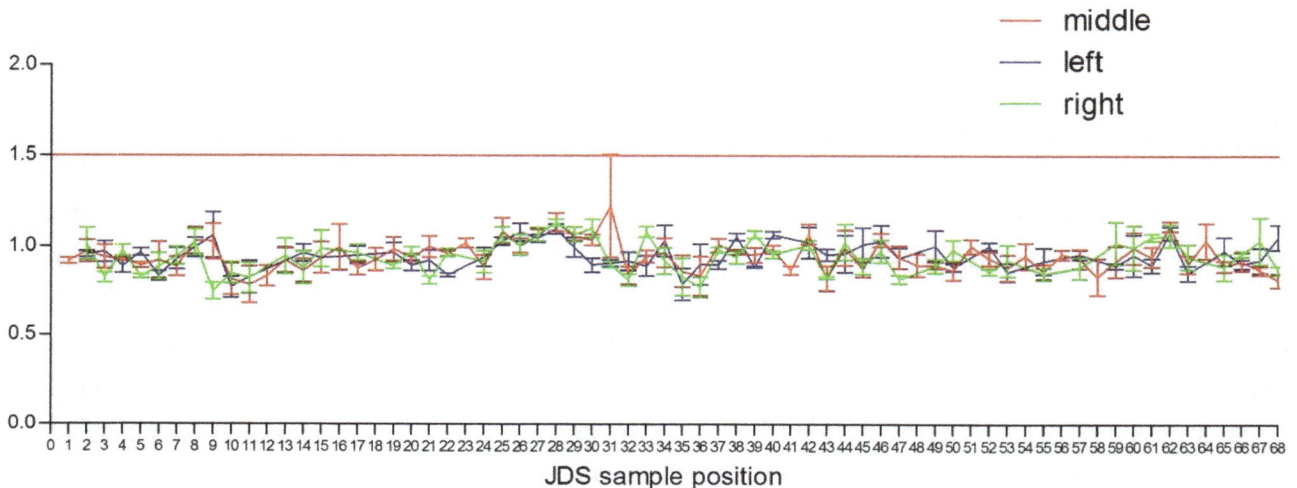

Figure 2. Results from umuC testing of the River Danube without enzymatic activation. The red line at 1.5 represents the limit value according to ISO 13829.

3.2. UmuC Results with Enzymatic S9 Activation

Investigation of the River Danube samples with enzymatic S9 activation showed exceedance of the limit value of 1.5 and elevated values before and after a few JDS sampling points (Figure 3). The values of all investigated sampling points had little standard deviations and were thus considered reliable. Values started to rise from JDS13 (Bratislava, SVK, river 1869 km) on until JDS28 (upstream Drava, HR, river 1632 km). The limiting value was exceeded at JDS15 (Medvedov, SVK, river 1806 km), JDS20 (Szob, HU, river 1707 km), JDS22 (downstream Budapest, HU, river 1632 km), JDS23 (Rackeve-Soroksar branch, HU, river 1586 km), JDS24 (Dunarföldvar, HU, river 1560 km) and JDS25 (Paks, HU, river 1533 km). Elevated values were also observed at JDS55 L (downstream Jantra, RO, river 532 km) but stayed below the limit of 1.5.

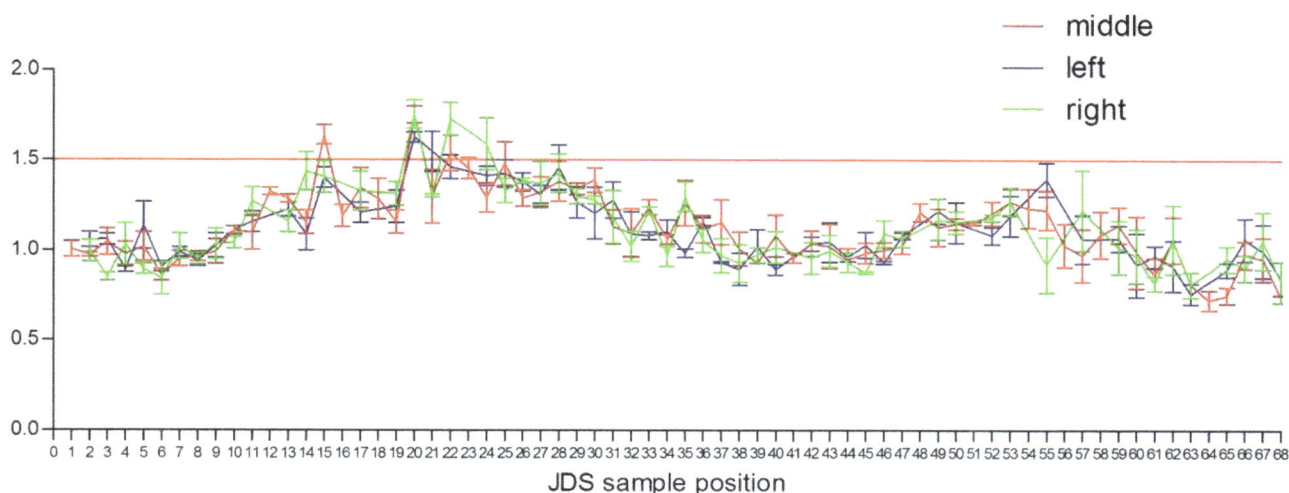

Figure 3. Results from umuC testing of the river Danube with enzymatic activation. The red line at 1.5 represents the limit value. Values at JDS15 M, JDS20 L,M,R, JDS21 L, and JDS22 R, clearly override the limit value. Values are also increased before and after JDS55 but do not exceed the limiting value. Growth of *Salmonella* is impaired beginning at JDS60 but does not fall below 75% compared to the growth control.

When elevated values were observed, they were mostly elevated at all three horizontal sampling point (e.g., JDS 20 left, middle and right). This leads to the conclusion, that the toxic signal has come from a point upstream as it has to be spread all over the whole width of the river. The definite source of the toxic signals is difficult to find, as the umuC is sensitive for at least 400 chemicals tested by Reifferscheid *et al.* [13]. One group of toxicants that need prior S9 activation and are known to be pollutant in surface waters are polychlorinated biphenyls (PCBs) [14,15] although they were found at very low levels in the River Danube [1].The possible sources are the large municipal waste water treatment plants, the outfall of large factories in these areas, and the agricultural land use of the watershed area for these sites.

The reduced growth rate from JDS60 to JDS68 triggered the values to around 0.80 to 0.85 which is close to the cytotoxic limit value according to ISO 13829 (Materials and Methods 2.2). The growth rate dropped by around 15%–20% which might be a reference for cytotoxicity in this stretch of the River Danube, but there was no parallel growth reduction found in the MTS test with eukaryotic cells (see below).

3.3. MTS Testing

For all investigated samples the MTS test did not show any toxic signals (Figure 4) and there were no differences all over the River Danube stretch. Although HepG2 liver cells are capable of phase one enzymatic liver modification and suitable for primary investigation [16] there was no detectable reduction of the cell viability. The values of the River Danube samples tend to be even a little bit elevated (10%–20%) compared to the control (deionized water), as they were only filtrated and contain still their natural salt concentration. The filtrated Danube water was osmotically better for the cells than the control and this must be the reason for the slightly elevated values. The MTS test did not lead to positive results with the applied cell line. Extending the tests to other cell lines (e.g., epithelial cell lines like IEC-18, fibroblastic cell lines like BALB/c 3T3 [17–19]) could bring further insights.

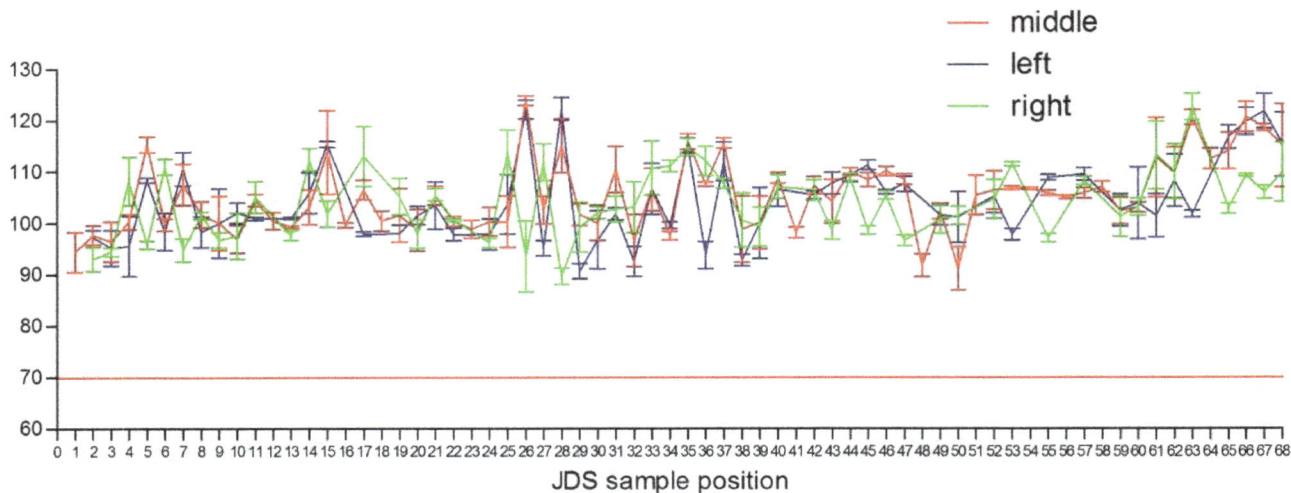

Figure 4. Values of the MTS results of the River Danube sampling points (x-axis). The y-axis represents percentage of viability compared to the control (deionized water, was set as 100%). The red line at 70% represents the limit value for an inhibition of growth caused by a toxic compound or a combination of compounds.

4. Conclusions

The examination of the JDS 3 River Danube samples provided a primary toxicological evaluation of the Danube and its major tributaries. The dense mesh of samples offered a unique chance for an assessment of this large transnational river system. Our data suggest that the Danube water in the river stretch between JDS13 and JDS 28 with elevated umuC values after S9 activation may carry a mutagenic burden. A direct comparison to the prior Danube surveys is not possible because toxicology was not investigated during JDS1 and only for sediment samples during JDS2. Further analysis at a high temporal resolution is needed to proof that our findings are consistent over time.

Acknowledgments

The Joint Danube Survey was organized by the International Commission for the Protection of the Danube River (ICPDR). The study was supported by the Austrian Science Fund (FWF), project nr P25817-B22. We further thank Georg Reischer, Stefan Jakwerth and Stoimir Kolarevic for their help in sampling.

Author Contributions

Clemens Kittinger and Gernot Zarfel had the original idea for the study and with Andreas H. Farnleitner and Andrea J. Grisold carried out the design. Rita Baumert, Bettina Folli, Michaela Lipp and Astrid Liebmann carried out the laboratory work. Clemens Kittinger was responsible for data cleaning. Clemens Kittinger and Alexander Kirschner drafted the manuscript, which was revised by all authors. All authors read and approved the final manuscript.

Conflicts of Interest

The authors declare to have no conflicts of interests and no financial relationships that might lead to a conflict of interests.

References

1. JDS 3. *Joint Danube Survey 3. A Comprehensive Analysis of Danube Water Quality*; ICPDR—International Commission for the Protection of the Danube River: Vienna, Austria, 2015.

2. Leusch, F.D.; Khan, S.J.; Gagnon, M.M.; Quayle, P.; Trinh, T.; Coleman, H.; Rawson, C.; Chapman, H.F.; Blair, P.; Nice, H.; *et al.* Assessment of Wastewater and Recycled Water Quality: A Comparison of Lines of Evidence from *in Vitro, in Vivo* and Chemical Analyses. *Water Res.* **2014**, *50*, 420–431.

3. Zegura, B.; Heath, E.; Cernosa, A.; Filipic, M. Combination of *in Vitro* Bioassays for the Determination of Cytotoxic and Genotoxic Potential of Wastewater, Surface Water and Drinking Water Samples. *Chemosphere* **2009**, *75*, 1453–1460.

4. Giuliani, F.; Koller, T.; Wurgler, F.E.; Widmer, R.M. Detection of Genotoxic Activity in Native Hospital Waste Water by the umuC Test. *Mutat. Res.* **1996**, *368*, 49–57.

5. Hamer, B.; Bihari, N.; Reifferscheid, G.; Zahn, R.K.; Muller, W.E.; Batel, R. Evaluation of the SOS/umu-Test Post-Treatment Assay for the Detection of Genotoxic Activities of Pure Compounds and Complex Environmental Mixtures. *Mutat. Res.* **2000**, *466*, 161–171.

6. Dizer, H.; Wittekindt, E.; Fischer, B.; Hansen, P.D. The Cytotoxic and Genotoxic Potential of Surface Water and Wastewater Effluents as Determined by Bioluminescence, Umu-Assays and Selected Biomarkers. *Chemosphere* **2002**, *46*, 225–233.

7. Kittinger, C.; Marth, E.; Reinthaler, F.F.; Zarfel, G.; Pichler-Semmelrock, F.; Mascher, W.; Mascher, G.; Mascher, F. Water Quality Assessment of a Central European River—Does the Directive 2000/60/EC Cover all the Needs for a Comprehensive Classification? *Sci. Total Environ.* **2013**, *447*, 424–429.

8. Hernando, M.D.; Heath, E.; Petrovic, M.; Barcelo, D. Trace-Level Determination of Pharmaceutical Residues by LC-MS/MS in Natural and Treated Waters. A Pilot-Survey Study. *Anal. Bioanal Chem.* **2006**, *385*, 985–991.

9. Macova, M.; Toze, S.; Hodgers, L.; Mueller, J.F.; Bartkow, M.; Escher, B.I. Bioanalytical Tools for the Evaluation of Organic Micropollutants during Sewage Treatment, Water Recycling and Drinking Water Generation. *Water Res.* **2011**, *45*, 4238–4247.

10. Armishaw, P.; Millar, R. A Natural Matrix (Pureed Tomato) Candidate Reference Material Containing Residue Concentrations of Pesticide Chemicals. *Fresenius J. Anal. Chem.* **2001**, *370*, 291–296.

11. Reifferscheid, G.; Heil, J.; Oda, Y.; Zahn, R.K. A Microplate Version of the SOS/umu-Test for Rapid Detection of Genotoxins and Genotoxic Potentials of Environmental Samples. *Mutat. Res.* **1991**, *253*, 215–222.

12. Knasmuller, S.; Mersch-Sundermann, V.; Kevekordes, S.; Darroudi, F.; Huber, W.W.; Hoelzl, C.; Bichler, J.; Majer, B.J. Use of Human-Derived Liver Cell Lines for the Detection of Environmental and Dietary Genotoxicants; Current State of Knowledge. *Toxicology* **2004**, *198*, 315–328.

13. Reifferscheid, G.; Heil, J. Validation of the SOS/umu Test using Test Results of 486 Chemicals and Comparison with the Ames Test and Carcinogenicity Data. *Mutat. Res.* **1996**, *369*, 129–145.

14. Flint, S.; Markle, T.; Thompson, S.; Wallace, E. Bisphenol A Exposure, Effects, and Policy: A Wildlife Perspective. *J. Environ. Manag.* **2012**, *104*, 19–34.

15. Haarstad, K.; Bavor, H.J.; Maehlum, T. Organic and Metallic Pollutants in Water Treatment and Natural Wetlands: A Review. *Water Sci. Technol.* **2012**, *65*, 76–99.

16. Baderna, D.; Colombo, A.; Romeo, M.; Cambria, F.; Teoldi, F.; Lodi, M.; Diomede, L.; Benfenati, E. Soil Quality in the Lomellina Area using in Vitro Models and Ecotoxicological Assays. *Environ. Res.* **2014**, *133*, 220–231.

17. Walum, E.; Hedander, J.; Garberg, P. Research Perspectives for Pre-Screening Alternatives to Animal Experimentation: On the Relevance of Cytotoxicity Measurements, Barrier Passage Determinations and High Throughput Screening *in Vitro* to Select Potentially Hazardous Compounds in Large Sets of Chemicals. *Toxicol. Appl. Pharmacol.* **2005**, *207*, 393–397.

18. Baderna, D.; Colombo, A.; Amodei, G.; Cantu, S.; Teoldi, F.; Cambria, F.; Rotella, G.; Natolino, F.; Lodi, M.; Benfenati, E. Chemical-Based Risk Assessment and *in Vitro* Models of Human Health Effects Induced by Organic Pollutants in Soils from the Olona Valley. *Sci. Total Environ.* **2013**, *463–464*, 790–801.

19. Kallweit, A.R.; Baird, C.H.; Stutzman, D.K.; Wischmeyer, P.E. Glutamine Prevents Apoptosis in Intestinal Epithelial Cells and Induces Differential Protective Pathways in Heat and Oxidant Injury Models. *JPEN J. Parenter. Enteral Nutr.* **2012**, *36*, 551–555.

5

Seasonal River Discharge Forecasting Using Support Vector Regression: A Case Study in the Italian Alps

Mattia Callegari [1,2], Paolo Mazzoli [3], Ludovica de Gregorio [1], Claudia Notarnicola [1], Luca Pasolli [4], Marcello Petitta [1] and Alberto Pistocchi [5,*]

[1] EURAC Research, European Academy of Bozen/Bolzano, Institute for Applied Remote Sensing, viale Druso, Bolzano 1-39100, Italy; E-Mails: mattia.callegari@eurac.edu (M.C.); ludovica.degregorio@eurac.edu (L.G.); claudia.notarnicola@eurac.edu (C.N.); marcello.petitta@eurac.edu (M.P.)

[2] Department of Earth and Environmental Science, University of Pavia, via Ferrata 1-27100 Pavia, Italy

[3] Research and development (R&D) Unit Suedtirol, Geographic Environmental COnsulting (GECO) Sistema—srl, via Maso della Pieve 60, Bolzano 39100, Italy; E-Mail: paolo.mazzoli@gecosistema.it

[4] Informatica Trentina Spa, via G. Gilli 2, Trento 38121, Italy; E-Mail: luca.pasolli@yahoo.it

[5] European Commission, Directorate-General Joint Research Centre (DG JRC), via E.Fermi 2749, Ispra (VA) 21027, Italy

* Author to whom correspondence should be addressed; E-Mail: alberto.pistocchi@jrc.ec.europa.eu;

Academic Editors: Roger Falconer and Clelia Marti

Abstract: In this contribution we analyze the performance of a monthly river discharge forecasting model with a Support Vector Regression (SVR) technique in a European alpine area. We considered as predictors the discharges of the antecedent months, snow-covered area (SCA), and meteorological and climatic variables for 14 catchments in South Tyrol (Northern Italy), as well as the long-term average discharge of the month of prediction, also regarded as a benchmark. Forecasts at a six-month lead time tend to perform no better than the benchmark, with an average 33% relative root mean square error (RMSE%) on test samples. However, at one month lead time, RMSE% was 22%, a non-negligible improvement over the benchmark; moreover, the SVR model reduces the frequency of higher errors associated with anomalous months. Predictions with a lead time of three months show an intermediate performance between those at one and six months lead time. Among the considered predictors,

SCA alone reduces RMSE% to 6% and 5% compared to using monthly discharges only, for a lead time equal to one and three months, respectively, whereas meteorological parameters bring only minor improvements. The model also outperformed a simpler linear autoregressive model, and yielded the lowest volume error in forecasting with one month lead time, while at longer lead times the differences compared to the benchmarks are negligible. Our results suggest that although an SVR model may deliver better forecasts than its simpler linear alternatives, long lead-time hydrological forecasting in Alpine catchments remains a challenge. Catchment state variables may play a bigger role than catchment input variables; hence a focus on characterizing seasonal catchment storage—Rather than seasonal weather forecasting—Could be key for improving our predictive capacity.

Keywords: seasonal hydrological forecast; snow cover area; support vector machine; regression; South Tyrol; Alps

1. Introduction

The prediction of water availability is a key element for effective water storage management [1]. This is particularly true when the available water storage capacity cannot compensate for reduced discharges over a full season, so that a more precise regulation needs to be planned.

While "quick response" hydrological events such as floods cannot be predicted, in the best cases, with a lead time longer than a few days as they critically depend on catchment input (daily or even hourly precipitation), available water volumes over extended periods (e.g., months) may depend on the rate of depletion of the catchments and are therefore driven by the catchment state, which is much more smoothly varying over time and can be arguably predicted with longer lead time.

In principle, it is possible to obtain monthly streamflow forecasts with long lead time based on catchment water balance models; however, these depend on the availability and reliability of long-term (at least one season [1–3]) forecasts of weather variables. Yang *et al.* [4] showed that the forecasting of streamflow is generally more accurate when catchment storage dominates over catchment input.

Moreover, the data collection, fieldwork, calibration, and validation for the setup of detailed, catchment-specific water balance models may be long and expensive; this poses practical constraints to application in many mountainous areas of the world, where hydrologic information is scarce and sparse, sites are difficult to access, and the operational management of a model for decision support may be difficult due to economic, organizational, and technical limitations (as discussed in e.g., [5]).

An alternative to water balance models is given by data-driven ("black box") models based on statistical relationships between time series of a target (e.g., monthly water discharge) and predictors.

Traditional black box models such as linear approaches [6] are being challenged recently by non-linear machine-learning methods such as the support vector regression (SVR) [7], reported by several authors to be more effective than previous methods in hydrologic time series forecasting [8]. The better performance of non-linear models may be related to the non-linear behavior shown by discharge as a function of catchment input and state variables.

Among the available retrieval algorithms, SVR is often adopted in the field of bio-physical parameter retrieval for the following characteristics: SVR is able to handle complex and non-linear problems, can manage different kinds of inputs, and can reach high level performance even in cases where few data are available [9,10].

The first two characteristics are in common with other retrieval strategies. Both non-parametric methods, such as Artificial Neural Network (ANN) and Genetic Algorithm, and parametric methods like Bayesian approaches can handle the non-linearity between input and output variables. However, the main limitation is related to the availability of a large training database to perform a robust retrieval. SVR overcomes this limitation because it is based on a geometrical concept. In this context, for ANN and statistical approaches, the main aim is to populate the feature space with as much data as possible, so that all the possible combinations of input and output features are available to build the mapping functions. On the other side, a retrieval-approach-based SVR identifies the margins of the tolerance tube around the input data. Once the margins have been identified, the mapping function can be identified without the need for more data [7].

In recent years, this type of model has been applied in several cases [11–15], particularly with reference to reservoirs [14], indicating a potential role in practical water resources management. Most applications, however, predict the streamflow at a given month on the basis of streamflows in previous months without using catchment state variables [16–18]. Dehghani *et al.* [19] highlighted the importance of snow water content for the prediction of streamflow in the Karun basin in Iran, but made use of basin air temperature as a proxy for snowmelt instead of referring specifically to snow-related variables. Wang and Fu [20] made use of soil moisture as a catchment variable for the improvement of adaptive monthly and sub-monthly streamflow predictions in the Yangtze River, and found that this variable appreciably improved the forecasts.

Examples of analyses focused on the catchment state, rather than the catchment input, have been published with reference to the summer response of spring to autumn precipitation [21,22], or to the summer runoff and vegetation indexes' response to groundwater storage [23], to some extent enabling long lead-time forecasting.

It has long been recognized that, in Alpine and similar settings, summer flows may be controlled by the winter accumulation of snow in the catchment [24]. The role of mountain snow and glacier water in sustaining runoff, even in the presence of trends of reduced precipitation, has been observed in many regions in the last years [25], and is increasingly included in hydrologic models applied to mountain areas [26,27].

This contribution examines the performance of SVR models for the forecasting of monthly streamflow in catchments of a European Alpine region with a lead time of one, three, or six months. Unlike many similar forecasting studies that make use of discharges only, we attempt to improve the model's predictive accuracy by introducing catchment input and state variables (climatic signals, and the snow-covered area) as predictors, as for a water balance model. Arguably, the SVR technique may be able to extract from the data the unknown, non-linear relationships between catchment input, state, and output, when a proper water balance model is difficult to identify. In the remainder of the paper, after introducing the study region and providing an intuitive introduction to the SVR technique, we discuss the results obtained in the prediction of streamflow with one, three, or six months lead time and the potentials and limitations of the approach, and provide some concluding perspectives.

2. Materials and Methods

2.1. Study Area and Available Discharge Data

The Alps, as well as other mountain regions in the world, are of strategic importance for water resources [28], given that they are a climate change hot spot and a key context where adaptation is required [29].

The study region considered here is the upper Adige catchment closed at Bronzolo/Branzoll, in the Autonomous Province of Bolzano/South Tyrol, Northern Italy (Figure 1). This is an alpine watershed with a drainage area of approximately 7400 km², with elevations ranging from about 200 m a.s.l. at the southern valley bottoms, to around 3900 m a.s.l. in the western upper ranges. The Adige is the second largest Italian river and is fed by a large tributary, the Isarco/Eisack, draining about half of the catchment. The area is covered by a relatively dense network of hydrographic and weather stations managed by the Province's Hydrographic Office (Figure 2). Among all the available water discharge stations, for the present study we considered just those in which the discharges have been continuously recorded for at least the last 20 years. Figure 1 shows the catchments closed at the selected stations, and Table 1 lists their main attributes.

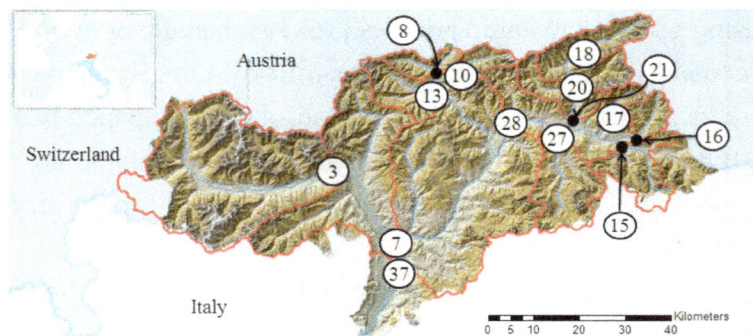

Figure 1. South Tyrol in northern Italy, location of the analyzed catchments.

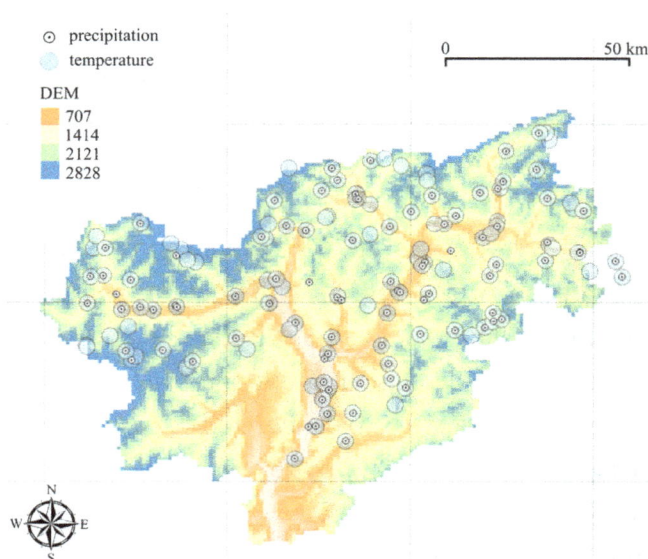

Figure 2. Available measurement stations in South Tyrol.

Table 1. Main characteristics of selected discharge measurement stations.

Station Name	Hydrographic Office Identifier	UTM 32N Coordinates (m)	Catchment Area (km²)	Min–Max–Mean Altitude (m a.s.l.)	Min–Max Discharge Q (m³/s)
Adige at Tel	3	$X = 659{,}075$ $Y = 5{,}170{,}831$	1,676	510–3,893–2,011	11.50–84.39
Adige at Ponte Adige	7	$X = 676{,}882$ $Y = 5{,}150{,}246$	2,705	240–3,893–1,907	22.68–114.96
Rio Fleres at Colle Isarco	8	$X = 685{,}584$ $Y = 5{,}201{,}188$	149	1,047–3,479–2,167	0.61–9.64
Rio Vizze at Novale	10	$X = 692{,}269$ $Y = 5{,}200{,}902$	108	1,375–3,500–2,186	0.76–13.03
Rio Ridanna at Vipiteno	13	$X = 685{,}246$ $Y = 5{,}195{,}138$	207	939–3,456–1,927	1.41–20.37
Rienza at Monguelfo	15	$X = 737{,}906$ $Y = 5{,}182{,}252$	264	1,096–3,217–1,827	2.16–12.50
Rio Casies at Colle	16	$X = 742{,}055$ $Y = 5{,}184{,}156$	117	1,198–2,825–1,960	0.86–6.22
Rio Anterselva at Bagni Salomone	17	$X = 735{,}133$ $Y = 5{,}190{,}077$	83	1,091–3,425–2,036	0.55–5.27
Aurino at Cadipietra	18	$X = 726{,}593$ $Y = 5{,}208{,}875$	149	1,047–3,485–2,167	1.46–27.06
Aurino at Caminata	20	$X = 725{,}134$ $Y = 5{,}198{,}678$	420	845–3,485–2,115	2.69–39.25
Aurino at San Giorgio	21	$X = 723{,}868$ $Y = 5{,}189{,}015$	613	819–3,485–2,038	5.40–56.18
Gadera at Mantana	27	$X = 719{,}819$ $Y = 5{,}184{,}323$	389	813–3,120–1,856	3.58–18.01
Rienza at Vandoies	28	$X = 706{,}592$ $Y = 5{,}188{,}228$	1,920	735–3,217–1,861	14.32–100.07
Adige at Bronzolo	37	$X = 677{,}940$ $Y = 5{,}142{,}627$	6,923	228–3,893–1,804	61.07–330.28

2.2. Introduction to Support Vector Regression (SVR)

2.2.1. Model Formulation

The statistical prediction of a variable $y(t + \Delta t)$ at an instant Δt from current time t may be framed as a particular case of regression problem, the goal of which is to identify a function $f(x)$ of a vector $x = (p_1, p_2, \ldots, p_m)$ of m (known) explanatory variables, so that $y(t + \Delta t) = f(x)$. The variables $p_1, \ldots, p_m$ forming vector **x** may include current and past data samples $[y(t), y(t - 1), \ldots, y(t - q)]$ for $q \leq m$, as well as additional numerical variables of any kind.

The function is identified on the basis of reference samples for which both $y(t + \Delta t)$ and **x** are known. Let us refer to a set of N samples $\{(x_1, y_1), (x_2, y_2), \ldots, (x_N, y_N)\}$ where each x_i and y_i, for $i = 1, \ldots,$ N, represent a particular value of the variable to be predicted, $y(t + \Delta t)$, and the corresponding vector of explanatory variables considered.

Traditional regression methods seek a function that minimizes prediction errors considering all N samples. The support vector regression (SVR) technique [30–34], instead, aims at finding the simplest function that can fit all the data while minimizing the sum of prediction errors above a predefined threshold. A review of the concepts and characteristics of these techniques with specific reference to hydrology is provided by [34]. For the sake of a quick illustration of the technique, let us first refer to the linear case, where the function to be estimated takes the form:

$$f(x) = \langle w, x \rangle + b, \tag{1}$$

w being a vector of weights of the same dimensionality of vector **x** of explanatory variables, and b a scalar. Here, the operator $\langle \cdot, \cdot \rangle$ refers to the scalar product of vectors. The "simplest function" is the function depending the least on input variables, and can be found by minimizing the norm $\|w\|^2 = \langle w, w \rangle$. As we want to minimize the sum of errors higher than a predefined threshold ε, we introduce the ε-insensitive loss function L_ε defined for each sample i as:

$$L_{\varepsilon i} := \begin{cases} 0 & \text{if } |f(x_i)-y_i| \le \varepsilon \\ |f(x_i) - y_i| - \varepsilon & \text{otherwise} \end{cases}. \tag{2}$$

From the loss function computed for each of the N reference sample, the SVR problem can be written as the following convex optimization problem:

$$minimize \ \tfrac{1}{2}\|w\|^2 + C\sum_{i=1}^{N} L_{\varepsilon i}, \tag{3}$$

where constant C is a model parameter, driving the tradeoff between the complexity of the function and tolerance of empirical errors. A graphical representation can be found in [31].

After appropriate mathematical manipulation, it can be shown that the weights can be written as a function of the reference sample vectors:

$$w = \sum_{i=1}^{N} \alpha_i x_i \tag{4}$$

and, consequently:

$$f(x) = \sum_{i=1}^{N} \alpha_i \langle x_i, x \rangle + b, \tag{5}$$

where α_i, $i = 1,\ldots,$ N, are constant coefficients.

Coefficient b can be estimated based on the Karush–Kuhn–Tucker (KKT) conditions [31], while the two parameters C and ε represent "tuning" parameters. In general, the SVR does not assume $f(x)$ to be linear, but stipulates it to be in the form:

$$f(x) = w \cdot \Phi(x) + b, \tag{6}$$

$\Phi(x)$ being a (non-linear) unknown function. In this case, it can be shown that the weights can be written as a function of the reference sample vectors:

$$w = \sum_{i=1}^{N} \alpha_i \Phi(x_i) \tag{7}$$

and, consequently:

$$f(x) = \sum_{i=1}^{N} \alpha_i \langle \Phi(x_i), \Phi(x) \rangle + b, \tag{8}$$

where α_i, $i = 1,\ldots,$ N, are constant coefficients. It should be noted that the identification of $f(x)$ does not require us to know the function $\Phi(x)$, but only the scalar products $\langle \Phi(x_i), \Phi(x) \rangle$ for , $i = 1,\ldots,$ N. Under certain assumptions [31], the scalar product $\langle \Phi(x_i), \Phi(x) \rangle$ can be rewritten using non-linear functions (called "kernels") $k(x_i, x)$. A commonly used kernel is the Gaussian radial basis function, uniquely identified by the single parameter represented by the kernel width γ, which represents an additional model parameter. In this way, the optimization follows procedures similar to those in the linear case, but function $f(x)$ is allowed to be non-linear and of arbitrary complexity. Further details on the method can be found in [31], while a more intuitive description of the approach specifically targeted to hydrological applications is given in [34].

2.2.2. Model Training and Feature Selection

No standard methods exist for calibrating C, ε, γ. The choice of these parameters is driven by the characteristics of the problem itself. Operationally, the SVR prediction process can be schematically divided into two parts, an off-line and an on-line procedure (Figure 3).

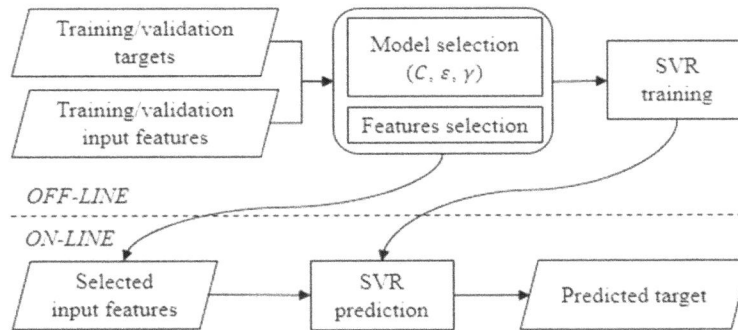

Figure 3. Support vector regression (SVR) general scheme separated between off-line and on-line phase. During the off-line phase the SVR is trained and the prediction function is estimated. The on-line phase refers to the operative part of the method, in which the prediction phase can accept a proper set of inputs for estimating the discharge in the target month.

The off-line part consists of SVR model training and supports the selection of relevant input variables. A common strategy for model training is to divide the available reference dataset into two subsets, one called "training set," which is used to find the optimal function for a fixed combination of parameters C, ε, γ, and one called "validation set," on which the performance in terms of a chosen empirical metric is computed (within this work, we used a set of metrics as explained in [35]). Instead of dividing the reference dataset into two subsets (*i.e.*, training and validation), we preferred to adopt a *t*-folds cross-validation procedure, which consists of randomly dividing the available reference samples into t subsets. Iteratively, $t - 1$ subsets are used for training, whereas the remaining subset is used for validation. In this way, all the samples are used for both training and validating the model. The cross-validation strategy may increase the representativeness of training and validation sets, especially in the case of limited samples being available [35].

The so-called "feature selection" operation aims at identifying those, among all explanatory variables ("features") considered, which are actually informative: The SVR model is trained iteratively with different combinations of features, and those combinations leading to the best model performances are eventually retained for further verification. Features that are found to be uninformative, or even worsen the performance of the model, are discarded.

2.2.3. Model Testing

Once the SVR model is properly trained with the selected input features, the SVR prediction can be used "on-line" for estimating a target variable using the set of selected input features. The estimates are compared to a set of known target variables (usually called "test set"), different from the training and validation ones, in order to independently assess the prediction accuracy of the method implemented.

The reference set can be built by exploring the whole time series, extracting at each instant the input feature vector by looking at the past and current values and the associated future target. For instance, monthly discharge at the instant $t + \Delta t$ may be predicted using monthly discharge at month $t, t-1$, etc., as well as the other explanatory variables at month $t, t-1$, etc.

2.3. Input Data Considered

We want to predict monthly discharges at month $t + \Delta t$, with Δt equal to one, three, or six months, using as input features the discharges measured at the current (t) and past ($t-1,..., t-q$) months (q being the number of months in the past that we want to consider), as well as other known variables that we assume may allow predicting discharges. These were chosen among the following:

- Catchment state variables: Snow storage represented by snow-covered area (SCA) as a proxy;
- Climate signals related to possible catchment input (temperature and precipitation): Indexes North Atlantic Oscillation (NAO), Wave Amplitude Index (WAI), Baroclinic Amplitude Index (BAI), Standardized Precipitation Index (SPI), and air temperature, briefly introduced and discussed hereafter;
- Catchment characteristics: Area, minimum, maximum, and mean elevation.

SCA has been extracted from snow maps obtained with MODIS images, which are freely provided by NASA (http://modis.gsfc.nasa.gov/) and acquired through a dedicated antenna available at EURAC Research, Bolzano. The snow products have been derived with a specific algorithm adapted to mountain areas [36,37]. This algorithm takes 250 m resolution, atmospherically corrected surface reflectance MOD09GQ-MYD09GQ (Terra-Aqua satellites) products as input. Cloud detection is carried out using 500 m resolution MOD09GA-MYD09GA products and 1 km resolution MOD021KM-MYD021KM products. MOD03-MYD03 are used for correct geolocation. Daily snow maps at 250-m resolution obtained with the algorithm described in [36] allow improved snow cover mapping with respect to the 500 m standard MODIS product, especially in mountainous terrain where snow cover shows high spatial variability due to highly irregular topography [37].

Climate signals announcing dry and wet periods can be derived from atmospheric circulation patterns. As shown in several studies, dry or wet conditions of a region can be connected to some particular large-scale climatic indicators, like the El-Nino Southern Oscillation (ENSO) or NAO [24,38–40], although a certain indicator may have different relevance in predicting climate in separate, but bordering, regions [24].

Other authors have shown a clear relationship between the phenomenology of meso-scale atmospheric conditions and extreme wet and dry events, especially in the mid-latitudes (see [41]). The SPI [42–44] has been proposed for application at different time scales [42,45,46], because it quantifies relative weather dryness and wetness, allowing comparison of climatic conditions in areas governed by different hydrological regimes.

It has been observed that the SPI computed over one month may correlate with unregulated stream flow, while the SPI over a longer period may correlate with reservoir storage [1,2]. SPI prediction based on statistical models has been conducted e.g., in Sicily [47] and Iran [48].

The WAI [49–51] and the BAI [52] are two additional well-established climatological indexes related to large-scale atmospheric circulation. The WAI is used to describe regimes of circulation in extra-tropical regions and can be used to identify atmospheric blocking events in Europe [49]. Atmospheric blocking leads to a stagnation of weather patterns, in turn possibly associated with floods, droughts, above- or below-normal temperatures, and other weather extremes. It is important to recognize a blocking pattern in its initial development. The BAI is related to heat fluxes and precipitation activity in the mid-latitudes [52], which may transport heat and momentum vertically and horizontally at synoptic scales in the mid-latitudes. These transfers may explain most of the high frequency variability in the extratropics [53]. The dominance of these climatic signals can be used to capture global trends that yield dry or wet periods [41].

Climatic signals may be related to catchment state and slow hydrological responses: Long lead-time forecasting has been successfully attempted on seasonal summer streamflows using ocean-atmospheric variables [54–56], suggesting that similar studies might be extended to a finer level of temporal detail (e.g., monthly). Studies on a two-century time series of discharges from the Po River, northern Italy, have highlighted a relationship between both winter precipitation and summer stream flow and the NAO [57].

In this case study, we considered SPI, WAI, BAI, NAO, and air temperature, the last deemed generally representative for snow melting.

Other variables may be in principle be added to the above list. For instance, cloud cover and catchment orientation (aspect) may be related to snowmelt, while vegetation and soil moisture indicators are representative of catchment state. In the catchments considered for the present analysis, the aspect varies at a relatively fine scale following the mountainous morphology, while the average aspect of each catchment is relatively constant. Vegetation and soil moisture were considered to be less representative than snow cover for the mountainous region of interest, as they account for relatively little water storage compared to snow. Finally, cloud cover was deemed a more indirect indicator than temperature. In a quest for balance between comprehensiveness of the model setup and complexity of model input preparation, we decided to limit the input variables to those discussed above.

In addition, a special input feature was also identified in the long-term average monthly discharge of the month of prediction. The latter was computed from the time series of discharges, considering the 10 years before the start of the prediction period. This choice, in general, puts an additional constraint on the selection of the hydrological stations, which need to be limited to those continuously monitored for at least 20 years.

2.4. Experimental Setup

In the study region, there are about 40 discharge gauging stations. However, many of these have not been monitored for a sufficiently long and continuous time period. In order to have a robust discharge prediction model, we chose to train a specialized SVR model only for the catchments with stations continuously monitored for at least 10 years, corresponding to a number of samples higher than 120; these correspond to the 14 measurement stations in Table 1.

The case of gauging stations with short or discontinuous monitoring is quite common in practice. Therefore, it is of interest to develop SVR models that may be applied to a catchment without specific calibration. A way to develop one such model is to consider all model training catchments together, while including catchment morphological descriptors as input features. In this way, we may hope to apply the model to any other catchment similar to the ones used for training. Such "regionalized" model calibration does not need to be repeated for each new catchment. For this reason, we complemented the catchment-specific models in the study region with a catchment-independent model, as explained in Section 2.4.2.

2.4.1. Data Preparation

The input dataset is formed by a time series of SCA, climatic signals (*i.e.*, NAO, WAI, BAI, SPI, and temperature), and measured monthly discharges.

The WAI and BAI indexes have been calculated from NCEP reanalysis. The NAO index has been taken by the Climate Prediction Center (CPC) of NOAA (National Oceanic and Atmospheric Administration) [58]. SPI and air temperature maps were obtained by interpolation of values measured at precipitation and temperature stations in the test area (Figure 2) over a 1-km resolution grid using ordinary kriging. A time series of the monthly mean SPI and temperature has been computed for each of the watersheds of the discharge stations by spatially averaging these maps (Figure 1).

We prepared the time series of all input features and target variables for each catchment for the period between November 2002 and May 2012, as the availability of the Aqua and Terra MODIS satellites, used to derive snow maps and thus extract the SCA, starts from 2002.

We selected three years to form the test set, while the remaining six years and seven months form the reference samples. This choice better ensures the independence between reference and test samples with respect to a random selection between the whole time series. Moreover, the choice of the reference set years as well as the choice of the test set years were made so as to ensure capturing as much inter-annual variability of discharges as possible. This is expected to yield a more reliable training of the SVR and a more realistic evaluation of the prediction accuracy on the test set.

The reference samples have been randomly divided into three subsets in order to set up the *t*-fold cross validation configuration explained in the above section. Then a multi-objective model selection procedure (as described in [35]) has been used to optimize with respect to more than one single figure of merit. In this approach both the percentage root mean square error (RMSE%) and mean square error (MSE) at the same time have been considered in the optimization.

2.4.2. Feature Selection

Feature selection is conducted together with model calibration during the off-line phase of SVR training (Figure 3). The feature selection procedure described hereafter was only applied to three representative catchments, a large one (n. 37, Adige at Bronzolo, 6923 km^2), a small one (n. 8, Rio Fleres at Colle Isarco, 149 km^2), and a medium-sized one (n. 28, Rienza at Vandoies, 1920 km^2), because of the time required for the examination of the high number of combinations of input features. With this choice, the procedure is faster and the representativeness of the selected features can be tested independently on the remaining 11 catchments of the study area.

The feature selection procedure was divided into three steps (Figure 4) and was based on the percentage root mean square error (RMSE%) of predicted *versus* observed discharges calculated on the validation samples.

The first step aims at selecting the input features related to catchment state, *i.e.*, the discharges of antecedent months and the snow accumulated in the catchment, described in our problem by the SCA time series. We tested as input several combinations of discharges of one to 12 antecedent months, and SCA from one to four months before the time of prediction. The same simulations have been repeated including and excluding the long-term average discharge of the month of prediction as input feature. An SVR was trained and tested for several combinations of input features, so as to identify their possibly optimal combination.

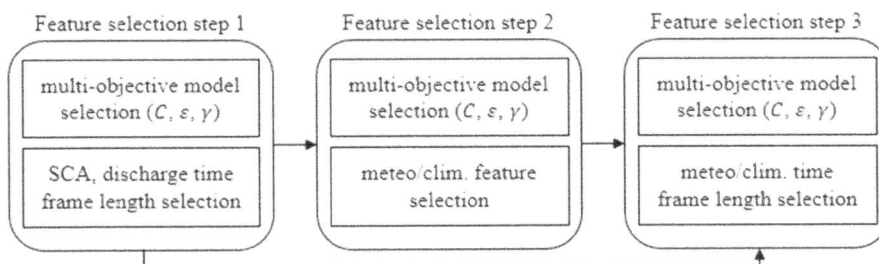

Figure 4. Feature selection scheme. The first step is relative to the selection of the time series length of the catchment state variables SCA and discharge. The second and the third step deal with the selection of the climatic and meteorological parameters that describe precipitation and temperature.

In the second step we select the climatic signals possibly relevant as input features. In order to better understand which parameter is actually informative, we added to the input feature selected in step 1 those variables that referred to the target month, thus mimicking an ideal forecast (*i.e.*, with no forecasting errors). The configuration that exhibits the highest performance on the validation set is selected.

In any operational prediction, the meteorological and climatic variables of the target month are not available, but they need to be predicted. Once we have identified those climatic signals relevant for the prediction, in step 3 of the feature selection process we replace the SVR input features that refer to the actual future value of the input features selected in step 2 with a combination of their values in a number of months before prediction; step 3 is therefore responsible for finding the optimal number of antecedent months for these variables.

2.4.3. Setup of a Catchment-Independent Support Vector Regression (SVR) Model

Following the approach described above, we have set up an SVR model for each individual catchment in the study region based on the features selected for three representative catchments and the parameters C, ε, and γ optimized individually for each catchment. In principle, any additional catchment in the area would require calibrating a specific SVR model. An SVR model calibrated on all catchments together on the basis of catchment features may be of practical interest as it would allow predictions on a generic catchment in the region. Therefore we have calibrated an SVR model for all catchments together, considering as target variables the normalized monthly discharges (discharges divided by watershed area), in order to have comparable quantities given the large variation of catchment sizes.

In the setup of an SVR model for all catchments together, catchment characteristics such as area and minimum, maximum, and average elevation were considered as input features along with the SCA and climate signals discussed above. On the contrary, the long-term average monthly discharge of the month of prediction has not been used as an input feature because its use implies the availability of a long discharge time series, in contrast with the goal of the approach—To also address poorly gauged catchments.

We iteratively simulated one of the 14 catchments as the one without a long time series available and thus we trained an SVR using the remaining 13 catchments considering the discharges normalized by catchment area, and SCA.

It is anticipated that an SVR for all catchments together may suffer from lower prediction accuracy with respect to a specialized SVR trained on a single catchment. For the setup of an SVR model for all catchments together, we followed the same procedure described above, by iteratively excluding one catchment at a time from the training phase and using it as a test set, so as to assess the performance obtained for the test samples of all the 14 catchments considered in this study.

3. Results and Discussion

The long-term average monthly discharge of the month of prediction, also used as an input feature, is the expected value of the discharge itself, and may be regarded as a benchmark for the performances obtained with different SVR configurations. Indeed, any prediction performing worse than this benchmark is basically of no use. Table 2 reports the RMSE% and the coefficient of determination (R^2) computed on the whole set of discharges for the 14 stations, by comparing observations with the long-term average discharge of the target month (*i.e.*, the benchmark predictor). The RMSE% is defined as:

$$RMSE\% = \sqrt{\frac{1}{N}\sum_{i=1}^{N}\left(\frac{A-P}{A}\right)^2} \times 100 \qquad (9$$

where A is the actual value and P the predicted value of the target. The "reference" RMSE% corresponding to the benchmark is equal to 33% ($R^2 = 0.7$) and sets the acceptance threshold for all other predictors. The RMSE% and R^2 of the long-term average predictor are obviously independent of the lead time. In the three-step feature selection procedure, any combination of features yielding a RMSE% higher than these values should be discarded.

Predictions at one month lead time with the features shown in Table 2 have a lower RMSE% with respect to the benchmark predictor (Figure 5). The mean RMSE% over the 14 catchments is consistently clearly reduced. With a prediction lead time of three months the improvement is less apparent, while with a lead time of six months the accuracy seems to be similar to the benchmark. The best performing SVR configuration for a lead time of one month reduces the RMSE% to 22% and increases R^2 to 0.76; for a lead time of three months, the corresponding figures are 28% and 0.68 (*i.e.*, a slight worsening), respectively; for a lead time of six months, we find RMSE% = 31% with no improvement on R^2.

The feature selection steps allow us to develop an understanding of the contribution of each single input feature of the SVR model. This is shown in Table 2. Step 1 of the feature selection procedure, in particular, allows us to identify the importance of antecedent monthly discharges, SCA, and the long-term average discharge of the target month.

SCA appears to be quite informative. Figure 6 shows the comparison between the RMSE% obtained on the individual catchments with SCA as input against the best configuration without SCA. As shown by the graph, the RMSE% decreases significantly when using SCA compared to using antecedent discharges only, except for the lead time of six months, in which case differences vanish. This is somehow expected as, at a longer lead time, the catchment state before prediction described by the SCA assumes less relevance.

Table 2. Performances of different SVR model configurations for single stations. The fourth and fifth column refer to the mean RMSE% and R^2 computed over the test samples of the 14 watersheds. In the third column "disch" stand for discharge, "dischAvg10" for the 10-year average discharge of the target month, and "temp" for air temperature. The notation—n: 0 stands for a time series of length $n - 1$ starting from the n-tested or the previous month to the current month from where the discharge prediction is done. As a benchmark, we considered the method based on the average discharge calculated over the 10 previous years, whose performances are shown in the first row. In addition, for a prediction lag of one and three months the results obtained with the AR model are shown.

Prediction Lag	Prediction Method	Input Features	Mean RMSE%	Mean R^2
All	10-year average	--	33	0.70
1	SVRsingle noMeteo	disch0, SCA-2:0, dischAvg10	22	0.76
	SVRsingle noMeteo noSCA	disch-11:0, dischAvg10	28	0.72
	SVRsingle noMeteo noDischAvg	disch-11:0, SCA-2:0	31	0.72
	SVRsingle meteoIdeal	disch0, SCA-2:0, dischAvg10, SPI, temp	21	0.79
	SVRsingle meteo	disch0, SCA-2:0, dischAvg10, SPI-1:0, temp-11:0	25	0.76
	AR model	disch-11:0	33	0.69
	AR model	disch-6:0	36	0.48
	AR model	disch-3:0	33	0.50
	AR model	disch0	33	0.49
3	SVRsingle noMeteo	disch0, SCA-1:0, dischAvg10	28	0.68
	SVRsingle noMeteo noSCA	disch-10:0, dischAvg10	33	0.69
	SVRsingle noMeteo noDischAvg	disch-9:0, SCA-2:0	33	0.68
	AR model	disch-11:0	35	0.70
	AR model	disch-6:0	69	0.03
	AR model	disch-3:0	56	0.01
	AR model	disch0	61	0.01
6	SVRsingle noMeteo	disch-10:0, SCA0, dischAvg10	31	0.70
	SVRsingle noMeteo noSCA	disch-11:0, dischAvg10	32	0.70
	SVRsingle noMeteo noDischAvg	disch-11:0, SCA-3:0	32	0.69

In addition, the long-term average discharge of the target month improves the prediction accuracy for all lead times (one, three, or six months): The mean improvement of the RMSE% over all the catchments analyzed, with respect to the corresponding model configuration without this input, is equal to 9%, 5%, and 1%, respectively, for a prediction lag equal to one, three, and six months (see Table 2). This indicates the capacity of the SVR model to better exploit this input at shorter lead times. At six months' lead time, the antecedent 12 months' discharges are shown to contain more or less the same information with respect to the long-term average discharge (see Table 2).

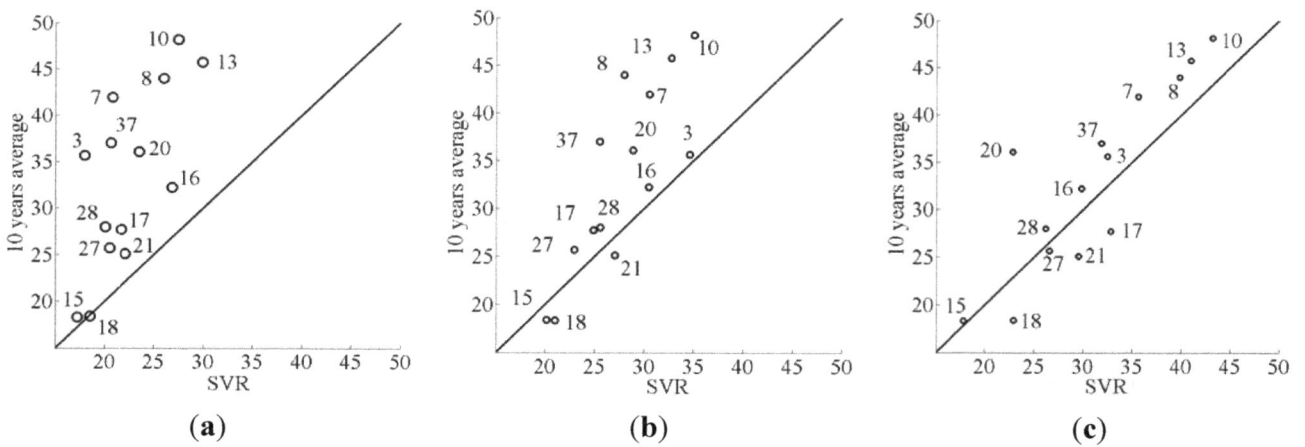

Figure 5. Comparison of the RMSE% obtained on the test set between the SVR method proposed (*x* axis) against the simple method based only on the average discharge of the target month computed over 10 previous years. In the scatterplots (**a**), (**b**), and (**c**) the prediction lag is equal to one, three, and six months, respectively.

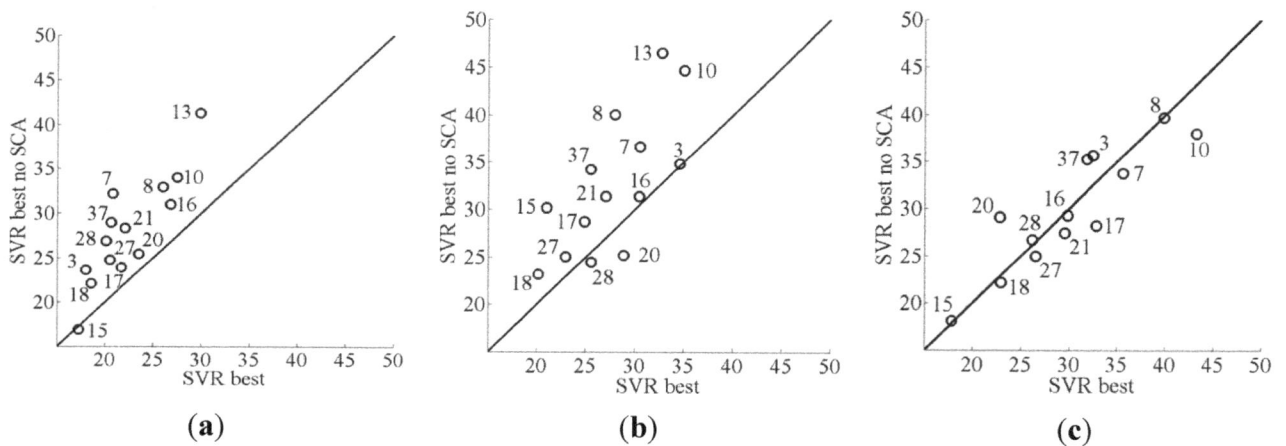

Figure 6. Comparison between the RMSE% obtained on the test set with (x axis) and without (y axis) SCA as SVR input feature. The scatterplots (**a**), (**b**), and (**c**) refer to the performances obtained by the SVR trained for the prediction with a lag equal to one, three, and six months, respectively.

SPI and temperature were the most significant input features resulting from the feature selection procedure on meteorological and climatic variables. However, their impact is appreciable only at one month lead time, in the case of an "ideal forecast," while using the values of antecedent months even increases the RMSE% (Figure 7). For a longer lead time, their inclusion as input features never improves the model. As an additional benchmark, we considered a linear autoregressive (AR) model set up with input ranging from one-month antecedent discharge to 12-month antecedent discharges. The statistical metrics of performance (RMSE% and R^2) of the AR model are shown in Table 2. These are only slightly better, and sometimes even worse, than the 10-year long-term average of the month of prediction. The

SVR is able to reduce the RMSE% with respect to the best AR method of 9% and 7% for a prediction lag equal to one and three months, respectively.

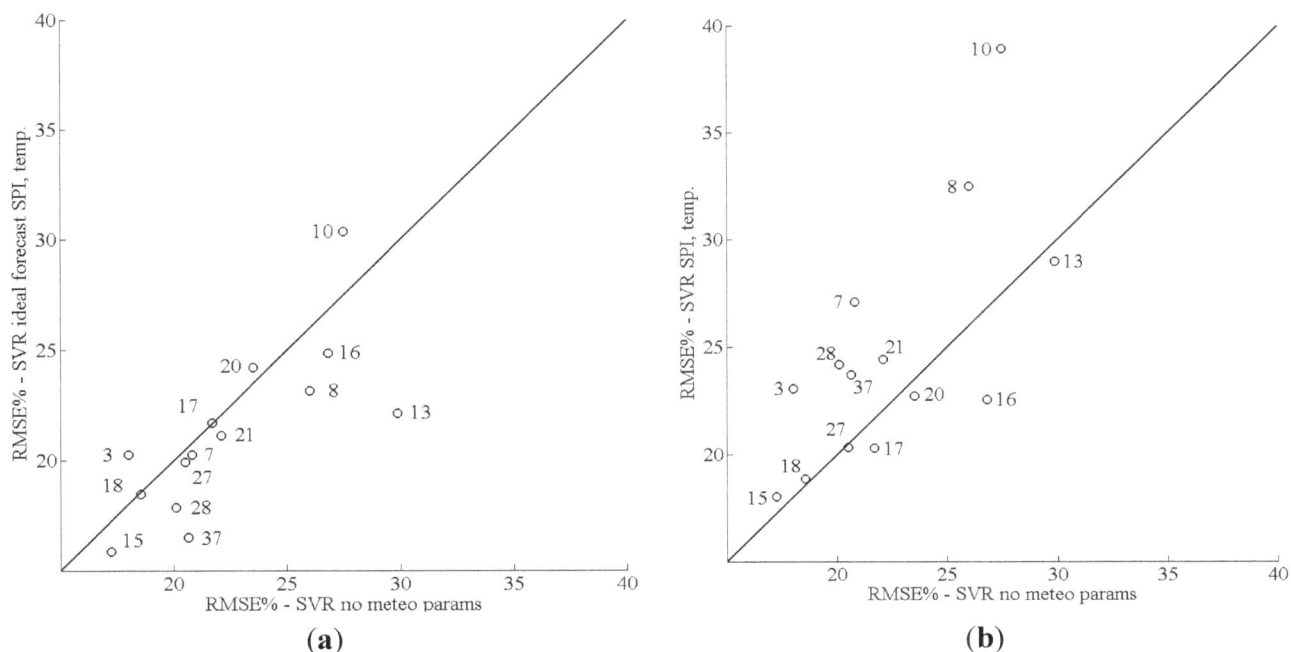

Figure 7. Comparison between the RMSE% obtained on the test set with (y axis) and without (x axis) SPI and temperature as SVR input feature. In (**a**) the actual target of SPI and temperature in the target month is employed (simulating an ideal forecast of this variable); In (**b**) the past time series of SPI and temperature is adopted. Both scatterplots refer to the performances obtained by the SVR trained for the prediction with a lag equal to one month.

The RMSE% is a metric of the general estimation accuracy for the different configurations tested and for the comparison with the benchmark method. However, a deeper error analysis on each catchment may yield useful additional insights. By looking at the error distribution of the Adige catchment closed at Ponte Adige (catchment identifier = 7 in Table 1), for instance, we have derived the error distribution on the test samples (shown in Figure 8a,b), for the most significant SVR configurations selected for predictions with one and three months lead time, together with the benchmarks. Besides yielding lower mean error values, the SVR models allow us to always keep error below 50%, while the benchmarks and SVR without SCA as input show a non-negligible fraction of higher errors for prediction lags equal to both one and three months.

Finally, it is of interest to inspect the cumulative volume errors: In many cases of practical interest, what matters is not necessarily the average error, but the total wrongly forecasted volume that may be associated with the monetary costs of water that was wrongly assumed to be available or lacking. Overestimation may, e.g., induce a hydropower plant manager to book unnecessary grid dispatching capacity, while underestimation may cause an irrigation manager to book unnecessary external supply. Figure 9 shows, as an example, a comparison of the volume forecasting errors for the benchmark of the long-term average discharge, for the SVR model with and without SCA as input, and for the best-performing AR model (the one with 12-month antecedent discharges), highlighting that the SVR model with SCA as input generates lower volume errors at a one-month lead time. At the same time, the

AR model and the SVR model without SCA introduce less appreciable advantages over the long-term average discharge. At three months' lead time, however, no advantage is shown by the SVR model, which performs no better than the benchmark (results not shown here for simplicity).

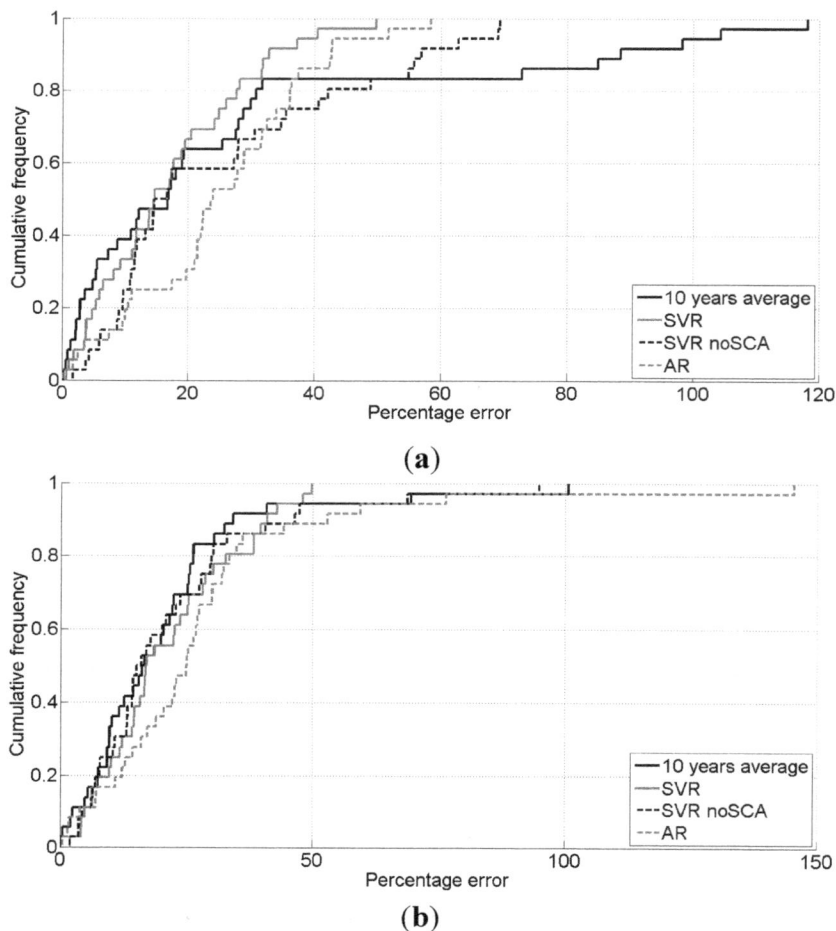

(a)

(b)

Figure 8. Cumulative frequency of the percentage error computed on the test set for watershed n. 7 (Adige at Ponte Adige), chosen as illustrative example. Each pdf shows the percentage error distribution for different input features configuration. The full gray line represents the selected SVR configuration. The benefit of the SVR approach with respect to the methods based on the 10-year average discharge (full black line) and to the AR method (dashed gray line) is clear as well as the improvement given by SCA (dashed black line). (a,b) show a prediction lag equal to one and three months, respectively.

When considering the SVR model calibrated for all catchments together, we observe that it generally yields less accurate predictions than a model set up for each individual catchment (see Table 3). At lead times of three and six months, none of the tested SVR configurations yields a better performance than the long-term catchment average, while at a lead time of one month the RMSE% may be reduced from 33% to 27%. Concerning the relevance of input features to prediction, the considerations expressed above for the SVR trained on individual catchments generally continue to hold: Also in this case the SCA and antecedent monthly discharges are retained as informative input features (step 1); the best performances are obtained using discharge data from the 12 months before prediction and the mean SCA from the two antecedent months as input features. On the other hand, none of the considered watershed attributes proved to be informative in our tests apart from catchment area, which is only used for normalizing the discharges. At step 2 of the feature selection procedure, climatic signals reveal a minor impact on prediction only in the case of an "ideal forecast."

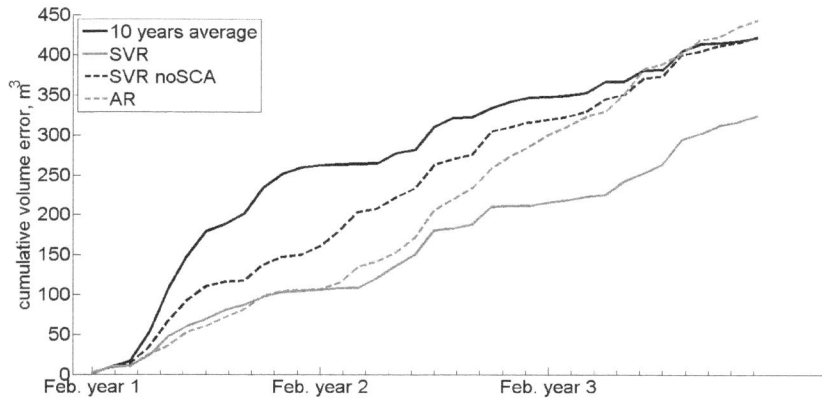

Figure 9. Cumulative volume errors of the one-month lead time forecast during the three test years (*i.e.*, 2005, 2010, 2011) relative to the catchment closed at Ponte Adige (*n.* 7).

Table 3. Performances of different SVR model configurations for all stations. The fourth and fifth column refer to the mean RMSE% and R square computed over the test samples of the 14 watersheds. In the third column "disch" stands for discharge and "temp" for air temperature. The notation—n: 0 stands for a time series of length $n - 1$ starting from the n-th previous month to the current month from where the discharge prediction is done. As a benchmark we considered the method based on the average discharge calculated over the 10 previous years, whose performances are shown in the first row.

Prediction Lag	Prediction Method	Input Features	Mean RMSE%	Mean R^2
All	10-year average	--	33	0.70
1	SVRall noMeteo	disch-11:0, SCA-1:0	27	0.75
	SVRall meteoIdeal	disch-11:0, SCA-1:0, SPI, temp	24	0.81
	SVRall meteo	disch-11:0, SCA-1:0, SPI-3:0, temp-10:0	28	0.78
3	SVRall noMeteo	disch-10:0, SCA-1:0, SPI, temp	36	0.70
	SVRall meteo	disch-11:0, SCA-1:0, SPI-3:0, temp-9:0	36	0.74

The results obtained for South Tyrol are in line with findings in other contexts using similar techniques [8,11–15,54].

Purely statistical predictions as presented here usually perform better than deterministic predictions based on seasonal weather or climate forecasts. For instance, Bastola *et al.* [59] highlighted the shortcomings of such deterministic predictions in the southeastern United States, while in general there is substantial agreement on the importance of catchment state characterization for effective streamflow prediction [60–62].

4. Conclusions

We have presented an application of SVR models for predicting the monthly mean discharge over 14 catchments in the Alpine region of South Tyrol (Northern Italy) using monthly discharge and SCA of antecedent months as input features. The comparison between the SVR models and the long-term monthly mean discharge, as well as with a simpler linear autoregressive model, is encouraging, with a mean improvement of the RMSE% on the 14 catchments selected in this study of 11%, 5%, and 2% for a prediction lag equal to one, three, and six months, respectively (Table 2), and a reduction in frequency of larger errors (Figures 8 and 9). An SVR model proves to be useful compared to the long-term mean and a linear model, when a penalty is associated with over- or underestimation of monthly streamflow, as it allows reducing volume errors when predicting at a one month lead time, compared to the benchmarks. This advantage, however, is no longer apparent when considering lead times of three months.

The analysis also highlights the usefulness of SCA as a catchment state variable in the study region. Through appropriate feature selection, we showed that the SVR models trained with the SCA show an RMSE% on average 6% lower for a prediction lag equal to one month compared to the SVR models that consider only antecedent discharges as input features. On the contrary, catchment input (meteorological/climatological) variables, of which SPI and temperature have been shown to be the most informative in explaining monthly discharges, have been shown to bring in little, if any, prediction improvement. For the mountainous region considered in this study, even a perfect knowledge of precipitation (as SPI) and temperature during the month of prediction would not substantially improve discharge predictions, which suggests the importance of catchment state controls on flows, a (partial) descriptor of which is identified in snow-covered areas. This corroborates the observation, general in catchment hydrology, that discharge may be a complex and nonlinear response to precipitation and evapotranspiration, and can be more often related to the antecedent storage than to the input of water in a catchment. On the other hand, in models based on the catchment water balance, precipitation (and evapotranspiration) is as essential for prediction as antecedent storage; black-box models in this case study definitely "learn" more from the latter than from the former.

In order to predict monthly discharges for catchments where a sufficiently long discharge time series is not available, a general SVR trained by using the data of similar catchments, after normalizing discharges with the watershed area, still leads to a slight improvement for predictions at a lead time of one month (RMSE% reduced of 5%) while at longer lead times it yields predictions no better than the benchmark, or even slightly worse. This suggests that case-specific setup of statistical models as applied in this study may be difficult to generalize.

Machine learning techniques such as SVR may help improve our capacity for seasonal hydrological forecasting. Monthly discharge predictions at one to three months with RMSE% below 30% and an appreciable reduction in the frequency of high errors may already enable better management of water resources through optimization of storage at (and release from) reservoirs. However, further research is required, particularly in the description of catchment state (e.g., by using better proxies of snow water equivalent than simple SCA), in order to improve prediction performance in view of more effective practical applications, as catchment input seems relatively irrelevant even in the case of an ideal prediction, hence even less relevant after considering errors from real seasonal weather forecasts.

Acknowledgments

The research was conducted with support awarded to GECOsistema, R&D Unit Suedtirol, by the Province of Bolzano (LP 14/2006, "Bando Innovazione" 2011), in collaboration with EURAC Research. Hydrological data were kindly provided by the Province of Bolzano, Hydrographic Office. Daniela Dellantonio (EURAC Research) is thankfully acknowledged for her proofreading of the manuscript.

Author Contributions

All authors contributed to writing different parts of the paper. Alberto Pistocchi ideated the research building on discussions with Claudia Notarnicola, Paolo Mazzoli and Marcello Petitta and coordinated the writing of the paper. Mattia Callegari analyzed the data, developed the SVR calculations, and contributed to writing large parts of the paper; Ludovica De Gregorio contributed to the analysis of data and SVR calculations; Paolo Mazzoli supervised the selection and interpretation of hydrological data and largely contributed to the discussion of the results; Claudia Notarnicola. supervised the snow cover remote sensing data processing; Luca Pasolli contributed to the selection and calibration of the SVR model as well as the interpretation and discussion of the results; and Marcello Petitta selected and investigated the climate parameters used in the paper.

Conflicts of Interest

The authors declare no conflict of interest.

References

1. Low Flow and Drought Research in Europe: Achievements and Future. Available online: http://webworld.unesco.org/water/ihp/pdf/conclusions_bratislava_lfdr_workshop08.pdf (accessed on 30 April 2015).
2. Mishra, A.K.; Desai, V.R. Drought forecasting using feed-forward recursive neural network. *Ecol. Model.* **2006**, *198*, 127–148.
3. Madadgar, S.; Moradkhani, H. A Bayesian framework for probabilistic seasonal drought forecasting. *J. Hydrometeorol.* **2013**, *14*, 1685–1705.
4. Yang, L.; Tian, F.; Sun, Y.; Hu, H. Attribution of hydrologic forecast uncertainty within scalable forecast windows. *Hydrol. Earth Syst. Sci.* **2014**, *18*, 775–786.
5. Chalise, S.R.; Kansakar, S.R.; Rees, G.; Croker, K.; Zaidman, M. Management of water resources and low flow estimation for the Himalayan basins of Nepal. *J. Hydrol.* **2003**, *282*, 25–35.
6. Pandit, S.M.; Wu, S.-M. *Time Series and System Analysis with Applications*; John Wiley & Sons, Inc.: Hpboken, NJ, USA, 1983.
7. Vapnik, V. *The Nature of Statistical Learning Theory*; Springer: New York, NY, USA, 1995.
8. Wang, W.; Chau, K.; Cheng, C.; Qiu, L. A comparison of performance of several artificial intelligence methods for forecasting monthly discharge time series. *J. Hydrol.* **2009**, *374*, 294–306.
9. Bruzzone, L.; Melgani, F. Robust multiple estimator system for the analysis of biophysical parameters from remotely sensed data. *IEEE Trans. Geosci. Remote Sens.* **2005**, *43*, 159–174.

10. Pasolli, L.; Notarnicola, C.; Bruzzone, L. Estimating soil moisture with the support vector regression technique. *IEEE Geosci. Remote Sens. Lett.* **2011**, *8*, 1080–1084.

11. Shiri, J.; Kisi, O. Short-term and long-term streamflow forecasting using a wavelet and neuro-fuzzy conjunction model. *J. Hydrol.* **2010**, *394*, 486–493.

12. Guo, J.; Zhou, J.; Qin, H.; Zou, Q.; Li, Q. Monthly streamflow forecasting based on improved support vector machine model. *Expert Syst. Appl.* **2011**, *38*, 13073–13081.

13. Shiri, J.; Cimen, M. A wavelet-support vector machine conjunction model for monthly streamflow forecasting. *J. Hydrol.* **2011**, *399*, 132–140.

14. Zhao, T.; Cai, X.; Yang, D. Effect of streamflow forecast uncertainty on real-time reservoir operation. *Adv. Water Resour.* **2011**, *34*, 495–504.

15. Lima, C.H.R.; Lall, U. Climate informed monthly streamflow forecasts for the Brazilian hydropower network using a periodic ridge regression model. *J. Hydrol.* **2010**, *380*, 438–449.

16. Liu, Z.; Zhou, P.; Chen, G.; Guo, L. Evaluating a coupled discrete wavelet transform and support vector regression for daily and monthly streamflow forecasting. *J. Hydrol.* **2014**, *519*, 2822–2831.

17. Huang, S.; Chang, J.; Huang, Q.; Chen, Y. Monthly streamflow prediction using modified EMD-based support vector machine. *J. Hydrol.* **2014**, *511*, 764–775.

18. Sun, A.Y.; Wang, D.; Xu, X. Monthly streamflow forecasting using Gaussian process regression. *J. Hydrol.* **2014**, *511*, 72–81.

19. Dehghani, M.; Saghafian, B.; Nasiri Saleh, F.; Farokhnia, A.; Noori, R. Uncertainty analysis of streamflow drought forecast using artificial neural networks and Monte Carlo simulation. *Int. J. Climatol.* **2014**, *34*, 1169–1180.

20. Wang, H.; Fu, X. Utilization of climate information and soil moisture estimates to provide monthly and sub-monthly streamflow forecasts. *Int. J. Climatol.* **2014**, *34*, 3515–3527.

21. Fiorillo, F.; Esposito, L.; Guadagno, F.M. Analysis and forecast of water resources in an ultra-centenarian spring discharge series from Serino (Southern Italy). *J. Hydrol.* **2007**, *336*, 125–138.

22. Fiorillo, F. Spring hydrographs as indicators of droughts in a karts environment. *J. Hydrol.* **2009**, *373*, 290–301.

23. Mendicino, G.; Senatore, A.; Versace, P. A Groundwater resource index (GRI) for drought monitoring and forecasting in a Mediterranean climate. *J. Hydrol.* **2008**, *357*, 282–302.

24. Bartolini, E.; Claps, P.; D'Odorico, P. Interannual variability of winter precipitation in the European Alps: Relations with the North Atlantic Oscillation. *Hydrol. Earth Syst. Sci.* **2009**, *13*, 17–25.

25. Singh, P.; Bengtsson, L. Hydrological sensitivity of a large Himalayan basin to climate change. *Hydrol. Process.* **2004**, *18*, 2363–2385.

26. Parajka, J.; Blöschl, G. The value of MODIS snow cover data in validating and calibrating conceptual hydrologic models. *J. Hydrol.* **2008**, *358*, 240–258.

27. Viviroli, D.; Gurtz, J.; Zappa, M. The Hydrological Modeling System PREVAH—Part II—Physical model description. In *Geographica Bernensia P40*; Institute of Geography, University of Berne: Berne, Switzerland, 2007, p. 86.

28. Viviroli, D.; Dürr, H.H.; Messerli, B.; Meybeck, M.; Weingartner, R. Mountains of the world, water towers for humanity: Typology, mapping and global significance. *Water. Resour. Res.* **2007**, *43*, doi:10.1029/2006WR005653.

29. European Environmental Agency. *Regional Climate Change and Adaptation: The Alps Facing the Challenge of Changing Water Resources*; EEA report No. 8/2009; European Environmental Agency: Copenhagen, Denmark, 2009.

30. Evgeniou, T.; Pontil, M.; Poggio, T. Regularization Network and support vector machines. *Adv. Comput. Math.* **2000**, *13*, 1–50.

31. Smola, A.J.; Sholkopf, B. A tutorial in support vector regressions. *Stat. Comput.* **2004**, *14*, 199–222.

32. Zahn, H.; Shi, P.; Chen, C. Retrieval of oceanic chlorophyll concentration using support vector machines. *IEEE Trans. Geosci. Remote Sens.* **2003**, *41*, 2947–2951.

33. Pai, P.F.; Lin, K.P.; Lin, C.S.; Chang, P.T. Time series forecasting by a seasonal support vector regression model. *Expert Syst. Appl.* **2010**, *37*, 4261–4265.

34. Raghavendra, N.S.; Deka, P.C. Support vector machine applications in the field of hydrology: A review. *Appl. Soft Comput.* **2014**, *19*, 372–386.

35. Pasolli, L.; Notarnicola, C.; Bruzzone, L. Multi-objective parameter optimization in support vector regression: General formulation and application to the retrieval of soil moisture from remote sensing data. *IEEE J. Sel. Top. Appl.* **2012**, *5*, 1495–1508.

36. Notarnicola, C.; Duguay, M.; Moelg, N.; Schellenberger, T.; Tetzlaff, A.; Monsorno, R.; Costa, A.; Steurer, C.; Zebisch, M. Snow cover maps from MODIS images at 250 m resolution, Part 1: Algorithm description. *Remote Sens.* **2013**, *5*, 110–126.

37. Notarnicola, C.; Duguay, M.; Moelg, N.; Schellenberger, T.; Tetzlaff, A.; Monsorno, R.; Costa, A.; Steurer, C.; Zebisch, M. Snow Cover Maps from MODIS Images at 250 m Resolution, Part 2: Validation. *Remote Sens.* **2013**, *5*, 1568–1587.

38. Piechota, T.C.; Dracup, J.A. Drought and regional hydrologic variation in the United States: Associations with ElNino-Southern Oscillation. *Water Resour. Res.* **1996**, *32*, 1359–1373.

39. Tadesse, T.; Wilhite, D.A.; Harms, S.K.; Hayes, M.J.; Goddard, S. Drought monitoring using data mining techniques: A case study for Nebraska, USA. *Nat. Hazards* **2004**, *33*, 137–159.

40. Barros, A.P.; Bowden, G.J. Toward long-lead operational forecasts of drought: An experimental study in the Murray-Darling River Basin. *J. Hydrol.* **2008**, *357*, 349–367.

41. Bordi, I.; Fraedrich, K.; Petitta, M.; Sutera, A. Extreme value analysis of wet and dry periods in Sicily. *Theor. Appl. Climatol.* **2007**, *87*, 61–71.

42. McKee, T.; Doesken, N.; Kleist, J. The relationship of drought frequency and duration to time scales. In Proceedings of the 8th Conference on Applied Climatology, Anaheim, CA, USA, 17–22 January 1993.

43. Hayes, M.J.; Svoboda, M.D.; Wilhite, D.A.; Vanyarkho, O.V. Monitoring the 1996 drought using the standardized precipitation index. *Bull. Am. Meteor. Soc.* **1999**, *80*, 429–438.

44. Vicente-Serrano, S.M.; Lopez-Moreno, J.I. Hydrological response to different time scales of climatological drought: An evaluation of the Standardized Precipitation Index in a mountainous Mediterranean basin. *Hydrol. Earth Syst. Sci.* **2005**, *9*, 523–533.

45. Keyantash, J.; Dracup, J.A. The quantification of drought: An evaluation of drought indices. *Bull. Am. Meteor. Soc.* **2002**, *83*, 1167–1180.

46. Bordi, I.; Sutera, A. Fifty years of precipitation: Some spatially remote teleconnections. *Water Resour. Manag.* **2001**, *15*, 247–280.

47. Bordi, I.; Fraedrich, K.; Petitta, M.; Sutera, A. Methods for predicting drought occurrences. In Proceedings of the EWRA (The European Water Resources Association) Conference, Menton, France, 7–10 September 2005.

48. Morid, S.; Smakhtin, V.; Bagherzadeh, K. Drought forecasting using artificial neural networks and time series of drought indices. *Int. J. Climatol.* **2007**, *27*, 2103–2111.

49. Sutera, A. Probability density distribution of large-scale atmospheric flow. *Adv. Geophys.* **1996**, *29*, 227–249.

50. Hansen, A.R.; Sutera, A. On the probability density distribution of planetary-scale atmospheric wave amplitude. *J. Atmos. Sci.* **1986**, *43*, 3250–3265.

51. Calmanti, S.; Dell'Aquila, A.; Lucarini, V.; Ruti, P.M.; Speranza, A. Statistical properties of mid-latitude atmospheric variability. In *Nonlinear Dynamics in Geosciences*; Springer: New York, NY, USA, 2007; pp. 369–391.

52. Dell'Aquila, A.; Ruti, P.M.; Sutera, A. Effects of the baroclinic adjustment on the tropopause in the NCEP-NCAR reanalysis. *Clim. Dyn.* **2007**, *28*, 2–3.

53. Fraedrich, K.; Bottger, H. A wavenumber frequency analysis of the 500 mb geopotential at 50°N. *J. Atmos. Sci.* **1978**, *35*, 745–750.

54. Soukup, T.L.; Oubeidillah, A.A.; Tootle, G.A.; Piechota, T.C.; Wulff, S.S. Long lead time streamflow forecasting of the North Platte River incorporating oceanic-atmospheric climate variability. *J. Hydrol.* **2009**, *368*, 131–142.

55. Chiew, F.H.S. Climate variability, seasonal streamflow forecast and water resources management. In *Climate Impact on Australia's Natural Resources: Current and Future Challenges*; Climate impact on Australia's Natural Resources: Queensland, Australia, 2003; pp. 21–23.

56. Chiew, F.H.S.; McMahon, T.A. Global ENSO-streamflow teleconnection, streamflow forecasting and interannual variability. *Hydrolog. Sci. J.* **2002**, *47*, 505–522.

57. Zanchettin, D.; Traverso, P.; Tomasino, M. Po river discharges: A preliminary analysis of a 200-year time series. *Clim. Chang.* **2008**, *89*, 411–433.

58. Nationa Weather Service (NWS) Climate Prediction Center: North Atlantic Oscillation. Available online: http://www.cpc.ncep.noaa.gov/products/precip/CWlink/pna/nao.shtml (accessed on 30 April 2015).

59. Satish, B.; Misra, V.; Li, H. Seasonal hydrological forecasts for watersheds over the southeastern United States for the boreal summer and fall seasons. *Earth Interact.* **2013**, *17*, 1–22.

60. Yossef, N.C.; Winsemius, H.; Weerts, A.; van Beek, R.; Bierkens, M.F.P. Skill of a global seasonal streamflow forecasting system, relative roles of initial conditions and meteorological forcing. *Water Resour. Res.* **2013**, *49*, 4687–4699.

61. Shukla, S.; Sheffield, J.; Wood, E.F.; Lettenmaier, D.P. On the sources of global land surface hydrologic predictability. *Hydrol. Earth Syst. Sci.* **2013**, *17*, 2781–2796.

62. Mahanama, S.; Livneh, B.; Koster, R.; Lettenmaier, D.; Reichle, R. Soil moisture, snow, and seasonal streamflow forecasts in the United States. *J. Hydrometeor.* **2012**, *13*, 189–203.

6

Modeling of Soil Water and Salt Dynamics and Its Effects on Root Water Uptake in Heihe Arid Wetland, Gansu, China

Huijie Li [1,†], Jun Yi [1,†], Jianguo Zhang [1], Ying Zhao [1,2,*], Bingcheng Si [2,3,*], Robert Lee Hill [4], Lele Cui [1,5] and Xiaoyu Liu [1]

[1] Key Laboratory of Plant Nutrition and the Agri-Environment in Northwest China, Ministry of Agriculture, Northwest Agriculture and Forestry University, Yangling 712100, China; E-Mails: lihuijie2015@gmail.com (H.L.); yijun_soil@yahoo.com (J.Y.); zhangjianguo21@nwsuaf.edu.cn (J.Z.); Soilhydrology@gmail.com (L.C.); liuxiaoyu417@163.com (X.L.)

[2] Department of Soil Science, University of Saskatchewan, Saskatoon, SK, S7N 5A8, Canada

[3] College of Water Resources and Architecture Engineering, Northwest Agriculture and Forestry University, Yangling 712100, China

[4] Department of Environmental Science and Technology, University of Maryland, College Park, MD 20742, USA; E-Mail: rlh@umd.edu

[5] Suide Test Station of Water and Soil Conservation, Yellow River Conservancy Committee of the Ministry of Water Resources, Yulin, Shanxi 719000, China

[†] These authors contributed equally to this work.

[*] Author to whom correspondence should be addressed; E-Mail: yzhaosoils@gmail.com (Y.Z.); Bing.Si@usask.ca (B.S.);

Academic Editor: Lutz Breuer

Abstract: In the Heihe River basin, China, increased salinity and water shortages present serious threats to the sustainability of arid wetlands. It is critical to understand the interactions between soil water and salts (from saline shallow groundwater and the river) and their effects on plant growth under the influence of shallow groundwater and irrigation. In this study, the Hydrus-1D model was used in an arid wetland of the Middle Heihe River to investigate the effects of the dynamics of soil water, soil salinization, and depth to water table (DWT) as well as groundwater salinity on *Chinese tamarisk* root water uptake. The modeled soil water and electrical conductivity of soil solution (EC_{sw}) are in good agreement with the observations, as indicated by *RMSE* values (0.031 and 0.046 $cm^3 \cdot cm^{-3}$ for soil water content, 0.037 and 0.035 $dS \cdot m^{-1}$ for EC_{sw}, during the model calibration and validation periods, respectively). The calibrated model was used in scenario analyses considering different DWTs, salinity levels and the introduction of preseason irrigation. The results showed that (I) *Chinese tamarisk* root distribution was greatly affected by soil water and salt distribution in the soil profile, with about 73.8% of the roots being distributed in the 20–60 cm layer; (II) root water uptake accounted for 91.0% of the potential maximal value when water stress was considered, and for 41.6% when both water and salt stress were considered; (III) root water uptake was very sensitive to fluctuations of the water table, and was greatly reduced when the DWT was either dropped or raised 60% of the 2012 reference depth; (IV) arid wetland vegetation exhibited a high level of groundwater dependence even though shallow groundwater resulted in increased soil salinization and (V) preseason irrigation could effectively increase root water uptake by leaching salts from the root zone. We concluded that a suitable water table and groundwater salinity coupled with proper irrigation are key factors to sustainable development of arid wetlands.

Keywords: arid wetland; water and salt dynamics; Hydrus-1D; root water uptake; *Chinese tamarisk*

---

1. Introduction

In arid and semi-arid wetlands, salinity and water scarcity are two serious and chronic environmental problems threatening the ecosystem [1]. In the Heihe River basin, China, wetlands are now often experiencing extended periods of high soil salinization levels and associated water availability problems due to the impacts of high evaporative conditions, poor surface drainage, human population pressures, and the associated changes in land use [2,3]. Soil salinity limits plant growth [4] and negatively influences soil quality [5,6], resulting in the change of community structure, density, and growth status. Wetland areas decreased by 42% from 1975 to 2010 within the middle Heihe River basin [7]. Reed marsh areas decreased from 597.8 ha in the late 1990s to the current 492 ha, and that the reed plant height and reed stem were reduced by 25% and 4%, respectively [8]. Owing to the changes in water availability, land desiccation, and salinization, the vegetation has shifted from hydrophytes towards halophytes, psammophytes, xerophytes and super-xerophytes [9]. Meanwhile, soil salinization has caused clogging of soil pores and channels for water flow that has resulted in a considerable reduction in soil permeability, soil porosity, and soil hydraulic conductivity [10]. Consequently, before developing reliable countermeasures, it is important to evaluate the interactions between soil water and salts and their impacts on plant water use, based on factors such as the groundwater quality, the water table, and the plant tolerance to salinity.

In an arid climate where rainfall is very limited, shallow groundwater plays a key role in ecosystem functions. Therefore, it is particularly important to understand the effects of the depth to the water table (DWT) and groundwater quality on root zone water contents, salinity, and plant water use. Jolly *et al.* [1] concluded that in arid/semiarid environments, where the surface water regime was vulnerable to rainfall variability, the persistence of wetlands can be dependent, either completely or partially, on the contributions from groundwater. Ayars *et al.* [11] reported that the potential for meeting crop water needs from shallow groundwater ranged up to 50% of the total irrigation. Crops like alfalfa and forage grasses exhibit more continuous water uptake patterns for their long growing seasons and robust root systems [12]. However, most of these previous studies have focused on farmland with few of them giving consideration to wetlands, especially in arid environments [1].

The interactions between the soil water, salt, shallow groundwater, and root water uptake are complex and influenced by numerous factors. Evaluating the interactions of these factors through field research is difficult and time-consuming. In addition, salt variation is often small and is not easily detectable in a single growth season. Simulation models that integrate the soil water movement, solute transport, and plant water uptake provide information that otherwise cannot be obtained from field experiment [13]. Since the 1970s, many numerical solutions were developed to describe water and solute transport [14–16]. Most of these models are based on the solutions to the Richards equation for water flow and the convection-dispersion equation (CDE) for solute transport [17]. However, accurate predictions of these models rely on precise measurement of hydraulic characteristics [18]. For some parameters (e.g., α, n and l in the van Genuchten-Mualem function [19,20]), however, it is difficult to measure hydraulic characteristics at the plot scale. Numerous studies have indicated that the laboratory measured hydraulic parameters and/or parameters derived from pedotransfer functions, in combination with inverse optimization algorithms are an effective approach for improving the description of unsaturated hydraulic functions [4,17,21].

In this study, a widely used model Hydrus-1D [22], which simulates one-dimensional transport of water, heat, and multiple solutes in variably saturated media, was adopted to simulate the soil water flow, solute transport and root water uptake in an arid wetland with shallow saline groundwater. Our objectives were: (I) to test the feasibility of the Hydrus-1D model approach in simulating water flow and solute transport with observed data; (II) to characterize the interactions between soil water, salt, and groundwater and their effects on *Chinese tamarisk* root water uptake and root distribution, and (III) to conduct scenario analyses for the soil water, salt dynamics and root water uptake under different conditions. In addition, the long term salinity trends under different DWTs and groundwater salinities (EC_{gw}) were investigated.

2. Materials and Methods

2.1. Study Area

The study area is located at Pingchuan town, in the Middle Heihe River basin, Gansu, China (39°20′ N, 100°06′ E). The landform is representative of a riparian wetland covered with *Chinese tamarisk*, which is the dominant plant community in this area and in the study area served as a shelter forest. Soil electrical conductivities (soil/water ratio of 1/5, $EC_{1:5}$) ranged between 1.51 and 26.72 dS·m^{-1}. The DWT ranged from 0.47 m in the rainy fall to 1.44 m in the dry winter. The salinity of the shallow groundwater varied between 2.0 and 4.0 dS·m^{-1}. The climate is a continental arid temperate zone with annual average precipitation of 116.7 mm from 1965 to 2010 with about 60% of the precipitation received during July–September. The annual average potential evaporation is 2366 mm·year^{-1}. The annual average temperature is 7.6 °C, and the lowest and highest temperatures are about −27 °C in January and 39.1 °C in July, respectively [23]. Soil was formed from alluvial deposits with a silty loam texture (Table 1).

Table 1. Soil physical properties and calibrated parameters at the study area.

Soil Layer (cm)	0–15	15–25	25–55	55–65	65–100	100–150
Texture	Silt Loam	Silt Loam	Silt Loam	Silt Loam	Silt Loam	Coarse Sand
Clay (%)	11	10	11	13	16	-
Silt (%)	54	65	66	64	56	-
Sand (%)	35	25	23	23	28	100
Bulk density(g·cm^{-3})	1.2	1.35	1.42	1.44	1.44	1.42
θ_r (cm^3·cm^{-3})	0.093	0.08	0.075	0.071	0.079	0.051
θ_s (cm^3·cm^{-3})	0.543	0.493	0.502	0.462	0.534	0.376
α (cm^{-1})	0.019	0.028	0.032	0.033	0.035	0.034
n	2.06	1.70	1.50	1.36	1.71	4.43
l	0.5	0.5	0.5	0.5	0.5	0.5
K_s (cm·d^{-1})	2.5	10.6	24.7	6.7	34.0	1428.0
λ (cm)	16.8	19.5	48.2	29.1	10.8	124.0

Note: The particle size limits were 0.05 to 2 mm for sand, 0.05–0.002 mm for silt and <0.002 mm for clay. θ_r, residual water content; θ_s, saturated water content; α, reciprocal value of air entry pressure; n, the smoothness of pore size distribution; l, pore connectivity parameter; K_s, saturated hydraulic conductivity; and λ, dispersivity.

2.2. Measurements

Field data were collected during two growing seasons of *Chinese tamarisk* (2 May to 27 October 2012 and 2013). Soil water content was measured using a Neutron Moisture Meter (NMM, L520-D, Nanjing, China) every 5 days with a depth interval of 10 cm down to 100 cm. Soil salinity based on the soil diluted extract method (soil/water ratio of 1/5, $EC_{1:5}$) was measured every 15 days with a measurement interval of 10 cm down to 90 cm. A shallow monitoring well was installed in the vicinity of the neutron probe to measure DWT every 5 days. The groundwater electrical conductivity (EC_{gw}) was measured every 15 days. At the beginning of the experiment, undisturbed soil samples (diameter, 5 cm, height 5 cm) from five representative layers were collected for the laboratory measurement of soil bulk density (BD), texture, saturated hydraulic conductivity (K_s) and water content (θ_s). The BD was calculated from the volume-mass relationship for each core sample. Soil texture was determined using the pipette sampling method [24]. K_s values of the undisturbed soil cores were determined using a falling head method [25]. The soil cores were first saturated from the bottom and then submerged in water for 24 h. After weighing, the saturated soil samples were dried at 105 °C to constant mass, and their mass-based saturated soil water content was determined. θ_s values were determined by multiplying saturated mass-based soil water content with BD. In addition, root distribution was measured using a root auger (Eijkelkamp, The Netherlands), and soil cores were sampled in 10 cm depth increments. Root biomass was obtained by washing away soil particles, oven-dried and weighed (Figure 1).

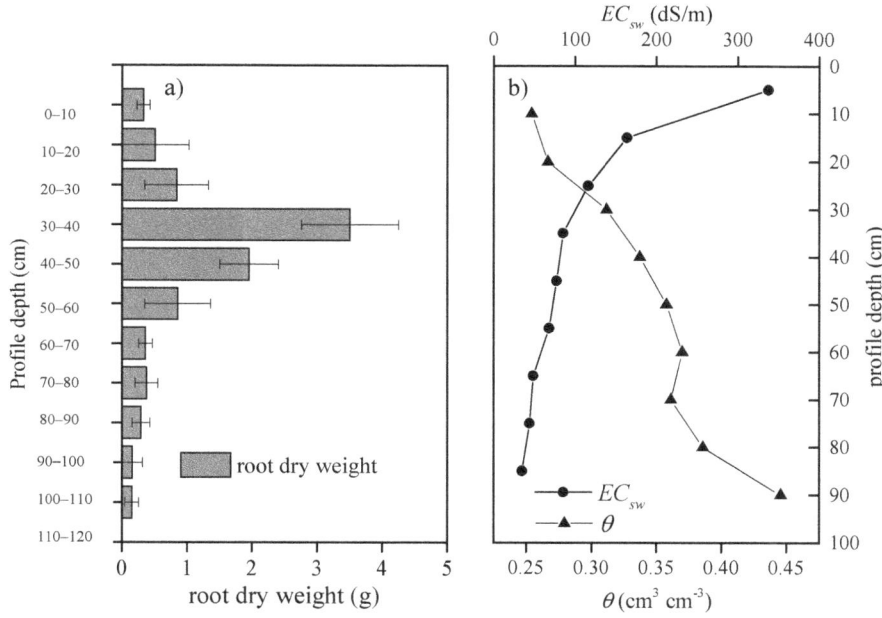

Figure 1. Graphical representation of (**a**) the root distribution of *Chinese tamarisk* measured at the start of the 2012 growing season and (**b**) the average value of soil water (θ) and EC_{sw} during the 2012 growing season.

2.3. Model Simulation

2.3.1. Soil Water Flow and Solute Transport

Simulations of soil water flow and solute transport were performed with Hydrus-1D [22]. This program numerically solves the Richards equation for water flow and uses advection-dispersion equations (CDE) for heat and solute transport in variably saturated porous media. Variably-saturated water flow is described using the Richards' equation:

$$\frac{\partial \theta}{\partial t} = \frac{\partial}{\partial z}\left[K(h)\left(\frac{\partial h}{\partial z} + 1\right)\right] - S(z, t) \tag{1}$$

where θ is soil water content ($L^3 \cdot L^{-3}$), t is time (T), z is the vertical space coordinate (L), K is the hydraulic conductivity ($L \cdot T^{-1}$), h is the pressure head (L), S is the sink term accounting for root water uptake ($L^3 \cdot L^{-3} \cdot T^{-1}$). The unsaturated soil hydraulic properties were described using the van Genuchten-Mualem functional relationships [19,20]:

$$\theta(h) = \begin{cases} \theta_r + \dfrac{\theta_s - \theta_r}{(1 + |\alpha h|^n)^m} & h < 0 \\ \theta_s & h \geq 0 \end{cases} \tag{2}$$

$$K(h) = K_s S_e^l [1 - (1 - S_e^{1/m})^m]^2 \tag{3}$$

$$S_e = \frac{\theta - \theta_r}{\theta_s - \theta_r} \tag{4}$$

where θ_r and θ_s are the residual and saturated water contents ($L^3 \cdot L^{-3}$), respectively. K_s ($L \cdot T^{-1}$) is the saturated hydraulic conductivity, α, (L^{-1}) and n represent the empirical shape parameters, $m = 1 - 1/n$; l is the pore connectivity parameter, which is taken as 0.5 [20]. S_e is the effective saturation.

2.3.2. Solute Transport

The partial differential equations governing equilibrium one-dimensional solute transport under transient flow in variably-saturated medium is defined in Hydrus-1D as:

$$\frac{\partial \theta C}{\partial t} = \frac{\partial}{\partial z}\left(\theta D \frac{\partial C}{\partial z}\right) - \frac{\partial v \theta C}{\partial Z} \tag{5}$$

where C is the solute concentration of the liquid phase ($M \cdot L^{-3}$). D is the dispersion coefficient ($L^2 \cdot T^{-1}$), and v is the average pore water velocity ($L \cdot T^{-1}$). The dispersion coefficient is defined as (ignoring molecular diffusion):

$$D = \lambda v \tag{6}$$

where λ is dispersivity (L). The dispersivity is viewed as a material constant independent of the flow rate. Since v is obtained from the numerical solution of the water flow model (the water flux q divided by θ), dispersivity is the only solute transport parameter needed for solving the CDE equation.

2.3.3. Root Water Uptake

The potential transpiration rate, T_p ($L \cdot T^{-1}$), is spread in the root zone according to the normalized root density distribution function, $\beta(z, t)$ (L^{-1}). The actual root water uptake, S, is obtained from the potential root water uptake (*i.e.*, potential transpiration) S_p, through multiplication with a stress response function $\alpha(h, h_\varphi, z, t)$ accounting for water and osmotic stresses [26,27] as follows:

$$S(h, h_\varphi, z, t) = \alpha(h, h_\varphi, z, t)S_p(z, t) = \alpha(h, h_\varphi, z, t)\beta(z, t)T_p(t) \tag{7}$$

where stress response function $\alpha(h, h_\varphi, z, t)$ is a dimensionless function of the soil water (h) and osmotic (h_φ) pressure heads ($0 \leq \alpha \leq 1$). $S_p(z, t)$ and $S(h, h_\varphi, z, t)$ are the potential and actual volumes of water removed from a unit volume of soil per unit of time ($L^3 \cdot L^{-3} \cdot T^{-1}$), respectively. The actual transpiration rate, T_a ($L \cdot T^{-1}$), is then obtained by integrating Equation (7) over the root domain L_R:

$$T_a = \int_{L_R} S(h, h_\varphi, z, t)dz = T_p \int_{L_R} \alpha(h, h_\varphi, z, t)\beta(z, t)\,dz \tag{8}$$

We further assumed that the effects of water and salinity were multiplicative [28]: $\alpha(h, h_\varphi) = \alpha(h)\,\alpha(h_\varphi)$, so that different stress response functions could be used. Root water uptake due to water stress was described using the model introduced by Feddes *et al.* [26]:

$$\alpha_h = \begin{cases} 0, & h > h_1, h \leq h_4 \\ \dfrac{h - h_1}{h_2 - h_1}, & h_2 < h \leq h_4 \\ 1, & h_3 < h \leq h_2 \\ \dfrac{h - h_4}{h_3 - h_4}, & h_4 < h \leq h_3 \end{cases} \tag{9}$$

where root water uptake is assumed to be zero close to saturation (*i.e.*, wetter than the "anaerobiosis point", h_1). For $h < h_4$ (the wilting point pressure head), root water uptake is also completely stressed. Root water uptake is considered to be at the potential rate when pressure heads range from h_2 to h_3, and when pressure head values are between h_1 and h_2 (or h_3 and h_4), root water uptake increases (or decreases) linearly with h.

Hydrus-1D assumes h_3 is a function of T_p and allows users to specify two different T_p (T_{p1} and T_{p2}) and h_3 ($h_{3\text{-}1}$ and $h_{3\text{-}2}$), respectively. The calculation equations are:

$$h_3 = \begin{cases} h_{3-1} + \dfrac{(h_{3-2} - h_{3-1})}{(T_{p1} - T_{P2})(T_{p1} - T_p)} & T_{p2} < T_p < T_{p1} \\ h_{3-2} & T_p < T_{p2} \\ h_{3-1} & T_p > T_{p1} \end{cases} \tag{10}$$

Root water uptake due to osmotic stress was described with an S-shaped function developed by van Genuchten, 1987 [28]:

$$\alpha_\varphi = \frac{1}{1 + \left(\dfrac{h_\varphi}{h_{\varphi 50}}\right)^p} \tag{11}$$

where p represents experimental constants. The exponent p was found to be approximately 3 when only salinity stress data was applied [28]. The parameter $h_{\varphi 50}$ represents the pressure head at which the water extraction rate is reduced by 50% during conditions of negligible water stress.

In Hydrus-1D, EC is expressed as electrical conductivity of soil solution (EC_{sw}). Alternatively, we measured $EC_{1:5}$ and converted it into the saturated paste extracts (EC_e) using the following relationship where $EC_e = (2.46 + 3.03\ \theta_{sp}^{-1})\ EC_{1:5}$, where θ_{sp} is the water content of the saturated paste ($\theta_{sp}\ \text{Kg}\cdot\text{Kg}^{-1}$) [29]. Then EC_e values are converted into EC_{sw} by the following equation [30]:

$$EC_{sw} = EC_e \frac{BD \cdot SP}{100 \cdot \theta} = EC_e \frac{\theta_s}{\theta} \tag{12}$$

where SP is the saturation percentage (the water content of the saturated soil-paste, expressed on a dry-weight basis), BD is the bulk density ($\text{g}\cdot\text{cm}^{-3}$) and θ_s is the saturated soil water content ($L^3 \cdot L^{-3}$).

Furthermore, Hydrus-1D uses the following relationship to convert EC_{sw} to osmotic pressure head (cm):

$$h_\varphi = -3.8106 EC_{sw} + 0.5072 \tag{13}$$

The equation is very similar to the relationship reported by the US Salinity Laboratory Staff (1954) [31] for estimating the osmotic pressure of soil solutions from EC measurements ($h_\varphi = -3.7188\ EC_{sw}$).

2.4. Initial and Boundary Condition

Initial conditions were set in the model with measured soil water contents and electrical conductivities on 2 May 2012. At the soil surface, an atmospheric boundary condition was specified using the daily data of precipitation and reference crop evapotranspiration (ET_0) obtained from the Linze Station (2 km away from the site). Daily values of the ET_p were calculated using the reference evapotranspiration (ET_0) via Penman-Monteith method [32] multiplying by a crop coefficient of the investigated *Chinese tamarisk*. The crop coefficient was estimated from fraction of ground cover and plant height [33], that is, crop coefficient for the middle season is 1.05, increased from 0.55 to 1.05 linearly in the first 10 days of plant development and decreased from 1.05 to 0.55 linearly in the last seven days of plant defoliation. Then, ET_p was divided into potential evaporation (E_p) and potential transpiration (T_p) according to Beer's Law:

$$E_p(t) = ET_p(t) \cdot exp^{-k \cdot LAI(t)} \tag{14}$$

$$T_p(t) = ET_p(t) - E_p(t) \tag{15}$$

where k is an extinction coefficient set to be 0.463 and LAI is the leaf area index. The model was used to directly calculate actual E and T given the soil moisture conditions and the root water uptake functions. LAI values were measured during different stages of the growing season using a LI-COR area meter (Model LI-3100C, LI-COR Environmental and Biotechnology Research Systems, Lincoln, Nebraska), and were linearly interpolated between the measurement dates. At the bottom, variable pressure head and concentration boundary conditions were specified for water flow and solute transport using the measured water table depths and groundwater EC, respectively. For solute transport, we assumed that the rain water was free of solutes and implemented a no flux boundary condition at the soil surface. The boundary conditions used in the calibration and validation processes are shown in Figure 2.

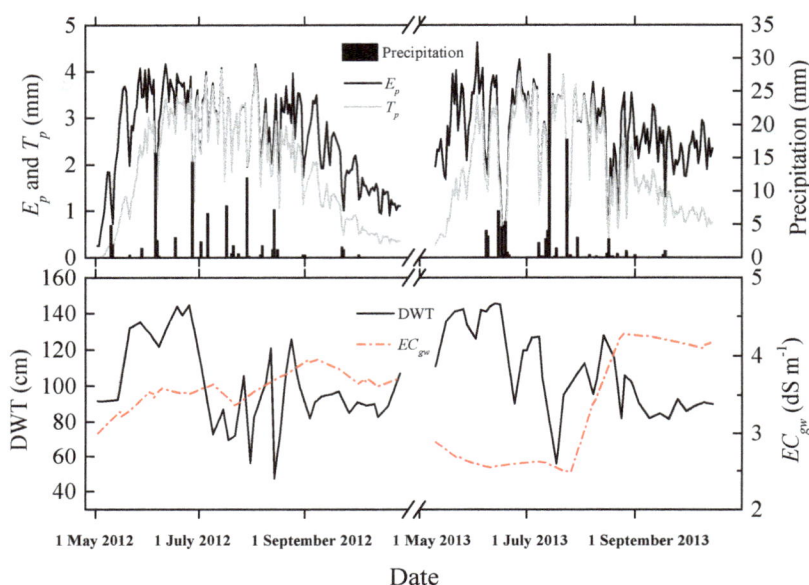

Figure 2. Dynamics of potential evapotranspiration, precipitation, depth to water table (DWT) and groundwater electric conductivity (ECgw).

2.5. Model Calibration and Validation

In this study, the Hydrus-1D model was calibrated using site-specific boundary conditions and measured water contents, θ, and electrical conductivities, EC_{sw} values during 2012. Saturated water content (θ_s) and hydraulic conductivity (K_s) were determined from the soil cores taken as stated above. The other van Genuchten-Mualem parameters were estimated via Rosetta pedotransfer functions [34] using the particle size distribution and bulk density dataset. For initial values of solute transport parameters in the root zone (0–100 cm), the dispersivity (λ) was set to an average value (8.9) based on 67 soils with silt loam textures according to Vanderborght and Vereecken [35]. Based on aquifer materials, thickness, and hydraulic conductivity, the dispersivity of the aquifer was fixed as 124 cm according to Gelhar et al. [36].

The parameters of the Feddes model were synthesized based on Moayyad [37] with Grinevskii [38]: $h_1 = -0.1$ cm, $h_2 = -2$ cm, $h_{3\text{-}1} = -80$ cm, $h_{3\text{-}2} = -250$ cm, $h_4 = -15,000$ cm. Without consideration of the water stress (*i.e.*, reset the Feddes model parameters to make water stress vanish), parameters of the S-shaped function developed by van Genuchten were fitted as: $h_{\varphi 50} = 326.4$ cm, $p = 3$. Root distribution was specified according to measured root dry weight distribution along the soil profile (Figure 1).

In Hydrus, inverse parameter estimation employed a relatively simple, gradient-based, local optimization approach based on the Marquardt-Levenberg method [39]. In this case, inverse solutions were used to optimize soil hydraulic and solute transport parameters simultaneously using the observed data, initial conditions, initial estimates, and boundary conditions. That is, α, n and λ in the five upper soil layers were fitted first since Hydrus could optimize 15 parameters at a time. θ_r was the last parameter estimated. Then, the model was validated with the observed data of the 2013 growing season without changing the calibrated parameters.

The agreement between the predicted and observed data was evaluated by root mean square error (*RMSE*) and coefficient of determination (R^2):

$$RMSE = \sqrt{\frac{\sum_{i=1}^{N}(O_i - P_i)^2}{N - 1}} \tag{16}$$

$$R^2 = 1 - \frac{\sum_{i=1}^{N}(P_i - O_i)^2}{\sum_{i-1}^{N}(O_i - \bar{O})^2} \tag{17}$$

where O_i and P_i are the *i*th values of observed and predicted values, respectively, and $\bar{O}$ is the average of observed values. N is the number of observations.

2.6. Simulation Scenarios

In order to understand the impacts of groundwater change on plant water use, we simulated root water uptake under different DWTs and EC_{gw} conditions. Taking the data of 2012 as the reference, eight DWT (*i.e.*, DWT would either raise or drop 15%, 30%, 45% and 60% based on the 2012 reference depth, respectively) and eight EC_{gw} (*i.e.*, EC_{gw} would either increase or decrease 15%, 30%, 45% and 60% based on the 2012 reference value, respectively) were assumed in this process (Table 2).

In addition, to evaluate long-term salinity trends, a long-term time series analysis was conducted considering the fluctuations of DWT and EC_{gw} in relation to 2012 base values. Firstly, using the LARS-WG weather generator [40] and historical meteorological data from 1954 to 2012, we generated 30 years of weather data which had the same statistical characteristics as the historical data. Then, soil salinization risks were assessed using the generated long-term time series data for three water levels (DWT is 60, 100 (average water level of 2012) and 140 cm) and three groundwater electrical conductivities (EC_{gw} is 1.75, 3.75 (average EC_{gw} of 2012) and 5.75 dS·m^{-1}). The initial value of EC_{sw} was taken as 3.75 dS·m^{-1} for different water tables and 107.25 dS·m^{-1} (average root zone EC_{sw} of 2012) for different EC_{gw}, respectively.

Salt normally accumulated before the plant germination in our studied area (*i.e.*, the dry and cold winter time). In order to elucidate the impacts of artificial watering on soil water, salt dynamics, and to determine how to alleviate salt stress on the arid wetlands, we evaluated the influence of a preseason irrigation event applied at the initial stages of plant growth starting on 2 May 2012. To account for irrigation amounts ranging from 1 to 80 cm, eighty Hydrus simulation scenarios were run. All simulation scenarios are listed in Table 2.

Table 2. Simulation scenarios performed in this study.

Main Scenarios	Scenarios in Detail
Root water uptake predictions	Depth to water table raise 15% of the 2012 reference depth, DWT+15%
	Depth to water table raise 30% of the 2012 reference depth, DWT+30%
	Depth to water table raise 45% of the 2012 reference depth, DWT+45%
	Depth to water table raise 60% of the 2012 reference depth, DWT+60%
	Depth to water table drop 15% of the 2012 reference depth, DWT-15%
	Depth to water table drop 30% of the 2012 reference depth, DWT-30%
	Depth to water table drop 45% of the 2012 reference depth, DWT-45%
	Depth to water table drop 60% of the 2012 reference depth, DWT-60%
	Groundwater electrical conductivity increase 15% of the 2012 reference value, $EC_{gw} + 15\%$
	Groundwater electrical conductivity increase 30% of the 2012 reference value, $EC_{gw} + 30\%$
	Groundwater electrical conductivity increase 45% of the 2012 reference value, $EC_{gw} + 45\%$
	Groundwater electrical conductivity increase 60% of the 2012 reference value, $EC_{gw} + 60\%$
	Groundwater electrical conductivity decrease 15% of the 2012 reference value, $EC_{gw} - 15\%$
	Groundwater electrical conductivity decrease 30% of the 2012 reference value, $EC_{gw} - 30\%$
	Groundwater electrical conductivity decrease 45% of the 2012 reference value, $EC_{gw} - 45\%$
	Groundwater electrical conductivity decrease 60% of the 2012 reference value, $EC_{gw} - 60\%$
Long term (30 years) salinity trends	Depth to water table is 60 cm, DWT = 60 cm
	Depth to water table is 100 cm, DWT = 100 cm
	Depth to water table is 140 cm, DWT = 140 cm
	Groundwater electrical conductivity is 1.75 dS/m, EC_{gw} = 1.75 dS/m
	Groundwater electrical conductivity is 3.75 dS/m, EC_{gw} = 3.75 dS/m
	Groundwater electrical conductivity is 5.75 dS/m, EC_{gw} = 5.75 dS/m
Preseason irrigation strategy	A human irrigation (irrigation amount range from 1 to 80 cm) applied at the initial stages of plant growth (2 May 2012).

3. Results and Discussion

3.1. Model Calibration and Validation

There was good agreement between observed and simulated soil water contents and salt contents as indicated by the smaller *RMSE* and higher R^2 values. Calibration periods resulted in *RMSE* = 0.031 $cm^3 \cdot cm^{-3}$ and R^2 = 0.88 for soil water and *RMSE* = 0.037 $dS \cdot m^{-1}$ and R^2 = 0.92 for EC_{sw}. Validation period resulted in *RMSE* = 0.046 $cm^3 \cdot cm^{-3}$ and R^2 = 0.82 for soil water and *RMSE* = 0.035 $dS \cdot m^{-1}$ and R^2 = 0.95 for EC_{sw}. These results demonstrated that despite the considerable demands on input data, Hydrus-1D was an effective tool for evaluating water and solute transport [41–43] and would be acceptable in performing scenario simulations. The calibrated parameters are shown in Table 1. There was generally very good agreement between the simulated and measured soil water contents, though there are some discrepancies. It is not possible to specifically identify the causes of the discrepancies, but they might be partially attributable to preferential flow caused by macropores and cracks [44,45], spatial heterogeneity and observation errors [46], and the locally occurring chemical processes, such as adsorption-desorption, and proportional root uptake [22], and precipitation/dissolution reactions in soils [47].

3.2. Soil Water and Salt Dynamics and Their Effects on Root Water Uptake of Tamarisk

The roots of *Chinese tamarisk* are primarily distributed in the 20–60 cm soil layer, accounting for 73.76% of total dry weight (Figure 1a) with the maximum values (35.55%) at the 30–40 cm soil depth. This distribution may be partially attributed to a large salt accumulation near the soil surface that is unfavorable to root growth (Figure 1b). The accumulation of salts has primarily been caused by high atmospheric demands that caused water movement towards the soil surface from the shallow saline groundwater. Since the groundwater has high salinity levels, salts are also transported with the water and accumulated in the root zone. In addition, scarce rainfall and poor surface drainage have also been shown to contribute to this process in arid regions [48]. Our results are consistent with the reports of Li *et al.* [49], who found that root growth of *tamarisk* seemed to be repressed when the salinity ($EC_{1:5}$) was greater than 6 dS·m^{-1}. Similarly, although the deeper soil layer contains little salt, shallow water table results in relatively high water contents and small values of aeration porosity. These conditions may limit root growth, respiration, and water uptake. Therefore, the optimum depth observed for plant growth was between 20 to 60 cm, because of the salt stress near the soil surface, and the saturation and anaerobic conditions below the 60 cm soil depth (Figure 3). Therefore, the long-term effects of water and salt stress caused *Chinese tamarisk* to develop its root system in the most suitable strata.

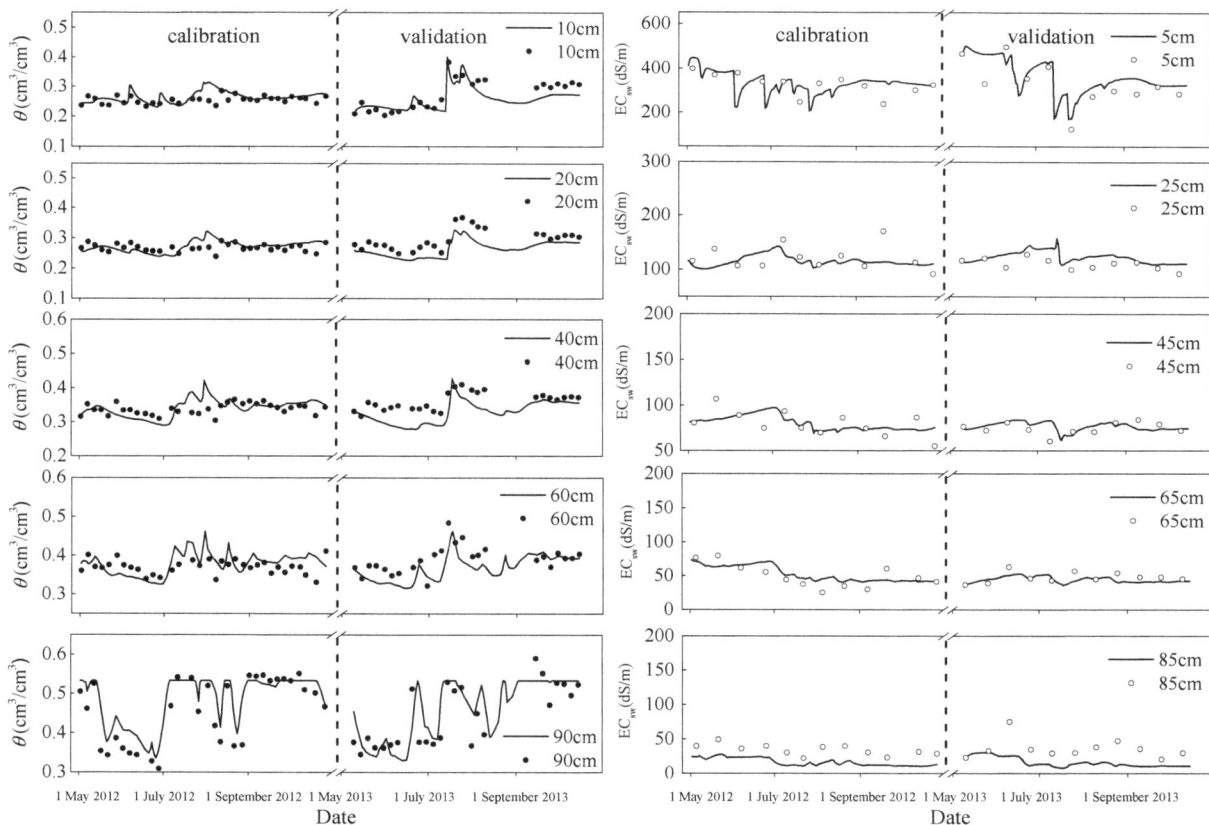

Figure 3. Measured and simulated soil water and EC_{sw} in both calibration and validation period during the growing season of *Chinese tamarisk*.

Dynamics of the profiled water contents were primarily attributable to the natural precipitation, evapotranspiration, and location of the water table. In general, soil water contents increased from the surface layer to the bottom layer because of a shallow water table. The soil moisture at the 20–60 cm strata remained relatively constant during the growing season and served as a soil moisture buffer layer (Figures 1 and 3). Because of the shallow groundwater tables, the soil moisture fluctuated dramatically for the 80–100 cm layer and ranged from 0.33 to 0.53 $cm^3 \cdot cm^{-3}$ (Figure 3). Similarly, due to the relatively limited precipitation and large quantity of evaporation during 2012, the EC fluctuated intensively with EC_{sw} values ranging from 207.1 to 448.4 $dS \cdot m^{-1}$ in the surface layer. Meanwhile, EC decreased from the surface layer to the bottom layer because of the intensive evaporation, poor surface drainage, and negligible precipitation (Figures 1 and 3).

Because of the sparse vegetative cover that was effected by water and salinity stresses, cumulative evaporation reached 149 mm during the growing season. Accordingly, the migration of salt with intensive evapotranspiration was thought to be the main cause for soil salinization in this area. Infiltration was only 91 mm during the 2012 growing season that was less than both evaporation and transpiration (Figure 4). Furthermore, rainfall infiltration could dissolve large quantities of soluble salts from the upper layer. Though precipitation in this region is unable to provide sufficient water for plant growth, cumulative upward soil water flux attributable to groundwater charge reached 216 mm during the growing season of 2012. Further, compared with the infiltration water from the upper boundary, the recharged water from the groundwater has a low salt concentration and can be easily utilized by plant roots. Therefore, the groundwater plays a critical role in the maintenance of *Chinese tamarisk* growth and water supplements. These observations are in agreement with Morris and Collopy [50], who reported that more than half the tree water uptake (*Eucalyptus Camaldulensis* and *Casuarina cunninghamiana*) was drawn from the groundwater. Satchithanantham *et al.* [51] found that during the dry mid-season, when the ET was at its peak, the groundwater supplied up to 92% of the water for consumptive use by potatoes. Ayars *et al.* [11] observed that almost 100% of the consumptive use by alfalfa was supplied by contributions from the shallow groundwater.

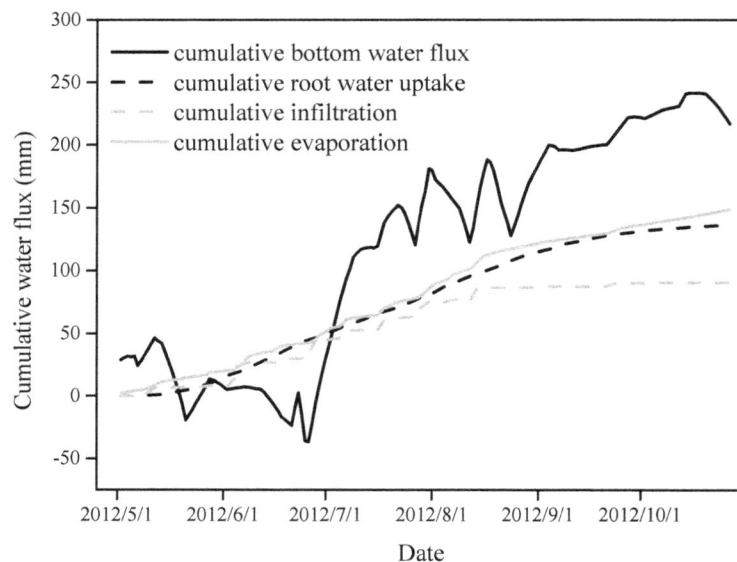

Figure 4. Cumulative water flux during the growing season of 2012.

Root water uptake reached 91.0% of its potential maximal value when water stress was considered and only 41.6% of that amount when both water and salt stress conditions were taken into consideration (Figure 5). This phenomenon has been attributed to the vast quantities of soluble salts that results in decreased solute potentials and increased ion toxicity [52]. These types of observations have resulted in assessments of salt stress being the dominating factor affecting root water uptake in arid riparian wetlands [1]. Therefore, there is a pressing need to develop appropriate management measures to reduce the impacts of water and salt stresses on *Chinese tamarisk*.

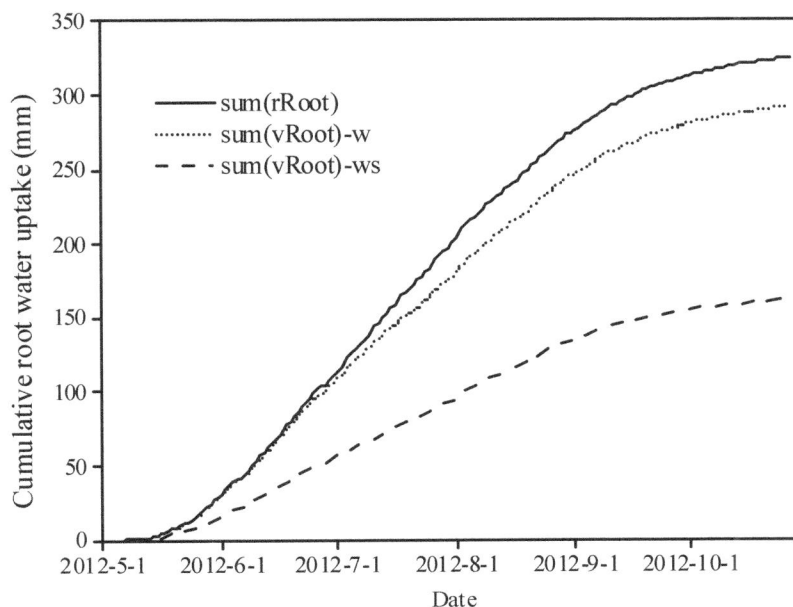

Figure 5. Cumulative root water uptake under different stress conditions. Note: sum(rRoot), potential cumulative root water uptake; sum(vRoot)-w, cumulative root water uptake under water stress only; sum(vRoot)-ws, cumulative root water uptake under coupled water and salt stress.

3.3. Scenario Simulations

3.3.1. Root Water Uptake Predictions

As indicated by Figure 6, cumulative root water uptake is more sensitive to fluctuations of water table than EC_{gw}. Root water uptake reached the maximum values of 136.6 mm when DWT was at the 2012 reference depth (CK), then decreased gradually as the water table rose. Cumulative root water uptake only reached 72.5 mm when DWT dropped 60% (DWT − 60%). This reduction in root water uptake was mainly because excessive shallow water table caused water stress in the root zone. Similarly, the cumulative root water uptake declined dramatically when the DWT increased from the 2012 reference depth to DWT + 60%. The cumulative root water uptake only reached 118.6 mm when DWT increased 60%. These simulation results suggest that either too shallow or too deep a water table will have dramatic impacts on the root water uptake. Contrary to the effects of groundwater table, root water uptake exhibited almost no change within the assumed range of EC_{gw} values. This relative lack of response may be partially attributed to a high degree of salt tolerance for this plant. For example, in many riparian systems of the southwestern United States, increased salinity caused by changes in water flow, have favored salt-tolerant *tamarisk* and greatly reduced the recruitment and growth of native salt-sensitive riparian species [53,54]. In addition, salt accumulation in the root zone is a slow process and the change of EC_{gw} may not result in obvious increases of root zone salinity within a single year which will be discussed in the following section (Section 3.3.2).

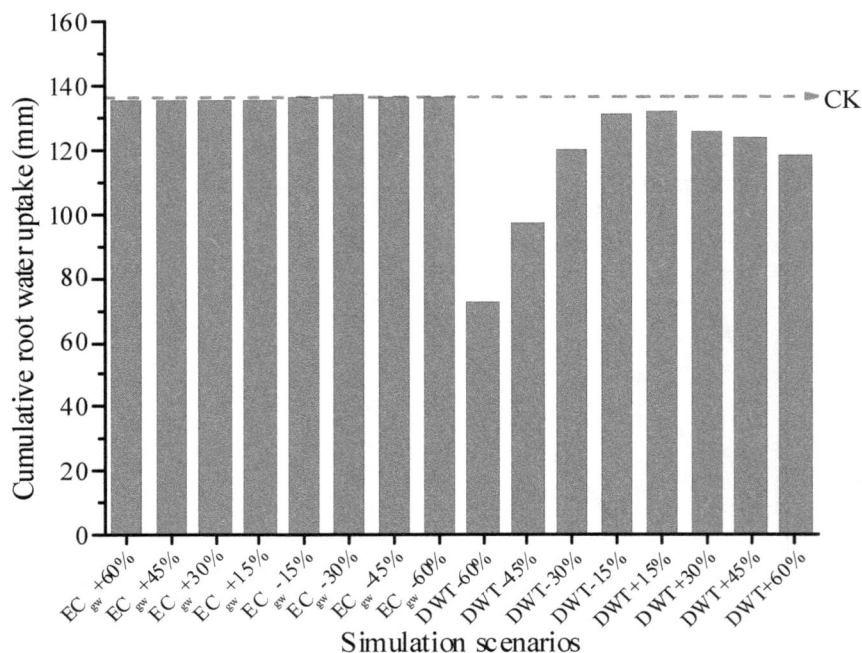

Figure 6. Cumulative root water uptake of *Chinese tamarisk* under different simulation conditions. CK is cumulative root water uptake of 2012 growing season.

3.3.2. The Long-Term Salinization Trends

To assess long-term salinity trends, Hydrus-1D was combined with a stochastic weather generator LARS-WG to evaluate the long term changes of soil salinity under different water tables and EC_{gw}. Figure 7a illustrates that soil salinization increased year by year with a saline shallow ground water (EC_{gw} = 3.75 dS·m^{-1}). In general, root zone EC_{sw} increased with the upward movement of water table whereas the amplitude of the EC_{sw} decreased with the elevation of the water table. Average root zone EC_{sw} after 30 years were 62.30, 57.34 and 47.15 dS·m^{-1} when DWT was at a depth of 60, 80 and 100 cm, respectively. These results indicated that a shallow water table contributed to increased soil salinization although the same conditions promoted root water uptake. Ibrakhimov *et al.* [55] found that elevated groundwater levels resulted in increased soil salinization by the annual addition of 3.5–14 t/ha of salts depending on groundwater salinity. Xie *et al.* [4] reported that there is a contradiction between available water, salt stress, and reed water uptake with variations in DWT. Similarly, root zone EC_{sw} increased with the increased EC_{gw}, values of 142.08, 177.53, and 210.55 dS·m^{-1} when accompanying EC_{gw} values were 1.75, 3.75, and 5.75 ds·m^{-1} after 30 years, respectively (Figure 7b). The results indicated that soil salinization conditions will deteriorate continuously without human intervention and highlighted the importance of preventing human induced EC_{gw} increases that would occur from subsurface irrigation drainage from farmland.

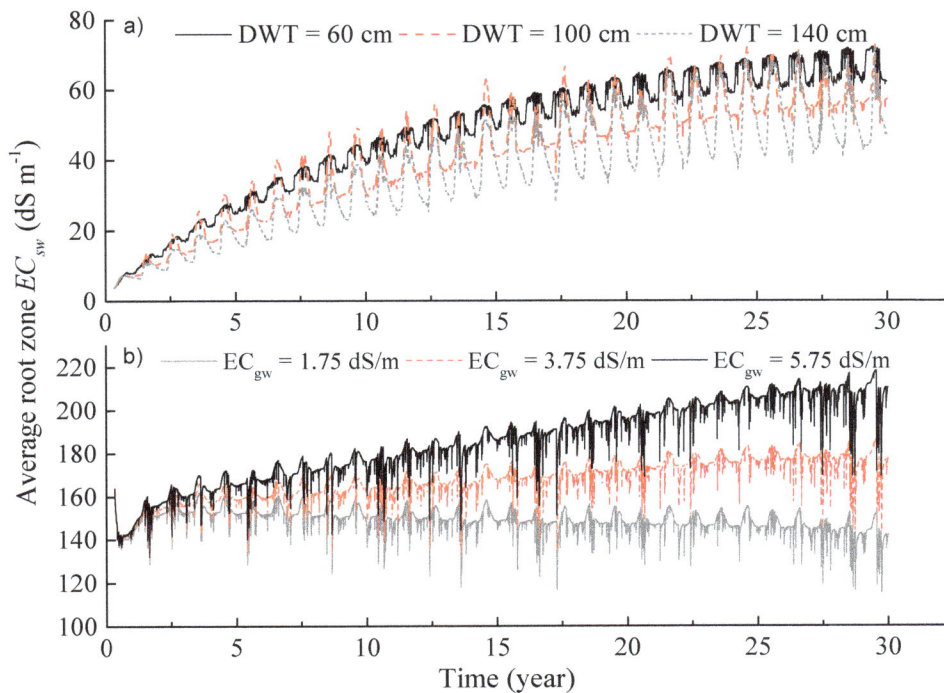

Figure 7. Temporal changes in soil salinity in averaged root zone as affected by (**a**) water table and (**b**) groundwater electrical conductivity.

3.3.3. Preseason Irrigation Strategy

Preseason irrigation increased root water uptake significantly. Compared with no irrigation, root water uptake increased 4.5%, 40.2%, 79.3%, 100.6% and 115.4% when the irrigation amounts were 10, 20, 30, 40 and 50 cm, respectively (Figure 8). These results indicated that root water uptake generally increased with increased irrigation quantities when irrigation amounts ranged from 10 to 50 cm. However, root water uptake increased only 2 mm (from 294 to 296 mm) when the irrigation quantities increased from 50 to 80 cm, and demonstrated that irrigation quantities less than 50 cm were sufficient to promote *tamarisk* root water uptake (Figure 9b). Note that cumulative root water uptake displayed a decreased trend when the irrigation quantities were less than 6 cm and indicated that although a small quantity of irrigation increased the root zone water content to some extent, it was not sufficient for the salt to be effectively leached out of the root zone. Furthermore, the irrigated water cannot be easily used by the plant roots because of the water contained a large amount of soluble salts that were dissolved from the upper soil layer. This observation implied that the precipitation in the region is not beneficial for plants and even threatens plant growth, since single precipitation events are normally less than 6 cm. The increased root water uptake resulting from fresh water irrigation was mainly because sufficient quantities of water from the upper boundary layers can leach salts below the root zone and effectively alleviate salt stress (Figure 9b). Because these salts can re-accumulate as a result of evaporation (Figure 9a), repeated irrigation is needed. The average root zone EC_{sw} dropped rapidly below 52.5, 27.7, and 13.7 dS·m^{-1} in 15 days when the irrigation quantities were 20, 40, and 60 cm, respectively. Therefore, irrigation before the growing season is essential, but too small or too large a quantity of irrigation is not advisable. In our case, 30–40 cm of preseason irrigation was reasonable.

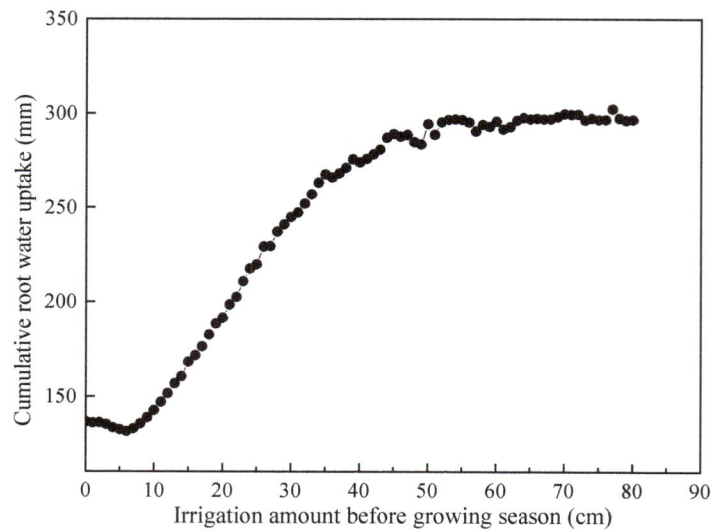

Figure 8. Cumulative root water uptake of *Chinese tamarisk* with different irrigation amounts.

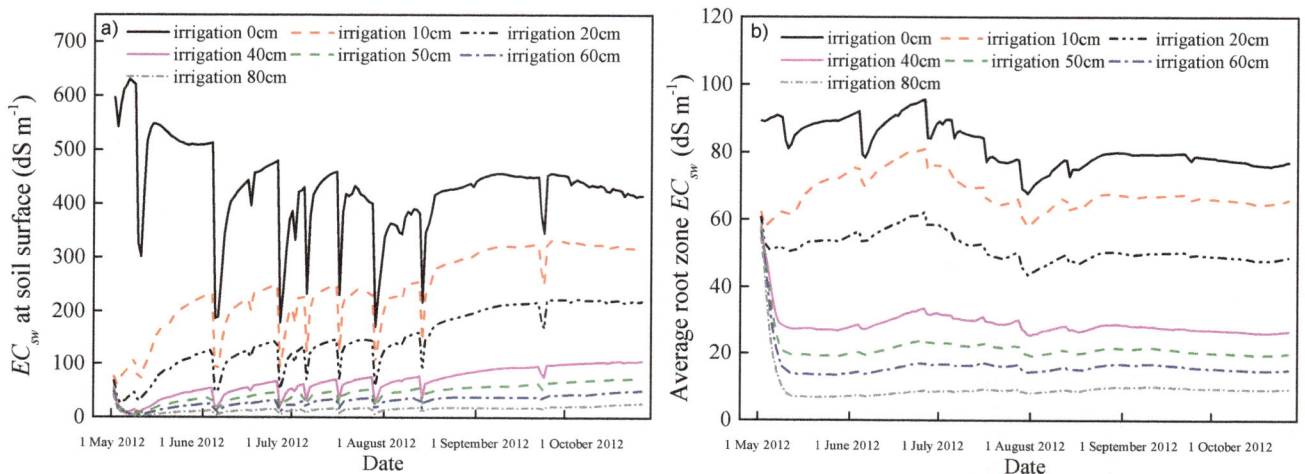

Figure 9. Temporal changes of soil salinity at the soil surface (**a**) and the averaged root zone electrical conductivity as affected by irrigation amount (**b**).

The advantages of artificial irrigation to maintain plant growth has also been addressed by other researchers. Holland *et al.* [56] observed that a two-fold to five-fold increase in plant water potential and a three-fold to six-fold increase in *Eucalyptus camaldulensis* water consumption in three to four months after watering in the riparian region. Xie *et al.* [4] reported that irrigation clearly increased reed water use, especially when irrigation quantities were higher than 3 $cm \cdot d^{-1}$. Askri *et al.* [57] demonstrated that in order to increase date palm water use, high irrigation frequencies and shallow groundwater are needed to maintain high water content and low salinity in the root zone. Therefore, we would suggest it would be beneficial to have artificial irrigation to maintain the sustainable development of the arid riparian wetlands.

4. Conclusions

In this study, soil water and salt dynamics and their effects on *Tamarisk* root water uptake were characterized by coupling measured data with simulation scenarios in the Heihe riparian wetland, China. The Hydrus-1D model simulations of soil water and salt dynamics matched the observed data fairly well during both the calibration and validation periods as indicated by smaller *RMSE* and higher R^2 values, which demonstrated the feasibility of using the model under different simulation scenarios.

Chinese tamarisk extends its root system into the most suitable strata with 73.6% of the total root system distributed in the 20–60 cm soil layer because of the long-term effects of water and salt stress. Groundwater is the main water source for *Chinese tamarisk* in the study area. Cumulative root water uptake only accounted for 41.6% of the potential value under the joint influences of water and salt stress. This result indicated the necessity of human interventions to alleviate water and salt stress. Furthermore, root water uptake was most sensitive to the fluctuations of water table levels. Too deep or too shallow a groundwater table was found to severely repress root water uptake. Shallow groundwater was found to result in increased soil salinization, especially when the groundwater contains a large amount of salts (high EC_{sw}). Preseason irrigation has the potential to leach salt out of the root zone and maintain the EC_{sw} at a reduced level during the growing season which would result in increased water uptake. Cumulative root water uptake increased when irrigation quantities were initially increased. Further increases in irrigation quantities diminished the increased rate of root water uptake.

Irrigation before the growing season is necessary, but the irrigation quantities should be taken into consideration. This study provided insights into soil water and salt redistribution and their effects on plant water use, and should help in the establishment of improved management practices for arid riparian wetlands.

Acknowledgments

This study was financially supported by National Natural Science Foundation of China (91025018, 41371233, 41371234, 41201279), Northwest Agriculture and Forestry University fund (Z111021308), and the "*111*" Project (B12007). We thank our colleagues for the joined field work and data analysis. Special thanks go to the staff of the Linze CERN station.

Author Contributions

All authors contributed to the design and development of this manuscript. Huijie Li and Jun Yi, carried out the data analysis and prepared the first draft of the manuscript; Ying Zhao is the graduate advisor of Huijie Li and contributed many ideas to the study; Jianguo Zhang, Bingcheng Si, Robert Hill, Lele Cui and Xiaoyu Liu provided some important advices on the concept of methodology and writing of the manuscript.

Conflicts of Interest

The authors declare no conflict of interest.

References

1. Jolly, I.D.; McEwan, K.L.; Holland, K.L. A review of groundwater-surface water interactions in arid/semi-arid wetlands and the consequences of salinity for wetland ecology. *Ecohydrology* **2008**, *1*, 43–58.

2. Niu, Y.; Liu, X.D.; Zhang, H.B.; Meng, H.J. The ecological function evaluation of wetland in upper and middle reaches of heihe basin. *Wetl. Sci.* **2007**, *3*, 215–220. (In Chinese)

3. Lei, K.; Zhang, M.X. The wetland resources in china and the conservation advices. *Wetl. Sci.* **2005**, *3*, 81–86. (In Chinese)

4. Xie, T.; Liu, X.; Sun, T. The effects of groundwater table and flood irrigation strategies on soil water and salt dynamics and reed water use in the yellow river delta, china. *Ecol. Model.* **2011**, *222*, 241–252.

5. Ahmad, M.U.D.; Bastiaanssen, W.G.M.; Feddes, R.A. Sustainable use of groundwater for irrigation: A numerical analysis of the subsoil water fluxes. *Irrig. Drain.* **2002**, *51*, 227–241.

6. Gowing, J.W.; Rose, D.A.; Ghamarnia, H. The effect of salinity on water productivity of wheat under deficit irrigation above shallow groundwater. *Agric. Water Manag.* **2009**, *96*, 517–524.

7. Jiang, P.H.; Zhao, R.F.; Zhao, H.L.; Lu, L.P.; Xie, Z.L. Relationships of wetland landscape fragmentation with climate change in middle reaches of heihe river, China. *J. Appl. Ecol.* **2013**, *24*, 1661–1668. (In Chinese)

8. Zhang, H.B.; Meng, H.J.; Liu, X.D.; Zhao, W.J.; Wang, X.P. Vegetation characteristics and ecological restoration technology of typical degradation wetlands in the middle of heihe river basin, zhangye city of gansu province. *Wetl. Sci.* **2012**, *10*, 194–199. (In Chinese)

9. Feng, Q.; Liu, W.; Su, Y.H.; Zhang, Y.W.; Si, J.H. Distribution and evolution of water chemistry in heihe river basin. *Environ. Geol.* **2004**, *45*, 947–956.

10. Tedeschi, A.; Dell'Aquila, R. Effects of irrigation with saline waters, at different concentrations, on soil physical and chemical characteristics. *Agric. Water Manag.* **2005**, *77*, 308–322.

11. Ayars, J.E.; Shouse, P.; Lesch, S.M. *In situ* use of groundwater by alfalfa. *Agric. Water Manag.* **2009**, *96*, 1579–1586.

12. Ayars, J.E.; Christen, E.W.; Soppe, R.W.; Meyer, W.S. The resource potential of *in-situ* shallow ground water use in irrigated agriculture: A review. *Irrig. Sci.* **2005**, *24*, 147–160.

13. Singh, R.; Singh, J. Irrigation planning in cotton through simulation modeling. *Irrig. Sci.* **1996**, *17*, 31–36.

14. Feddes, R.A.; Kabat, P.; van Bakel, P.; Bronswijk, J.J.B.; Halbertsma, J. Modelling soil water dynamics in the unsaturated zone—State of the art. *J. Hydrol.* **1988**, *100*, 69–111.

15. Jong, R.d.; Bootsma, A. Review of recent developments in soil water simulation models. *Can. J. Soil Sci.* **1996**, *76*, 263–273.

16. Saito, H.; Šimůnek, J.; Mohanty, B.P. Numerical analysis of coupled water, vapor, and heat transport in the vadose zone. *Vadose Zone J.* **2006**, *5*, 784–800.

17. Jacques, D.; Šimůnek, J.; Timmerman, A.; Feyen, J. Calibration of richards' and convection-dispersion equations to field-scale water flow and solute transport under rainfall conditions. *J. Hydrol.* **2002**, *259*, 15–31.

18. Wang, T.; Franz, T.E.; Zlotnik, V.A. Controls of soil hydraulic characteristics on modeling groundwater recharge under different climatic conditions. *J. Hydrol.* **2015**, *521*, 470–481.

19. Van Genuchten, M.T. A closed-form equation for predicting the hydraulic conductivity of unsaturated soils. *Soil Sci. Soc. Am. J.* **1980**, *44*, 892–898.

20. Mualem, Y. A new model for predicting the hydraulic conductivity of unsaturated porous media. *Water Resour. Res.* **1976**, *12*, 513–522.

21. Šimůnek, J.; van Genuchten, M.T.; Šejna, M. Hydrus: Model use, calibration, and validation. *Tansac Asabe* **2012**, *55*, 1261–1274.

22. Šimůnek, J.; van Genuchten, M.T.; Sejna, M. The hydrus-1d software package for simulating the one-dimensional movement of water, heat, and multiple solutes in variably-saturated media. *Univ. Calif. -Riverside Res. Rep.* **2005**, *3*, 1–240.

23. Zhao, W.; Liu, B.; Zhang, Z. Water requirements of maize in the middle heihe river basin, china. *Agric. Water Manag.* **2010**, *97*, 215–223.

24. Gee, G.W.; Or, D. 2.4 particle-size analysis. *Methods Soil Anal. Part* **2002**, *4*, 255–293.

25. Klute, A.; Dirksen, C. Hydraulic conductivity and diffusivity: Laboratory methods. In *Methods of Soil Analysis: Part 1—Physical and Mineralogical Methods*; American Society of Agronomy: Madison, WI, USA, 1986; pp. 687–734.

26. Feddes, R.A.; Kowalik, P.J.; Zaradny, H.X. *Simulation of Field Water Use and Crop Yield*; Centre for Agricultural Publishing and Documentation: Wageningen, The Netherlands, 1978.

27. Šimůnek, J.; Hopmans, J.W. Modeling compensated root water and nutrient uptake. *Ecol. Model.* **2009**, *220*, 505–521.

28. Van Genuchten, M.T. *A Numerical Model for Water and Solute Movement in and below the Root Zone*; United States Department of Agriculture Agricultural Research Service U.S. Salinity Laboratory: Riverside, CA, USA, 1987.

29. Slavich, P.G.; Petterson, G.H. Estimating the electrical conductivity of saturated paste extracts from 1: 5 soil, water suspensions and texture. *Soil Res.* **1993**, *31*, 73–81.

30. Corwin, D.L.; Lesch, S.M. Application of soil electrical conductivity to precision agriculture. *Agron. J.* **2003**, *95*, 455–471.

31. Richards, L.A. U.S. salinity laboratory staff. In *USDA Handbook No. 60. Diagnosis and Improvement of Saline and Alkali Soils*; USDA: Washington, DC, USA, 1954; p. 13.

32. Allen, R.G.; Pereira, L.S.; Raes, D.; Smith, M. *Crop Evapotranspiration-Guidelines for Computing Crop Water Requirements-FAO Irrigation and Drainage Paper 56*; FAO: Rome, Italy, 1998; Volume 300, p. 6541.

33. Allen, R.G.; Pereira, L.S. Estimating crop coefficients from fraction of ground cover and height. *Irrig. Sci.* **2009**, *28*, 17–34.

34. Schaap, M.G.; Leij, F.J.; van Genuchten, M.T. Rosetta: A computer program for estimating soil hydraulic parameters with hierarchical pedotransfer functions. *J. Hydrol.* **2001**, *251*, 163–176.

35. Vanderborght, J.; Vereecken, H. Review of dispersivities for transport modeling in soils. *Vadose Zone J.* **2007**, *6*, 29–52.

36. Gelhar, L.W.; Welty, C.; Rehfeldt, R.K. A critical review of data on field-scale dispersion in aquifers. *Water Resour. Res.* **1992**, *28*, 1955–1974.

37. Moayyad, B. *Importance of Groundwater Depth, Soil Texture and Rooting Depth on Arid Riparian Evapotranspiration*; New Mexico Institute of Mining and Technology: Socorro, NM, USA, 2001.

38. Grinevskii, S.O. Modeling root water uptake when calculating unsaturated flow in the vadose zone and groundwater recharge. *Mosc. Univ. Geol. Bull.* **2011**, *66*, 189–201.

39. Marquardt, D.W. An algorithm for least-squares estimation of nonlinear parameters. *J. Soc. Ind. Appl. Math.* **1963**, *11*, 431–441.

40. Semenov, M.A.; Barrow, E.M.; Lars-Wg, A. *A Stochastic Weather Generator for Use in Climate Impact Studies*; User Manual: Hertfordshire, UK, 2002.

41. Kandelous, M.M.; Kamai, T.; Vrugt, J.A.; Šimůnek, J.; Hanson, B.; Hopmans, J.W. Evaluation of subsurface drip irrigation design and management parameters for alfalfa. *Agric. Water Manag.* **2012**, *109*, 81–93.

42. Ramos, T.B.; Šimůnek, J.; Gonçalves, M.C.; Martins, J.C.; Prazeres, A.; Pereira, L.S. Two-dimensional modeling of water and nitrogen fate from sweet sorghum irrigated with fresh and blended saline waters. *Agric. Water Manag.* **2012**, *111*, 87–104.

43. Siyal, A.A.; Bristow, K.L.; Šimůnek, J. Minimizing nitrogen leaching from furrow irrigation through novel fertilizer placement and soil surface management strategies. *Agric. Water Manag.* **2012**, *115*, 242–251.

44. Garg, K.K.; Das, B.S.; Safeeq, M.; Bhadoria, P.B.S. Measurement and modeling of soil water regime in a lowland paddy field showing preferential transport. *Agric. Water Manag.* **2009**, *96*, 1705–1714.

45. Patil, M.D.; Das, B.S.; Bhadoria, P.B.S. A simple bund plugging technique for improving water productivity in wetland rice. *Soil Tillage Res.* **2011**, *112*, 66–75.

46. Vazifedoust, M.; van Dam, J.C.; Feddes, R.A.; Feizi, M. Increasing water productivity of irrigated crops under limited water supply at field scale. *Agric. Water Manag.* **2008**, *95*, 89–102.

47. Robarge, W.P. *Precipitation/Dissolution Reactions in Soils*; CRC Press: Boca Raton, FL, USA, 1999; p. 2.

48. Herczeg, A.L.; Dogramaci, S.S.; Leaney, F.W.J. Origin of dissolved salts in a large, semi-arid groundwater system: Murray basin, australia. *Mar. Freshw. Res.* **2001**, *52*, 41–52.

49. Li, C.; Lei, J.; Zhao, Y.; Xu, X.; Li, S. Effect of saline water irrigation on soil development and plant growth in the taklimakan desert highway shelterbelt. *Soil Tillage Res.* **2015**, *146*, 99–107.

50. Morris, J.D.; Collopy, J.J. Water use and salt accumulation by *Eucalyptus camaldulensis* and *Casuarina cunninghamiana* on a site with shallow saline groundwater. *Agric. Water Manag.* **1999**, *39*, 205–227.

51. Satchithanantham, S.; Krahn, V.; Sri Ranjan, R.; Sager, S. Shallow groundwater uptake and irrigation water redistribution within the potato root zone. *Agric. Water Manag.* **2014**, *132*, 101–110.

52. Bastias, E.; Alcaraz-Lopez, C.; Bonilla, I.; Martinez-Ballesta, M.C.; Bolanos, L.; Carvajal, M. Interactions between salinity and boron toxicity in tomato plants involve apoplastic calcium. *J. Plant Physiol.* **2010**, *167*, 54–60.

53. Brotherson, J.D.; Field, D. Tamarix: Impacts of a successful weed. *Ra ngelands* **1987**, *9*, 110–112.

54. Sala, A.; Smith, S.D.; Devitt, D.A. Water use by tamarix ramosissima and associated phreatophytes in a mojave desert floodplain. *Ecol. Appl.* **1996**, 888–898.

55. Ibrakhimov, M.; Khamzina, A.; Forkutsa, I.; Paluasheva, G.; Lamers, J.P.A.; Tischbein, B.; Vlek, P.L.G.; Martius, C. Groundwater table and salinity: Spatial and temporal distribution and influence on soil salinization in khorezm region (uzbekistan, aral sea basin). *Irrig. Drain. Syst.* **2007**, *21*, 219–236.

56. Holland, K.L.; Charles, A.H.; Jolly, I.D.; Overton, I.C.; Gehrig, S.; Simmons, C.T. Effectiveness of artificial watering of a semi-arid saline wetland for managing riparian vegetation health. *Hydrol. Process.* **2009**, *23*, 3474–3484.

57. Askri, B.; Ahmed, A.T.; Abichou, T.; Bouhlila, R. Effects of shallow water table, salinity and frequency of irrigation water on the date palm water use. *J. Hydrol.* **2014**, *513*, 81–90.

7

Actors' Perceptions of Issues in the Implementation of the First Round of the Water Framework Directive: Examples from the Water Management and Forestry Sectors in Southern Sweden

E. Carina H. Keskitalo

Geography and Economic History, Umeå University, Umeå 901 87, Sweden;
E-Mail: Carina.Keskitalo@geography.umu.se;

Academic Editor: Miklas Scholz

Abstract: The EU Water Framework Directive exerts a major impact on water management structure and aims, and water use activities in the member states. This paper reviews the perceptions of the early WFD implementation in a case study area in southern Sweden. The focus is on the perceptions of both water management and forestry actors, the latter as a potential diffuse source impact on water quality. This study highlights the considerable complexity of reorienting or rescaling governance given the complex existing systems particular to the area, the multi-interpretable early policies on implementation and the complexity of interpreting the regionally-focused WFD approach in the largely locally-focused Swedish system. While the first phase of implementation is now long past, conclusions on the complexity of reorienting systems remain relevant, particularly with regard to non-point sources.

Keywords: water framework directive; governance; Västerhavet; Sweden; forest

1. Introduction and Aim

The EU Water Framework Directive (WFD) is currently being implemented across the EU with the aim of bringing all water bodies up to "good ecological and chemical status" by 2015 with full enforcement of the directive by 2027 [1]. The supra-national delegation of authority on water issues introduced with the EU WFD has been seen as being a potentially radical policy innovation [2], but also includes a number of issues for its implementation. These include the designation of river basin districts as well as the appointment of a suitable authority to apply the rules of the Directive (Art. 3.1) with significant local involvement [3]. All of these requirements are likely to play out differently in different national contexts due to varying prerequisites in terms of administrative structure, established interests and policy priorities. The WFD may contribute to a rescaling of governance, whereby very different impacts in terms of actors" losses or gains of influence may now result as compared with the previous structure in each country [4,5]. There are also indications that changes in actor roles will depend on how they are able to deal with the rescaling, *i.e.*, their ability to work across scales and with the participatory and integrative requirements of the WFD.

Sweden is an interesting example for WFD implementation given Sweden's major focus on municipal self-government and the historically strong importance of resource-based interests such as forestry, which has a diffuse source impact on water quality [6–8]. This paper studies the perceptions of the actors in the first round of the WFD directive implementation within the Swedish Västerhavet Water District. In particular the study asks how participants perceive the changing governance structure, including authority distribution, due to the WFD implementation. This study thereby adds a governance-oriented case study in southern Sweden to assessments of WFD implementation in other case study areas in Sweden (e.g., [6,7,9]). The following sections describe the study's background with regard to the case study in terms of changes caused by the WFD, based on a multi-level governance perspective. The paper then outlines the case study method and the results, particularly in relation to the impact at regional and local levels.

2. Background

Multi-level governance is regularly defined as the integration of decision-making in government on several levels and the integration of the supranational, in particular the EU-level as well as private parties and NGOs [10]. Multi-level governance is also regularly seen to include changes in the distribution of authority and the inclusion of potentially competing administrative requirements (*cf.* [3]). As a result, the implementation of new tools, such as the Water Framework Directive, could be expected to impact the distribution of authority. Given the framework design of the WFD, and the relatively large degree of freedom this allows the individual state concerning the implementation of the directive [5], the perceptions of challenges and requirements for reorientation of practices amongst those involved in the implementation are important to understand and include the impact on new governance structures. The WFD can be seen as a multi-level governance tool [4] that results in the rescaling of water governance: it changes the distribution of authority across levels and may result in some actors losing or gaining influence, although this also depends on their own capabilities to work across scales [4]. Such changes will impact the national level, which will also develop specific ways to implement the Directive that serve as guides at the regional and local level. Regional and local levels within a state will thus face considerable demands as concerns regarding the integration and implementation of new organisational tools that originate, and are driven by, logics external to the state organisation [3]. In addition, other land use or industrial sectors with a potential impact on water quality will need to comply with the WFD and its requirements in both a stakeholder role and as impacted by potential measures under the WFD.

The general requirements of the WFD include (Art. 13) that regional water authorities must be developed on a river basin scale, and each of these are required to develop a river basin Management Plan and Programmes of Measures, including environmental quality standards along with identification of cost-effective measures to fulfil these. The Programmes should also describe which authorities are responsible for which measures, they should coordinate with different stakeholders when relevant, and define the timeframe for accomplishing these measures. Development is carried out on a cyclical basis, with the first cycle of implementation of the Water Framework Directive lasting until 2015, with upcoming cycles ending in 2021 and 2027, respectively [11].

With regard to implementation in the Swedish system, the greatest changes resulting from the WFD are that water issues are treated by the river basin rather than by the county or municipality [12]. A specific process has also been introduced which concerns decision-making on the (binding or non-binding) nature of environmental quality standards or norms for water in Swedish legislation at large. This process is not dealt with here, but see for instance [3]. Sweden has thus been divided into five river basin districts, each of which consist of river basin areas composed of several county administrative boards, *i.e.*, including several regional denominations as these are divided in the general Swedish system of delineation of authority. Each river basin district is administered by a Water Authority, which, in the Swedish system, is identified as one of the county administrative board in the area. Each Water Authority is responsible for developing water management plans and related programmes of measures for its river basin area, including the requisite cooperation with other county administrative boards, municipalities and others in local bodies [13]. Due to the requirement for consultation, the water authorities in two of the districts included in this study (Västerhavet) have suggested that the local water councils should be developed to include local stakeholders impacted by water management. Such water councils are based on local water organisations (water management councils) that have been in place, particularly in southern Sweden, since the 1950s but have now been developed into a format which is compliant under the WFD in order to monitor water management plans and potential measures [12]. While it has been emphasised in some literature that water councils do not hold decision-making authority (e.g., [12]), the structure for management under the WFD at the local and regional levels was initially somewhat inconsistent, for example discussing the role of proposals for potential measures from such bodies [13] which potentially led to confusion amongst stakeholders.

Changes to the existing governance system, particularly in southern Sweden and Västerhavet, in response to WFD implementation in Sweden were thus relatively elaborate, and include:

- Regional delineation of water, based on river basins, which have been established in Sweden by including several existing regional delineations (counties) together that make-up the water district.
- Designation of a water authority to develop water management plans and programmes of measures, which in Sweden consist of one selected county administrative board from the counties that make up the water district.
- In southern Sweden, development of local water councils to serve as collaboration bodies for the regional water authority, rather than in their previous format of collaboration bodies for the municipality.

The above changes to the Swedish system of water governance therefore mean that the regional level, which is generally de-emphasised in the Swedish system as a non-elected, mainly implementing, arm of the state, is instead emphasised, although now with new regional delineations. The changes in water council areas also result in local collaboration bodies (apart from the municipality itself) becoming more emphasised than previously. This is a shift from the pre-existing local coordination bodies, which were largely working in relation to the municipality and have not been judged to have exerted a major impact on measures taken. Rather the municipality is the emphasised sub-national decision-making unit in the Swedish system, *cf.* e.g., [7]. Both of these changes may, to some extent, impact what is often called the *planning monopoly* of the Swedish municipality, where the municipality is responsible for land use and planning decisions at the local level [14]. Jurisdictional mismatch may arise, particularly between previous systems of management and those newly introduced, posing significant implementation difficulties on both regional and local levels as well as in relation to land use. One example is the local level of the municipality which was formally responsible for the long-term protection of the water supply [13].

Authority distribution, or more broadly the governance of water, could thus be expected to result in both discussions on specific decision-making rights as well as discussions on coordination, in particular for the first implementation round of this system. Impacts affect not only in water management but also amongst stakeholders who may be impacted by the implementation of the WFD. In Sweden, forestry is a large-scale land use and thereby a potential diffuse source of impact on water quality (e.g., through fertilization of forest plantations where runoff may impact water systems). Forestry is, however, also particular in that, historically, forestry planning as a highly important industry traditionally functions separate from land use planning, whereas aspects of the WFD now require some integration [15]. As a result, concerns in the forestry sector over implications of the WFD were, from the outset, relatively extensive and as a result were dealt with in this study.

3. Case Study and Methodology

The case study targets the Västerhavet Water District, located in an area where water management issues have traditionally been pronounced. The area includes Sweden's second largest city, Gothenburg, and is the site of part of the County Administrative Board of Västra Götaland, which is the water authority for the Västerhavet Water District (See Figure 1).

Firstly this study draws upon a literature review on the Västerhavet water management system (for example water authority reports, water management plan and programmes of measures). Secondly, the study conducted a total of seven semi-structured interviews undertaken in 2010 during the first round of the implementation of the WFD. Interviews targeted the main administrative actors, as defined in the above background section, who were identified based on a strategic selection. These interviewees represented the Regional Water Authority and County Administrative Board (the water authority located at the Västra Götaland County Administrative Board and its Secretariat); the Göta River Water Council as the water council for the largest tributary in the area; the Göta River Water Management Council as a pre-existing stakeholder body of this water council, and Gothenburg City as the largest municipality in the area. In order to discuss opportunities for water management which would take into account industry in other sectors (potentially impacting water quality), actors in forestry have also been interviewed. Forestry was selected because it is a major industry which constitutes a potential diffuse source, as contamination from e.g., fertilizers used in forestry are dispersed across large areas and with considerable impact, particularly to small streams crossing the forested areas. Interviews with actors in forestry targeted a selection of state and small-scale forestry interests: the Swedish Forest Agency, the state forest company Sveaskog, and the association for small-scale forest owners in the area (Södra skogsägarna). Interview questions in all cases targeted their experience of the implementation of the WFD, including the development of river basin management plans, consultation and participation. The interviews further targeted interviewees' perception of changes in the distribution of authority resulting from the WFD and its impact on existing administrative systems. All interviews were undertaken in person, recorded and transcribed in full. Interviews typically lasted about an hour.

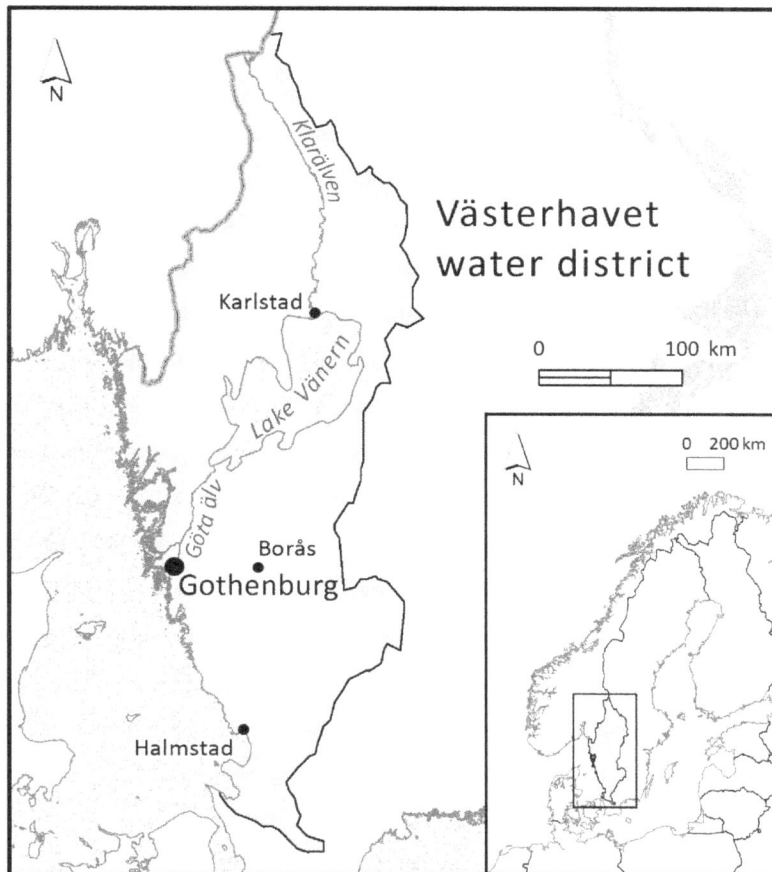

Figure 1. The Västerhavet Water District (revised from Västerhavet Water Authority 2011, GIS graphics courtesy of Magnus Strömgren).

Thirdly, the study draws upon a review of the Minutes and consultation statements. With regard to the Minutes, the study selected Minutes of the Göta River Water Council (Göta Älvs vattenråd) meetings ranging from its establishment in 2006 to 2009, which include some discussion of the stakeholders' perceptions of the process and implementation of the Directive (public material available on the council website). The study further draws on statements from the consultation received by Västerhavet Water Authority on the Management Plan and Programmes of Measures. Comments—Often in document format with lengths varying from a page to a couple of pages or longer—Were submitted during the March–September 2009 period of referral for these documents. The review of statements has targeted: (a) statements submitted by the main actors in administration as defined in interviewee selection above (Gothenburg Municipality and Göta River Water Council, as well as the summary of statements by the Västra Götaland County Administrative Board); and (b) statements submitted by stakeholders in forestry (a broader selection including all state forest organisations, forest owner or management bodies submitting statements during the consultation: the Swedish Forest Agency, general statements by the Federation of Swedish Farmers LRF which is also the national private forest owners' association, the forest industry association Skogsindustrierna, private forest owner organizations Skogsägarna Mellanskog and Södra Skogsägarna, forestry management industry Bergvik Skog and state forest company Sveaskog). This selection does not include the forest products industry.

All statements, protocols and interviews were coded according to categories of perceptions on the development of the programmes of measures, participation and impacts on authority of different actors within the Swedish structure. All Swedish to English translations in the results section from interviews and other types of documents were done by the author.

4. Results

4.1. Perceptions of the Development of Programmes of Measures: The Role of the National and Regional Level

In general, interviewees noted that there were complications with the participation and involvement in the WFD in the first phase of implementation. This was due to very extensive and generally-oriented material and to the relatively short period for referrals onto the programmes as well as the water council structure for participation. However, most actors who were interviewed also noted that the WFD has had a greater impact in terms of focusing attention on water issues. A typical comment in interviews was that "We have more focus on water issues then we did before". Also interviewees within the forestry sector typically noted that the water directive has brought "Water issues [in forestry] ... more significantly into focus".

Organisation on national level in order to support the process, on regional (water authority *vs.* broader county administrative board structure) and on local (water council *vs.* municipality) level was impacted by the WFD process. On each of these levels, the role of the political control of the process was discussed and, to some extent, used as an explanation for some of the implementation problems so far. The national level was particularly criticised for its limitation in developing the guidance on implementation in a timely manner. In the interview the Swedish Forest Agency noted that its work on the Water Framework Directive was delayed by the time that was spent establishing the water authorities, as well as by the delay in guidance from the Swedish EPA on the process of work in the water authorities: guidance to what was, at the time, regarded as a potentially large impacting factor in terms of diffuse sources of pollution was thereby perceived as limited. Actors generally noted that political control either at national or lower levels—e.g., in providing advice on how to balance conflicting interests—Had been limited and that most of the work with developing the implementation process for the WFD had fallen to civil servants. For example as an interviewee noted: "Political control ... has been very weak ... the Ministry of the Environment, overarching water management but also the ... now... Ministry for Rural Affairs ... have maintained a very low profile".

This was seen to result in problems with regard to the role of the regional level. For example, interviewees noted that: "Water authorities themselves have had to do a lot of work in the development (of water management in relation to the Water Framework Directive) ... as guidance for their organisation was lacking from the start". Statements from the consultation on water management plans also note that the Water Authorities are expert administrative organisations (staffed by civil servants) and thus unable to make the political decisions required by weighing environmental and other societal goals, which results in limited means for decision-making. As a result, authorities may have had to move towards more general and assessment-based measures.

On the basis of work on environmental quality assessment, suggestions for programmes of measures and management plans were finalised for the first cycle of WFD implementation in December 2008, and, according to the schedule of implementation, were required to come into force in December 2009. The management plan summarised the work so far with regard to the areas of mapping of water bodies, development of environmental quality norms and programmes of measures, and participation in water management assessment. The water authority concluded that it had "[e]stablished that extensive and broad measures are required to reach the environmental quality norms" ([13] author's translation). Rather than suggesting detailed measures, the authority identified 38 measures that constitute the basis for physical and other direct measures, which were given its mandate to direct measures to authorities and municipalities. One example is No. 21 on forestry that: "The Swedish Forest Agency should, in cooperation with the Swedish EPA and the Swedish Board of Fisheries, determine ordinance that imposes requirements for … protection zones and other measures close to water so that good chemical status and good or high level ecological status is maintained or achieved" (Measure No. 21 of the Management Plan [13] (p. 132) and the Programme of Measures [16] (p. 11), author's translation). The Management Plan notes state that given the complex nature of environmental quality norms, authorities and municipalities will, in many cases, need to undertake supplementary studies to clarify the measures necessary to achieve these ([13] author's translation).

The Management Plan thereby concluded that setting environmental quality norms require not only technical but also administrative, legal and practical coordination. Thus, it advocated stepwise implementation which includes knowledge building and the development of supporting administrative measures [13]. Both the role of formal administrative actors and the water councils are highlighted, as well as funding issues: "Our goal is to have the water councils as a central part of the development of the Programme of Measures, where the water councils themselves develop suggestions for measures that lead to improved water quality. This local knowledge and local acceptance of processes is important in order to implement the right measures at the right place at the right cost. Formally there is no hindrance to active water councils pursuing their own management work" ([13], author's translation). Statements on water councils as potentially undertaking management—And thereby potentially conflicting with municipal steering—Were not clear on the roles of different actors. Similarly, it was noted that "[T]he municipalities and county administrative boards will have a central role in the realisation of the programmes of measures. It is important that the experience with supervision and testing, among other matters, is utilised in order to continuously improve work" ([13] (p. 174); *cf.* [16]). On the other hand, "[F]or measures that lack a clear responsible operational actor to be undertaken in practice, a financial support system to finance or encourage actors to implement cost-effective measures is needed. In order to finance this support system and develop measures in different areas, financial instruments may be necessary" (ibid.). Clarification of the roles and funding has thus remained unsolved at this early stage of implementation.

Many bodies commented on the preliminary character of the Management Plan and Programmes of Measures, as well as noting problems in the documentation and consultation for these plans. Amongst other issues, actors discussed the limitations to the consultation process in that the documents were too large and complex to properly involve the citizenry. Additionally suggestions for measures were developed under a severe time limit. An interviewee in administration also noted that the implementation in a state such as Sweden would have to take longer: "Many parts have been based on expert judgement ... progress isn't quick when you have all these watercourses, which is not a situation for the rest of Europe".

Therefore the WFD resulted in having relatively large-scale discussions on how to implement the system, largely at regional level and below with perceived limited state guidance. In addition, discussions on what extent different sectors should integrate considerations and what potential considerations were needed in the process were undertaken. As a result of these concerns, some actors provided statements to the consultation that this first water management cycle should, given the limited time and material, be regarded as a trial period to provide the basis for follow-up and assessment of proposed measures. However, while management programmes were criticised for not being detailed enough, it was noted that having too detailed and developed programmes would have limited local influence, knowledge and application by water councils and other actors in the future [17]. An interview with the Water Authority noted: "Detailed programmes of measures ... I do not think it matters how many people we have had, they could not have been made in any case, because they need to be developed through cooperation ... and cooperation takes time".

4.2. Perceptions at Local Level

At the local level, the water councils were a new format for stakeholder participation. The Göta Älv Water Council was funded by the water authority administered by Göteborgsregionen (GR), which also administered the preceding format of local water cooperation stakeholder body since the 1980s. For example, from the basis of the established participatory structure, the Water Council also contacted other potentially interested bodies and allowed e.g., Kungälv Municipality to join the Council. Locally, WFD requirements were thus interpreted and implemented within the particular context of the area, and added to by consultations with additional bodies that were not part of the previous stakeholder body. A view that typified the understanding of interviewees in administration was: "When it starts to concern ... concrete measures we should have more people in... from the planning side, from the technical side".

While the Management Plan noted that the water council, county administrative boards and the municipalities would play central roles in developing measures [13], interviewees noted difficulties with the somewhat unclear definition of roles and authority. Concerns particularly targeted the role of the Water Councils. For example, an interviewee at the local level noted that: "Proposing measures is a rather far-reaching step. Who do you represent when you suggest measures in a water council? ... There [should be] democratically elected bodies that should determine these kinds of things". Water councils, another interviewee judged, "are also a somewhat odd constellation ...different water councils differ as to who joins and [who] is judged to be important. If representatives from fishing, fisheries councils and so on take part and play a dominant role in the water councils their issues are given a large scope".

On the forestry side, concerns were also related to the local character of participation envisioned within the Water Framework Directive. Interviewees noted that while the forest owners' organisation Södra has taken the stand that it should be represented in all water councils, the Swedish Forest Agency and Sveaskog, for example, noted that they have had no such opportunities. The interviewee at Sveaskog said: "A forest owners' association … [has the] opportunity to let its members represent it in the water councils, but for us who have a total of perhaps 600–700 employees, if we join every water council then that will be a very heavy burden".

As a result of these concerns, forest sector participation was also developed through setting up an informal forestry water council group, based on an initiative by the Federation of Swedish Farmers to join with participants from the forest industry sector (forest owners and forest owners' associations) to discuss the implications of the Water Framework Directive. This group, amongst other matters, discussed the potential impact of the WFD on forestry and potential ways of responding to any such requirements. For instance, the interviewee at Sveaskog noted that Sveaskog had been involved in the dialogue with water authorities, which differed depending on water authority, with the industry being a primary party: "My experience is that this dialogue is the result of a forest sector initiative, it was not … the communication path intended from the start, rather as the issue was raised the forest sector has taken this initiative". Therefore, to some extent the higher-level organisation of forestry was developed as a result of what was seen as a local focus in the WFD and which was perceived as not sufficiently including more large-scale actors from forestry, especially as the sector may become impacted in relation to management of non-point or diffuse sources such as forest fertilization, which is a potential issue along large areas close to water.

Many issues thus remained to be developed, largely with regard to decision-making and, crucially, financing and potential reimbursement to actors who might be impacted by these measures. One interviewee in administration noted, in a rather typical comment, that "[t]he crunch issue is how to finance measures". As most monitoring is carried out by public bodies, one issue was whether municipalities should fund further demands imposed by the Water Directive, or whether funding from the state level should be provided (e.g., [18]). The Water Authority noted that while funding had so far been supplied to water authorities supporting their work with the systematisation of knowledge, and to water councils for administration: "One concern is that the lack of financial resources will limit the opportunities to undertake concrete measures within water management activities. It is feared that the lack of means could also impact on the possibilities for everyone to participate, or participate on equal terms, in the coordination work" ([13], author's translation). For example, Gothenburg Municipality, in their comments in the consultation, stated that the municipalities will play an important role in implementing the Directive and that the financing of the programmes of measures in this respect has to be made clear, as at the time it was difficult to assess which costs different actors needed to bear from the general Programme of Measures and its general suggestions on costs for different measures. Coordination between different actors would be needed for further development in the current and future implementation cycles.

5. Discussion and Conclusions

The implementation of the WFD during its first phase has shown the large-scale challenge of setting up new regional and local structures for decision-making on water quality. While the early stages resulted in a greater awareness of the importance of water management among participants, a number of issues were noted. Concerns regarding relationships between elected and participatory bodies in clarifying requirements and developing and determining measures were noted, as a clear structure for making political decisions on priorities did not exist on the water authority level at the time (*cf.* [14]). These results concur with studies in other water districts in Sweden. In the case of a more northern Swedish water district, Hammer *et al.*, observed that challenges included coordinating monitoring within the district, that all stakeholders in the WFD were able to understand the technical information, and to manage diffuse sources of pollution [6]. Similar to this study's results, Andersson *et al.*, observed that: "Necessary municipal involvement in WFD-related measures for improving water quality will be difficult to achieve without associated guidance and financial support from the higher levels" [9] (p. 81). Both studies also noted that local actors at municipal level and in water councils listed unclear roles and responsibilities and funding issues as impediments to their participation. In addition, a number of potential measures also subject to the potential willingness of landowners (such as agricultural or forest land) to undertake them (*cf.* [9,19]). Considerations found in this study are not dissimilar to studies on WFD implementation in other areas, which have also noted that it may not be clear to all local actors to what extent certain issues are technical or may determine further decision-making ([4]; *cf.* [2,20]). It has also been noted that that there are problems in decision-making when dealing with the complexity involved in WFD implementation (*cf.* [21]) and in "learning to work with new directives" [22] (p. 65). In the case of non-point sources such as forestry, and other large-scale land uses, such consideration may also relate to the scale at which implementation takes place; for instance, in this case, with regards to how to participate in a discussion of measures at a local level.

The WFD was built on a governance logic that is not particular to the specific state systems and cannot necessarily take issues into account such as the potential diverging role of different source emissions or varying structures which are of nationally important sectors in different countries. This may thereby be regarded as an illustrative case of the difficulty of integrating new governance principles and systems, such as those which are EU-based (*cf.* [10]). As highlighted in the Västerhavet Case, this process is affected by existing organisations on a sectoral as well as regional and national scale. As a result, "Whereas much multi-level governance research has focused on the relationships of EU central institutions to member states, it is the regional factors that seem to be of equal impact ... a core challenge of EU policy will be to recognize these different scales" ([23] (p. 244); *cf.* [24–27]). This type of novel multi-level considerations posed by implementation that takes place in context of different institutional settings—Where there exists an "institutional ambiguity" with relation to what rules and norms to apply and how—Has been highlighted by Hajer [28] and van Leeuwen *et al.* [29]. Van Leeuwen *et al.*, showed that the implementation of the EU Marine Strategy Framework Directive was considerably unclear as a result of interplay between national, regional and EU levels ([29]; *cf.* [27]).

While many of these problems in the Swedish process were compounded by the short period for participation, and to some extent with having to keep up with the EU schedule for implementation given a late start, they do highlight structural aspects of some of the implementation problems observed. While the formalisation of policy advice and a process for dealing with the challenges for implementation in these cases may require adjusting EU structures to the established national system [10], the case illustrates the extensive complexity in reorienting or rescaling governance based on a regional approach for the largely locally-focused Swedish system, and given its local government planning monopoly as well as the roles of established sectoral practices.

Acknowledgment

The research was funded through Future Forests, a multi-disciplinary research programme supported by the Foundation for Strategic Environmental Research (MISTRA), the Swedish Forestry Industry, the Swedish University of Agricultural Sciences (SLU), Umeå University, and the Forestry Research Institute of Sweden.

Conflicts of Interest

The author declares no conflict of interest.

References

1. *Directive 2006/60/EC of the European Parliament and of the Council of 23 October 2000 Establishing a Framework for Community Action in the Field of Water Policy (the Water Framework Directive)*; European Union: Brussels, Belgium, 2006.

2. Moss, B. The Water Framework Directive: Total environment or political compromise? *Sci. Total Environ.* **2008**, *400*, 32–41.

3. Keskitalo, E.C.H.; Pettersson, M. Implementing multi-level governance? The legal basis and implementation of the EU Water Framework Directive for forestry in Sweden. *Environ. Policy Gov.* **2012**, *22*, 90–103.

4. Moss, T.; Newig, J. Multilevel water governance and problems of scale: Setting the stage for a broader debate. *Environ. Manag.* **2010**, *46*, 1–6.

5. Liefferink, D.; Wiering, M.; Uitenboogaart, Y. The EU Water Framework Directive: A multi-dimensional analysis of implementation and domestic impact. *Land Use Policy* **2011**, *28*, 712–722.

6. Hammer, M.; Balfors, B.; Mörtberg, U.; Petersson, M.; Quin, A. Governance of water resources in the phase of change: A case study of the implementation of the EU Water Framework Directive. *Ambio* **2011**, *40*, 210–220.

7. Lundqvist, J.L. Integrating Swedish Water Resource Management: A multi-level governance trilemma. *Local Environ.* **2004**, *9*, 413–424.

8. Bratt, A. Municipal officers on implementing the EU water framework directive in Sweden regarding agricultural nutrient flows. *Local Environ.* **2004**, *9*, 65–79.

9. Andersson, I.; Pettersson, M.; Jarsjö, J. Impact of the European Water Framework Directive on local-level water management: Case study Oxunda Catchment, Sweden. *Land Use Policy* **2012**, *29*, 73–82.

10. Marks, G.; Hooghe, L. Contrasting Visions of Multi-level Governance. In *Multi-level Governance*; Bache, I., Flinders, M., Eds.; Oxford University Press: Oxford, UK, 2004.

11. Vattenmyndigheterna. Hur vattenarbetet planeras. Available online: http://www.vattenmyndigheterna.se/vattenmyndigheten/Om+vattenforvaltning/Vattenplanering/ (accessed on 13 November 2013).

12. Frantzén, F. Berörda intressenters deltagande i vattenförvaltning—Förändring, förnyelse och arv (Stakeholder participation in water management—Change, renewal and legacy). *VATTEN J. Water Manag. Res.* **2012**, *68*, 275–283. (In Swedish)

13. *Förvaltningsplan Västerhavets Vattendistrikt. Vattenmyndigheten i Västerhavets Vattendistrikt vid Länsstyrelsen i Västra Götalands län*; Vattenmyndigheten Västerhavet: Göteborg, Sweden, 2009. (In Swedish)

14. Hedelin, B. Potential Implications of the EU Water Framework Directive in Sweden. Available online: http://www.nordregio.se/Global/EJSD/Refereed%20articles/refereed14.pdf (accessed on 13 November 2013).

15. Stjernström, O.; Karlsson, S.; Pettersson, Ö. Skogen och den kommunala planeringen. *Plan* **2013**, *1*, 42–45.

16. *Åtgärdsprogram Västerhavet. Vattenmyndigheten i Västerhavets Vattendistrikt vid Länsstyrelsen i Västra Götalands län*; Vattenmyndigheten Västerhavet: Göteborg, Sweden, 2009. (In Swedish)

17. *Göta älvs Vattenråds Remissvar på Samrådsmaterial 2009 från Västerhavets Vattendistrikt*; Göta älvs Vattenvårdsförbund/Göta River Water Council: Göteborg, Swenden, 31 August 2009. (In Swedish)

18. Swedish Government. *Prissatt Vatten? Betänkande av Vattenprisutredningen*; Statens offentliga utredningar (SOU): Stockholm, Sweden, 2010.

19. Benner, T.; Reinicke, W.H.; Witte, J.M. Multisectoral networks in global governance: Towards a pluralistic system of accountability. *Gov. Oppos.* **2004**, *39*, 191–210.

20. *Common Implementation Strategy for the Water Framework Directive*; European Union: Brussels, Belgium, 2004.

21. Hatton-Ellis, T. Editorial: The Hitchhiker's Guide to the Water Framework Directive. *Aquat. Conserv: Mar. Freshw. Ecosyst.* **2008**, *18*, 111–116.

22. Beunen, R.; van der Knaap, W.G.M.; Biesbroek, G.R. Implementation and integration of EU environmental directives. Experiences from the Netherlands. *Environ. Policy Gov.* **2009**, *19*, 57–69.

23. Kastens, B.; Newig, J. The Water Framework Directive and agricultural nitrate pollution: Will great expectations in Brussels be dashed in Lower Saxony? *Eur. Environ.* **2007**, *17*, 231–246.

24. Glachant, M. The need for adaptability in EU environmental policy design and implementation. *Eur. Environ.* **2001**, *11*, 239–249.

25. Orr, P.; Colvin, J.; King, D. Involving stakeholders in integrated river basin planning in England and Wales. *Water Resour. Manag.* **2007**, *21*, 331–349.

26. Koontz, T.M.; Newig, J. Cross-level information and influence in mandated participatory planning: Alternative pathways to sustainable water management in Germany's implementation of the EU Water Framework Directive. *Land Use Policy* **2014**, *38*, 594–604.

27. Albrecht, J. The Europeanization of water law by the Water Framework Directive: A second chance for water planning in Germany. *Land Use Policy* **2013**, *30*, 381–391.

28. Hajer, M. Policy without Polity? Policy Analysis and the Institutional Void. *Policy Sci.* **2003**, *36*, 175–195.

29. Van Leeuwen, J.; van Hoof, L.; van Tatenhove, J. Institutional ambiguity in implementing the European Union Marine Strategy Framework Directive. *Mar. Policy* **2012**, *36*, 636–643.

8

The Influence of a Eutrophic Lake to the River Downstream: Spatiotemporal Algal Composition Changes and the Driving Factors

Qian Yu [1], Yongcan Chen [1], Zhaowei Liu [1,*], Nick van de Giesen [2] and Dejun Zhu [1]

[1] State Key Laboratory of Hydroscience and Engineering, Tsinghua University, Beijing 100084, China; E-Mails: yqcherie@126.com (Q.Y.); chenyc@mail.tsinghua.edu.cn (Y.C.); zhudejun@tsinghua.edu.cn (D.Z.)

[2] Department of Water Management, Faculty of Civil Engineering and Geosciences, Delft University of Technology, PO Box 5048, 2600 GA Delft, The Netherlands; E-Mail: n.c.vandegiesen@tudelft.nl

* Author to whom correspondence should be addressed; E-Mail: liuzhw@tsinghua.edu.cn;

Academic Editor: Miklas Scholz

Abstract: Algal blooms have been frequently found at the upper reaches of the Tanglang River, which is downstream from the eutrophic Dianchi Lake. The eutrophic lake upstream is considered to be a potential source of phytoplankton, which contributes to the development of harmful algal blooms in the river downstream and can cause many serious problems for the river ecology. However, few studies focused on these kinds of rivers. Therefore, a field observation and laboratory analysis were conducted in this study. The results showed that the Tanglang River was obviously spatially heterogeneous due to the eutrophic Dianchi Lake upstream. The toxic *Microcystis* from the Dianchi Lake dominated the phytoplankton at the upper reaches, but these were gradually, rather than immediately, replaced by centric diatoms and chlorococalean green algae in the middle and lower reaches. The results of correlation analysis indicated that the changes in hydrodynamic conditions and underwater light intensity accounted for the spatial variations. The differences in the adaptability of different algae to changing aquatic environments explained the spatial variations of phytoplankton abundance. The dominant algae, most of which was from the Dianchi Lake upstream, determined the characteristics of the total abundance at the Tanglang River.

Keywords: eutrophication; phytoplankton structures; *Microcystis*; hydrodynamic conditions; turbidity

---

1. Introduction

Algal blooms have become a frequent sight in the upper reaches of the Tanglang River during the past few years [1]. In high-bloom seasons, the surface blue-green scums can stretch a long distance in the river downstream [1]. This phenomenon is rare elsewhere in rivers because the initial inoculum of phytoplankton at the beginning of rivers is usually small [2]; as a result, most often, blooms occur only in the middle or lower reaches of large, nutrient-rich rivers [3–5]. The cause of this unusual phenomenon is the upstream eutrophic Dianchi Lake, which may supply a large amount of dominant lacustrine algae to the river downstream. Thus, the Dianchi Lake and the Tanglang River constitute a special case of a river downstream from a eutrophic lake. Nevertheless, the Dianchi Lake-to-the Tanglang River system is not unique because all eutrophic lakes and their outflows fall into this category [6], and the water qualities in these rivers downstream are usually poor. However, few studies have been conducted, especially on this special case [2,6].

Eutrophic lake-to-river systems can cause many problems for society as well as the downstream environment. First, the smelly surface scums might decrease the recreational use of rivers [7]. Moreover, some Cyanobacteria can be toxic, and this includes *Microcystis*, which is also the main constituent algae found in the Dianchi Lake. The contaminated river water cannot be used directly. This can lead to drinking water crises [8] and water shortages for surrounding communities and for industrial and agricultural production [1,9]. In order to prevent and control these aforementioned problems, it is important to estimate the impact of eutrophic lakes on rivers downstream and to identify the driving factors of different algae's growth and loss. Prygiel and Leitão [6] found that Cyanophycean from the reservoir of Val Joly contaminated the River Helpe Majeutr downstream and disappeared within 30 km. However, they did not systematically analyze the reason of these spatial variations in phytoplankton species.

After analyzing the different characteristics of bloom-forming algae in lakes and rivers [10,11], it is found that some Cyanobacteria, *i.e.*, *Microcystis*, are not adapted to turbulent flows and might be replaced by some fluvial algae, *i.e.*, centric diatoms and chlorococalean green algae [9,12,13] in fast-flowing rivers. Many existing cases have proven that intensified turbulence of water may lead to algae species shift from buoyant Cyanobacteria to sinking Bacillariophyta [14–17]. However, these experiments were conducted in the same aquatic systems (either in lakes or in rivers) rather than in the lake-to-river coupled systems. There is a need to test that if the spatial variations in phytoplankton species exist in the Tanglang River downstream from the highly eutrophic Dianchi Lake. In addition, it is also important to detect the driving factors accounting for these spatial variations.

In the present study, we hypothesized that the abundance of the lacustrine algae from the Dianchi Lake upstream would decrease in the Tanglang River downstream, and they would sustain a long distance rather than disappear immediately. In addition, we also hypothesized that the hydrodynamic condition rather than nutrient levels or water temperature impacted the spatial variations in the phytoplankton abundances and compositions in the river downstream from the eutrophic lake. To test the hypotheses, two datasets were included in this study: Dataset1 included information about the entire Pudu Catchment and Dataset2 focused on the upper reaches of the Pudu River, namely the Tanglang River which is directly influenced by the Dianchi Lake upstream. The objectives of this research were: (1) to evaluate the impact of an upstream eutrophic lake on the downstream river in time and space; (2) to distinguish between the rivers found downstream from eutrophic lakes and other rivers without upstream algae sources; and (3) to examine the driving factors of total phytoplankton abundance in the river downstream from a eutrophic lake.

2. Materials and Methods

2.1. Study Area

The study area consists of the mouth of the Dianchi Lake and the downstream Pudu River (Figure 1). The Dianchi Lake is located on Yunnan–Guizhou Plateau in Southwest China. It covers 308.6 km^2 and the average depth is 4.4 m. The lake is one of the most severely polluted lakes in China and is frequently disturbed by *Microcystis* blooms [18]. The Pudu River downstream (294 km) is mainly composed of two sections: the Tanglang River (120 km) and the lower Pudu River (174 km). The Tanglang River is the only downstream outlet for the Dianchi Lake with an average slope of 2‰. The lower Pudu River is downstream in the Tanglang River with an average slope of about 6.1‰. There are two main tributaries flowing into the lower Pudu River, namely the Zhangjiu River and the Xima River. Another large tributary, the Jiulong River first merges with the Xima River and then the rivers flow into the lower Pudu River together. There is a gate between the Dianchi Lake and the Pudu River, called Zhongtan Gate. However, this gate is mainly for controlling water level but not discharges. Hence, it will not influence the outflow from the Dianchi Lake to the Pudu River. The annual average discharge of the Pudu River is 91.2 m^3/s. The subtropical monsoon climate provides the study area relatively constant temperatures throughout the year with an annual mean temperature of about 15 °C [19].

The Dianchi Lake has been widely investigated before due to the serious nature of the annual *Microcystis* blooms [18,20,21]. However, the Tanglang River was only sampled once in 1982 for phytoplankton identification by Qian *et al.* [22], and the entire Pudu River was once studied for pollutant sources identification by Yu *et al.* [23]. This is the first time that the Dianchi Lake and the Pudu River are being studied together as a whole system.

2.2. Datasets and Sampling

A total of nine sampling stations over the entire Pudu River were included in Dataset1 (Figure 1, S1–S9). Among them, three stations were located on the Tanglang River; the other three were located on the lower Pudu River, and the rest were located at the three main tributaries, respectively. The samples were collected on three consecutive days in three representative seasons, namely the rainy season (August to September in 2010), the normal season (November in 2010) and the dry season (March in 2011).

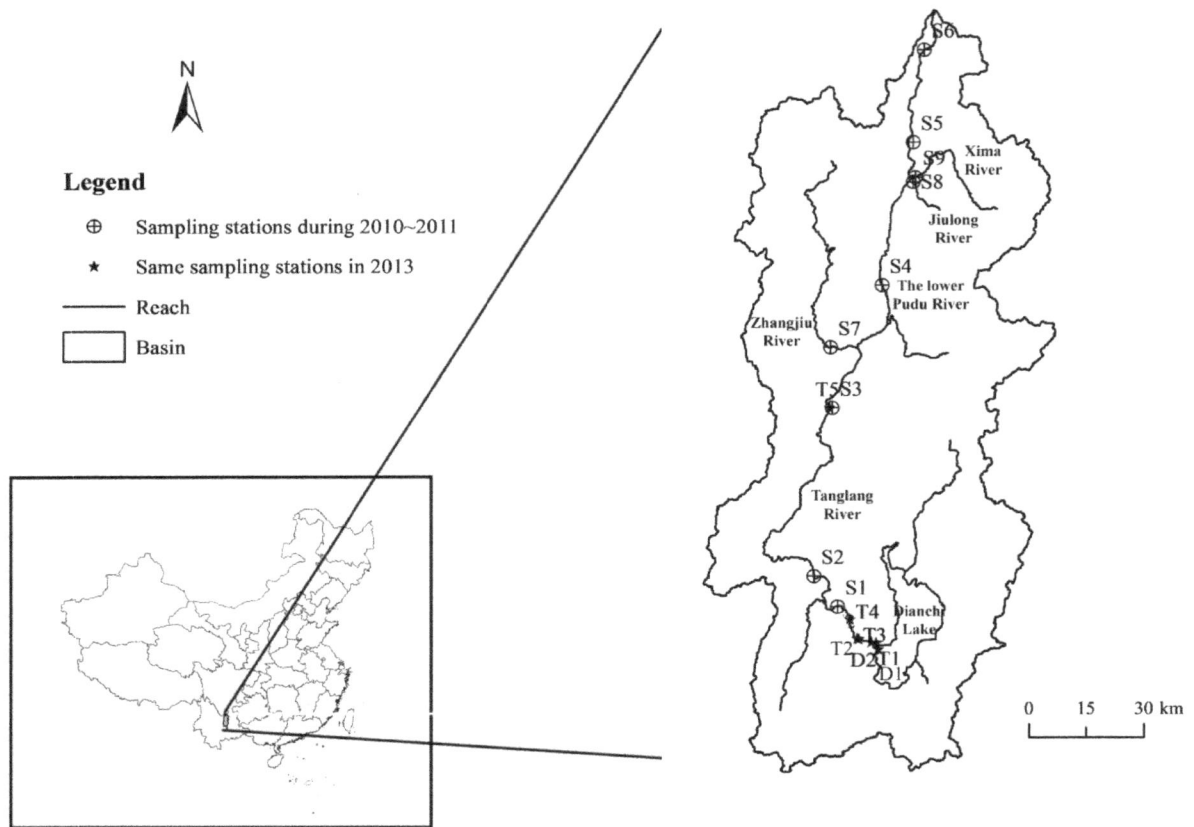

Figure 1. Map of study area and sampling stations in the two datasets.

The samplings included in Dataset2 were carried out at nine monitoring stations within one day in June 2013 (the rainy season) and at eleven monitoring stations on two consecutive days in September 2013 (the rainy season), respectively. Seven monitoring stations were situated in the same places for both the June and September measurements (Figure 1, D1–D2, and T1–T5). These seven stations included two stations at the mouth of the Dianchi Lake: −2.04 km (D1), 0 km (D2, reference station), and five others along the Tanglang River: 1.47 km (T1), 4.83 km (T2), 5.46 km (T3), 11.38 km (T4), and 114.60 km (T5). The samples of surface water (0.5 m in depth) were taken from the shore of the lake or the middle of the river. They were preserved *in situ* with Lugol's iodine solution. The 1 L water samples taken from each station were transported to the laboratory in darkened bottles. After 24 h standing in a laboratory, the supernate was siphoned from the samples and 30 mL of the remaining residue was prepared for further phytoplankton identification.

2.3. Phytoplankton Data and Environmental Variables

In Dataset1, eleven environmental parameters were determined according to State Environmental Protection Administration (SEPA) standard methods [24]. They were water temperature (WT), pH, suspended solids (SS), dissolved oxygen (DO), chemical oxygen demand by the potassium permanganate method (COD_{Mn}), chemical oxygen demand (COD), 5-day biochemical oxygen demand (BOD_5), ammonia nitrogen (NH_3–N), total phosphorus (TP), total nitrogen (TN) and chlorophyll-a (Chla). In this dataset, Chla was regarded as a proxy of the phytoplankton biomass. In Dataset2, eight environmental variables and phytoplankton species were determined. The environmental parameters were only detected in September. WT and turbidity were measured *in situ* at all eleven stations in September 2013 using YSI 6600 (Yellow Spring Instruments, Yellow Springs, OH, USA), a multiparameter water quality sonde. The DO was detected *in situ* by YSI PRO-ODO, a dissolved oxygen meter. TN, TP, orthophosphate (PO_4^{3-}) and dissolved silicon (Si) were examined in the laboratory followed SEPA [24]. The time of travel (TT) was also introduced to characterize the movement of phytoplankton in the Tanglang River. Different from completely mixed in lakes, phytoplankton pass through rivers as a plug [2]. In this study, we simply represented TT as the ratio of the reach lengths and the average velocities of the reach in the Tanglang River. The surface velocity of water was measured *in situ* by LS300 (Nanjing Sheng Rong equipment limited company, Nanjing, China), a portable flow meter. Phytoplankton species were identified and counted at species level by using a Nikon E100 (Sendai, Japan) biological microscope (10×40) according to phytoplankton morphology [25]. Species with filamentous and colonial forms were counted at the cell level.

2.4. Statistical Analysis

We performed the Q-style hierarchical Cluster Analysis (CA) and Pearson correlation analysis using SPSS 20.0 to determine the spatial differences of sampling stations and the driving factors of different algae, respectively. The Q-style hierarchical CA is a multivariate method to cluster similar samples into one cluster based on the characteristics of objects [23,26]. The similarity is measured by Euclidean distance using Ward's method based on normalized data [27]. In these cases, by means of the Q-style hierarchical CA, nine spatial stations located along the Pudu River were divided into specific-numbered groups based on the similarity of 11 environmental variables in Dataset1. The results of the Q-style CA can help to discriminate the spatial differences. The Pearson correlation was first used to detect the differences among site clusters in Dataset1 and then used in Dataset2 to study the driving factors of different phytoplankton phyla.

3. Results

3.1. Physical and Chemical Variables

In Dataset1 (Table 1), the highest WT in the Pudu River was observed in August (September) 2010 with an average temperature of 22.16 °C and this temperature was 6.8 and 7.4 °C higher than the temperatures measured in March 2011 and November 2010, respectively. The temperature stayed stable in the space. The pH was always higher than 7 and lower than 9. The highest concentration of TN appeared in March 2011. The concentration variations of NH_3–N were not consistent with TN. The concentrations of TP were much higher in November than in March or August (September).

In Dataset2 (Table 1), the WT in the Dianchi Lake varied little from the WT in the Tanglang River downstream. This was similar with the findings in Dataset1. The average temperature in June 2013 (26.79 °C) was about 4.22 °C higher than that in September 2013 (22.57 °C). Nutrient concentrations were very high at all stations in September. The concentrations of TN ranged from 4.31 to 31.85 mg/L, while those of TP ranged from 0.22 to 2.92 mg/L. Both the levels of TP and PO_4^{3-} increased rapidly in the middle reaches of the Tanglang River after a long period of stabilization in the upper reaches of the river. However, these then dramatically decreased in the lower reaches of the Tanglang River. PO_4^{3-}

was significantly correlated with TP ($R = 0.772$, $P < 0.01$). The concentrations of Si ranged from 0.2 to 4.06 mg/L. The lowest concentration was found in the mouth of the Dianchi Lake with the concentrations increasing along the river. The turbidity maintained at high values (>53.7 NTU+) and showed no obvious trend in the longitudinal. The velocity (Figure 2) and the TT increased significantly in the river downstream and reached the high values at T5, 1.5 m/s and 1.557 days (d), respectively. The spatial increasing variations in velocities were the most obvious among all the factors.

Table 1. Mean and range of environmental parameters in Dataset1 and Dataset2.

Dataset	Parameters	Unit	Mean	Range	No.
Dataset1 (9 stations spread over the entire Pudu River in August (September) and November 2010, March 2011)	WT	°C	17.49	12.8 (Mar.)–24.0 (Sept.)	81
	pH	-	7.56	6.4 (Nov.)–8.6 (Mar.)	81
	SS	mg/L	71.85	35 (Nov.)–120 (Mar.)	81
	DO	mg/L	7.60	3.6 (Aug.)–13.0 (Sept.)	81
	COD$_{Mn}$	mg/L	3.98	1.05 (Aug.)–8.48 (Mar.)	81
	COD$_{Cr}$	mg/L	27.39	7.2 (Mar.)–57.0 (Mar.)	81
	BOD$_5$	mg/L	4.29	0.5 (Sept.)–13.0 (Sept.)	81
	NH$_3$–N	mg/L	1.73	0.137 (Sept.)–5.360 (Mar.)	81
	TP	mg/L	0.49	0.017 (Mar.)–4.008 (Nov.)	81
	TN	mg/L	3.70	0.516 (Sept.)–10.110 (Mar.)	81
	Chla	mg/L	4.64	0.3 (Nov.)–28.5 (Mar.)	81
Dataset2 (11 stations spread over the Tanglang River in September 2013) [c]	WT	°C	22.04	20.66 (T2–8)–23.32 (D2)	11
	Length	km	35.80	−2.04 (D1)–114.60 (T5)	11
	Turbidity	NTU+	80.45	53.7 (T2–4)–185.4 (T4)	11
	Velocity	m/s	0.54	0.05 (T1)–1.556 (T2–6)	10 [a]
	TT	d	0.42	0.339 (T1)–1.557 (T5)	9 [b]
	DO	mg/L	8.08	4.13 (T2–6)–13.22 (D2)	11
	TP	mg/L	1.05	0.22 (D2)–2.92 (T4)	11
	TN	mg/L	12.10	4.31 (T2–6)–31.85 (T2–7)	11
	PO$_4^{3-}$	mg/L	0.41	0.084 (T2)–1.44 (T2–7)	11
	Si	mg/L	1.74	0.2 (D1)–4.06 (T2–8)	11

Notes: The month of the lowest (highest) value in Dataset1 is indicated in brackets. The sampling station in September 2013 of the lowest (highest) value in Dataset2 is indicated in the brackets as well. T2–4, T2–6, T2–7 and T2–8 were the other four sampling stations in September 2013, which are not discussed in this paper because they are not located in the same places as the sampling stations in June. The increasing number indicates the longer distance from D2. [a] Velocity was detected *in situ* from D2 to T9 in September 2013; [b] The time of travel at the mouth of the Dianchi Lake (D1 and D2) was not calculated due to the calculation method of TT being only applicable to plug flow in the river; [c] The environmental variables at all eleven sampling stations in September 2013 were determined. Among the eleven stations, seven of them were located in the same places with the stations in June 2013. Hence, we only compared the temporal and spatial variations in phytoplankton abundance at the same seven stations. However, when the relationships between the abundance and environmental variables in September were studied, we considered all eleven pieces of data.

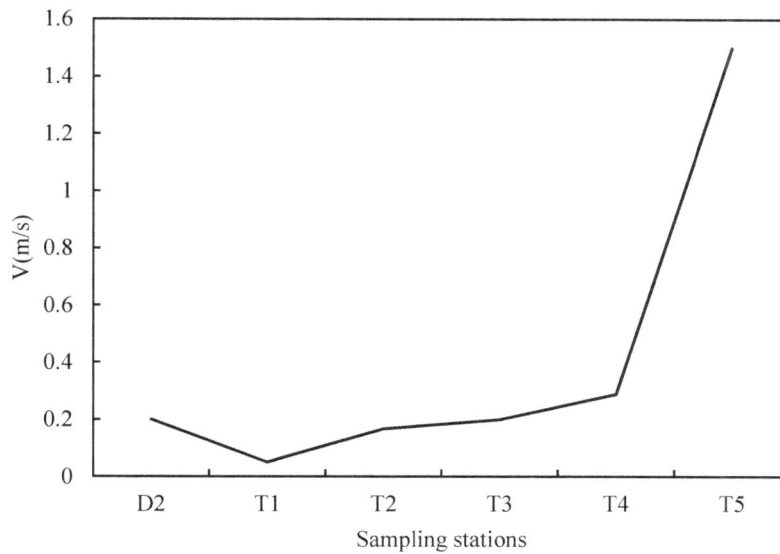

Figure 2. The velocities in the same seven sampling stations in September 2013.

3.2. Spatial Site Grouping Based on Environmental Variables

According to Dataset1, we partitioned the nine stations into three and five groups based on eleven physicochemical parameters, respectively. As shown in Table 2 (three clusters), the classifications were not consistent throughout the three typical seasons, but they showed similar trends. In general, the three stations situated along the Tanglang River (S1–S3) belonged to one group (C1), the three stations located on the lower Pudu River (S4–S6) were clustered into the same spatial cluster (C2) and the three monitoring stations at the three tributaries (S7–S9) were identified as another group (C3). We found that these three spatial clusters were representative of three different river types. C1 represents a river downstream from a eutrophic lake that is directly influenced by the lacustrine algae flowing from the lake. C2 represents a river influenced by the flow of the river upstream together with other tributaries. C3 represents rivers without upstream eutrophic lakes. However, when the nine stations were classified into five groups (Table 2), the three stations within C1 were distributed into three different groups for both the dry season and the rainy season. In contrast, the other stations within C2 and C3 remained in the same clusters without changes. This classification revealed more distinct differences among three stations within C1 compared to other stations within C2 and C3. Moreover, the correlations between TP and Chla in C1, C2 and C3 were 0.074, −0.462 ($P < 0.05$) and 0.518 ($P < 0.01$), respectively. These different correlations showed the different characteristics of the three various spatial clusters to some extent. TP was found to be significantly positively related to the total biomass in tributaries (C3) while weakly correlated with the biomass in the Tanglang River (C1).

Table 2. Site grouping according to Q-style Cluster Analysis (CA) based on Dataset1 (The sites with the same number were gathered into the same cluster).

Sites	Three Clusters			Five Clusters		
	Dry Season	Normal Season	Rainy Season	Dry Season	Normal Season	Rainy Season
S1	1	1	1	1	1	1
S2	1	1	1	2	1	2
S3	1	1	2	3	2	3
S4	2	1	3	4	2	4
S5	2	1	3	4	3	4
S6	2	1	3	4	2	4
S7	3	2	2	5	4	3
S8	3	3	3	5	5	5
S9	3	3	2	5	5	3

3.3. Spatial and Temporal Phytoplankton Compositions

All of the samples involved in Dataset2 in this study were characterized by high abundances (above 10^8 cells/L). The highest abundance was observed at the monitoring stations situated in the Dianchi Lake with the level of abundance decreasing further downstream in the river (Figure 3). The compositions of phytoplankton in the Dianchi Lake-to-the Tanglang River system were similar in June and September 2013. The spatial variations were more obvious than temporal variations. A total of 90+ species were identified in June and 100+ species were identified in September, which belonged to seven different phyla. The symbol of + was used to show that more than one species were found, but could not been identified. Bacillariophyta (diatoms) contributed the most species (36 in June and 39 in September), followed by Chlorophyta (green algae, 24+ and 25+), Cyanobacteria (19+ and 26), Xanthophyta (3 and 3), Euglenophyta (5 and 2), Dinophyta (dinoflagellates, 1 and 4) and Cryptophyta (2 and 1). Among them, Cyanobacteria contributed the most to the total abundance, followed by Chlorophyta and Bacillariophyta. These three phyla made up more than 98% of the total abundance at all stations in June and September, except for station T5 where the phyla made up 86% of the total abundance in June (Figure 3). Cyanobacteria dominated the Dianchi Lake and then entrained within the outflow directly entered the Tanglang River. They made up 85% of the total abundance at all sampling stations in September and almost all stations in June except T5 (Figure 3). *Microcystis* was the main genus of Cyanobacteria at almost all stations. The correlations between the abundance of *Microcystis* and that of Cyanobacteria in June and September were 0.726 ($P < 0.05$) and 0.993 ($P < 0.01$), respectively. Cyanobacteria lost their leading position in relation to the disappearance of *Microcystis* at T5 in June. Due to the low percentage of Cyanobacteria at T5 in June (3.3%), Chlorophyta (56.3%) and Bacillariophyta (27.1%) dominated this station. Along the Tanglang River, the percentage of Cyanobacteria decreased quickly within 28 km away from the Dianchi Lake in June and then declined more gradually. The percentage of Chlorophyta increased dramatically within 28 km away from the Dianchi Lake and then softened, while the proportion of Bacillariophyta continued rising (Figure 4). The abundance of *Microcystis* in the Tanglang River decreased exponentially while the abundances of centric diatoms and chlorococcales green algae increased both in June and September.

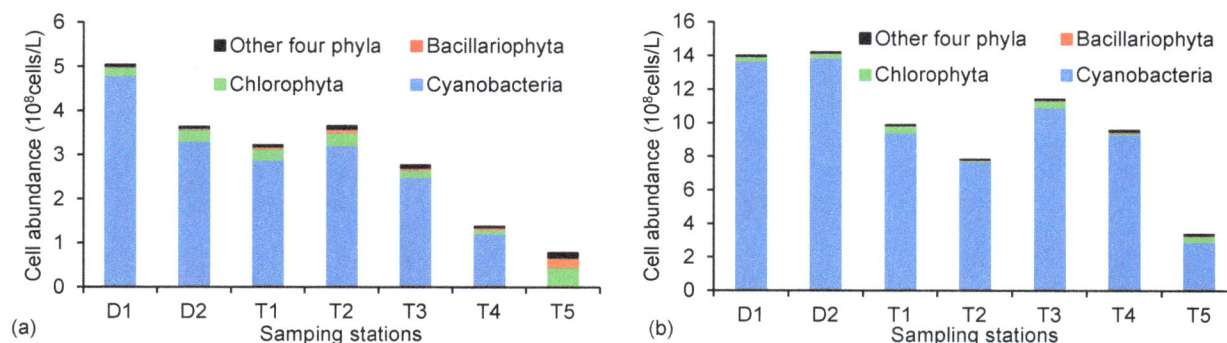

Figure 3. The phytoplankton compositions at the same seven stations (**a**) in June (**b**) in September (Other four phyla: Xanthophyta, Euglenophyta, Dinophyta and Cryptophyta).

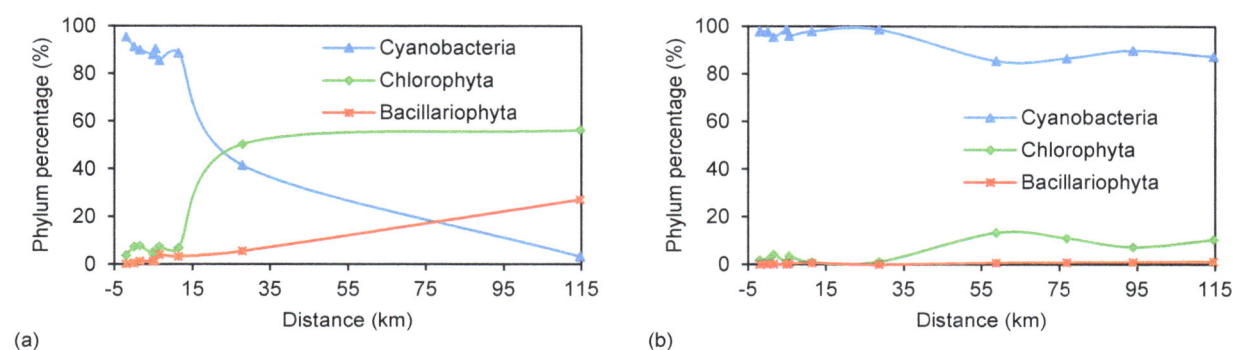

Figure 4. The trends of three main phytoplankton phyla in the Dianchi Lake-to-the Tanglang River (**a**) in June according to samplings at nine sampling stations (**b**) in September according to samplings at eleven sampling stations.

Seasonal variations in the abundances were distinct in this study (Figure 5). The total abundance recorded in September was almost two times higher than that in June. The abundance of Cyanobacteria entering the Tanglang River at T1 was 9.44×10^8 cells/L in September, which was two times greater than the corresponding measurement of 2.89×10^8 cells/L in June (Figure 5). In June, the percentage of *Microcystis* decreased from 84.3% at D2 to 64.6% at T3 to 0 at T5. In September, the percentage of *Microcystis* at D2 was 80.66%, similar to that in June. However, the percentage increased a little at T3 after a mild decrease at T1. Moreover, the percentage still remained around 77.5% at T5, which was different from the measurements in June. Unlike Cyanobacteria, the total abundance of other algae at T1 was 4.47×10^7 cells/L in September, which was similar to the 3.22×10^7 cells/L recorded in June (Figure 5). However, the spatial variations in the abundances of three main phyla were not as obvious in September as they were in June (Figure 4). Nevertheless, we still found that the percentage of Cyanobacteria decreased downstream in the river, while that of Chlorophyta and Bacillariophyta increased (Figure 4). In June, the Cyanobacteria abundance at T5 decreased by 99.2% in comparison to that at D2, the Chlorophyta abundance at T5 increased by 64.8% compared to that at D2 and the Bacillariophyta abundance at T5 increased by 378.7% compared with that at D2 (Figure 4). In September, the corresponding values were 78.8%, 31.0% and 73.5%, respectively (Figure 4).

In spite of differences in abundances, the phytoplankton compositions in the Tanglang River were similar in June and September.

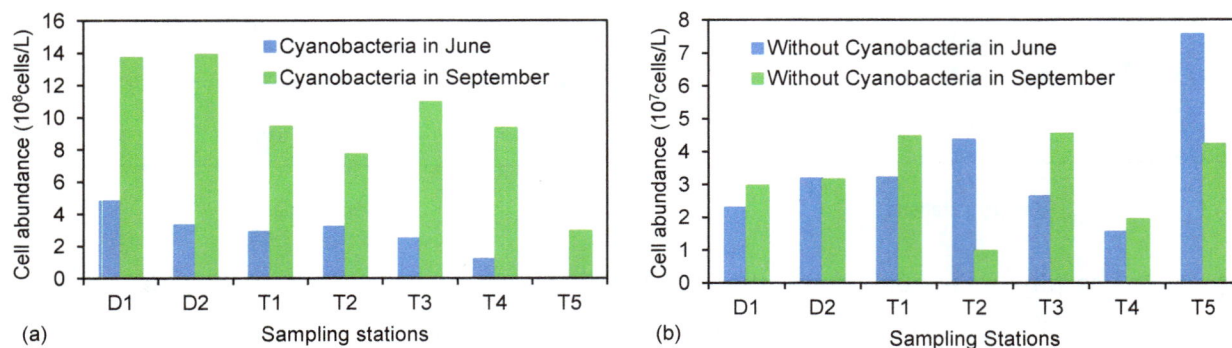

(a)

(b)

Figure 5. Comparisons between (**a**) the abundance of Cyanobacteria in June and in September and (**b**) the abundance without Cyanobacteria in June and in September.

3.4. Relationships between Phytoplankton Abundance and Environmental Variables in the Dianchi Lake-to-the Tanglang River in September 2013

The most influential factors on the total abundance and three main phyla in September were different. Due to the high correlation ($R = 0.999$, $P < 0.01$) between Cyanobacteria abundance and the total abundance, the driving factors of the total abundance were similar to those of the Cyanobacteria. According to Table 3, Cyanobacteria abundance was negatively correlated with all the nutrients, including TN, TP, PO_4^{3-} and Si. The abundances of green algae and diatoms were weakly related with these nutrients. In addition, the abundances of main phytoplankton were not significantly related with WT. The abundance of Cyanobacteria was significantly negatively related with the reach lengths ($R = -0.696$, $P < 0.05$) and the TT ($R = -0.617$, $P < 0.05$). Although the correlations were not significant, both Chlorophyta abundance ($R = 0.248$, $R = 0.044$) and Bacillariophyta abundance ($R = 0.370$, $R = 0.331$) were positively correlated with the reach lengths and the TT. The velocity inhibited the abundance of Cyanobacteria ($R = -0.186$) but encouraged the abundances of green algae ($R = 0.564$) and diatoms ($R = 0.297$). Moreover, Cyanobacteria abundance was positively related with the turbidity ($R = 0.638$), while Bacillariophyta were negatively related to the turbidity ($R = -0.412$).

Table 3. Pearson correlation coefficients between variables and phytoplankton abundances in September 2013.

Abundance	Length	Velocity	TT	WT	Turbidity	DO	TN	TP	PO_4^{3-}	Si
Total abundance	−0.690 *	−0.163	−0.619 *	0.544	0.652	0.273	−0.165	0.008	−0.502	−0.586
Cyanobacteria	−0.696 *	−0.186	−0.617 *	0.556	0.638	0.294	−0.161	−0.003	−0.516	−0.593
Chlorophyta	0.248	0.564	0.044	−0.366	0.229	−0.521	−0.043	0.254	0.418	0.253
Bacillariophyta	0.370	0.297	0.331	−0.362	−0.412	−0.229	−0.262	−0.053	0.235	0.350
Microcystis	−0.677 *	−0.141	−0.585	0.510	0.654 *	−0.234	−0.164	0.023	−0.509	−0.573

Note: * $P < 0.05$.

4. Discussion

4.1. The Influence of an Upstream Eutrophic Lake on Temporal and Spatial Phytoplankton Composition Variations in the River Downstream

Shallow lakes at low latitudes usually succumb to single species blooms, especially *Microcystis* (one of the toxic Cyanobacteria) [28]. These blooms are not only tolerant to higher insolation, but they also grow better at higher temperatures [29,30]. Therefore, Cyanobacterial blooms threaten George Lake in Uganda [29], as well as Taihu Lake [8] and Dianchi Lake [31] in China on a frequent basis. The year-round warm temperatures at the Dianchi Lake have lengthened the optimal periods for *Microcystis* blooms compared to other lakes. The stable aquatic conditions, suitable temperatures and sufficient nutrients create the proper environment for *Microcystis* to grow [32] in the Dianchi Lake. The main difference between the measurements in June and September was the quantity of the *Microcystis*. Since most of the phytoplankton in the upper reaches of the Tanglang River came from the Dianchi Lake upstream, the phytoplankton structures were similar in June and September. The big difference in the abundance of *Microcystis* from the Dianchi Lake upstream determined the variations downstream. Considering that the abundance of the *Microcystis* in June in the Dianchi Lake was much smaller than that in September, the abundance of the *Microcystis* in June (2.47×10^8 cells/L at T1) in the Tanglang River was also much smaller than that in September (7.95×10^8 cells/L at T1). However, the decrease rate of the *Microcystis* in June was larger than that in September. The cause of this phenomenon might be the discharges. The average discharge in June is about 112 m^3/s while the average discharge in September is about 200 m^3/s. The smaller discharge in June means that the TT in the Tanglang River might be longer than that in September. In this case, a large sum of the *Microcystis* in June would be lost due to longer transport time. However, a large amount of the *Microcystis* in September would be transported downstream rather than be lost otherwise. Therefore, the *Microcystis* would decrease much more in June within the same distance in the Tanglang River, as compared to September.

We regarded *Microcystis* as representative of the Dianchi Lake in this study due to their rare occurrence in rivers [1]. The percentage of the *Microcystis* abundance recorded at the different stations may reflect the impact of the Dianchi Lake. Unlike in the rivers without eutrophic lakes upstream where the maximum abundance occurs in the middle reaches [33] with centric diatoms dominating [3] or in the lower reaches with potamoplankton such as chlorococcalean colonial greens dominating [4], *Microcystis* dominated the upper and middle reaches of the Tanglang River with the total abundance declining downstream in the river. The influence of the Dianchi Lake was significant on the upper reaches of the Tanglang River, but the impact lessened further away from the lake as *Microcystis* abundance decreased. The spatial variations of the decrease in the *Microcystis* and the increase in green algae and diatoms accounted for the distinct differences among the three stations in site cluster C1 over space. In addition, the characteristics of C3 (tributaries) and C1 (rivers affected by the eutrophic lake upstream directly) were completely different. The total abundance was significantly positively related with TP in C3 because the growth of algae in C3 required nutrients and the level of TP increased as the algae abundance increased [34]. However, the total abundance was not correlated with TP at all in C1 because most of the algae in C1 were from the Dianchi Lake upstream. Even though some fluvial algae in C1 needed nutrients to grow, the decay of lacustrine algae does release nutrients [34]. Hence, the relationship between the total abundance and nutrients in C1 (the Tanglang River) was more complicated. With more turbulent flow in the middle and the lower reaches of the Tanglang River, the *Microcystis* from the Dianchi Lake were not able to adapt to the river's aquatic environment [15] and thus the abundance of *Microcystis* obviously declined. Consequently, the other competitor algae could obtain the extra resources necessary to grow. Hence, Cyanobacteria were gradually replaced by Chlorophyta and Bacillariophyta in the lower Tanglang River.

4.2. Impacts of Environmental Variables on the Spatial Phytoplankton Variations

When we consider the large number of studies on eutrophication in lakes and the distinct characteristics between rivers and lakes, we find several possible driving factors that can influence algae growth in rivers including the level of nutrients, the temperature level, incident light intensities and hydrological conditions [1,2,35]. In the river downstream from a lake, the biggest difference between the lacustrine ecosystem and the riverine ecosystem might be the main reason for the phytoplankton variations in the space. In this case, the significant change in hydrodynamic conditions was an important factor [36] as velocities had the most apparent variations from the Dianchi Lake to the Tanglang River downstream in comparison to other physical or chemical factors. Levels of nutrients and WT were not responsible for spatial phytoplankton variations in nutrient-rich lake-to-river systems. Moreover, unlike in lakes, the grazing pressure in rivers is not that important [1] due to the limited number of survivable predators. Only fast-growing aquatic organisms (such as rotifers) that commonly have low filtering rates can live in turbulent rivers [11,37]. The characteristics of the dominant algae may determine the characteristics of the total abundance. Therefore, in the Tanglang River, the total abundance was influenced by the driving factors of Cyanobacteria.

The level of nutrients was not a limiting factor in this study. P as a limiting factor in fresh lakes is widely accepted by researchers [38] while its role as a limiting nutrient in rivers is still unsettled [2]. In this combined system, the limiting nutrient found in upstream lakes may significantly affect rivers downstream. However, the discipline of P-limited does not work in the Dianchi Lake. Chen *et al.* [39] considers TP to be one of main limiting factors in the Dianchi Lake, while Wan *et al.* [20] argues that TN has replaced TP as the limiting nutrient. Wei *et al.* [21] believes that the limiting nutrients vary with regard to space in the Dianchi Lake. In view of these contrary opinions, it may be much more difficult to determine the limiting nutrient in the Tanglang River. Nevertheless, in light of the requirements for algae growth, we found that the phytoplankton in the Tanglang River was neither P-limited nor N-limited, as in many other rivers [40], because the concentrations of TP and TN were much greater than the limiting standards 3–6 µg/L and 100 µg/L [10], respectively. The weak correlations we found between TN, TP and the abundance also supported this point (Table 3). In addition, due to the high concentration (>60 µg/L) [35], PO_4^{3-} had no significant impact on the phytoplankton abundances in the Tanglang River, either.

We also found that the slight change in WT in the space was not responsible for the longitudinal variations of phytoplankton. Although WT is a crucial factor for temporal succession in phytoplankton [30,41], the correlation between WT and the abundance in the Tanglang River revealed that the river temperature did not significantly influence the growth or the loss of algae in the space. High temperatures (above 20 °C) favour the growth of Cyanobacteria [29]. The WT in the Tanglang

River were suitable for Cyanobacteria. In addition to flourishing in higher temperatures more than other competitor algae, Cyanobacteria blooms can also increase local water surface temperatures by light absorption [29]. We found that the WT at T5 without surface assemblages was around 1.2 °C lower than that at the upper stations in September 2013.

Hydrological and hydrodynamic conditions, including reach lengths, the time of travel and velocities, were among the main regulatory factors responsible for spatial variations in the phytoplankton abundances. The total abundances were significantly negatively related to reach lengths and the TT because Cyanobacteria (mainly *Microcystis*), originated from the Dianchi Lake, are not adapted to turbulent flow [6,15,16]. On the contrary, they tend to be favoured under scenarios of weak mixing, reduced discharge and slow flows [11,42–45]. Hence, *Microcystis* blooms frequently disturb shallow and relatively stable lakes [1,29,46]. In the case, the Dianchi Lake is a permanent source of the *Microcystis* inoculum for the Tanglang River downstream. However, the importance of this inoculum depends on the turbulence in the rivers downstream. In the slow-flowing rivers, the high inoculum from the lakes will continue to grow [45,46]. Otherwise, in the fast-flowing rivers, it will only add to the turbidity [2]. The Tanglang River is a turbid, fast-flowing river (about 0.54 m/s of average flow velocity in September and 1.5 m/s for the largest flow velocity at T5 in September). The short TT (0.339–1.557 d) was unable to maintain *Microcystis* in the Tanglang River. According to Reynolds' maximum replicate rate equation [10], the minimum doubling time of *Microcystis aeruginosa* is about 2.74 d, longer than TT of the Tanglang River. The *Microcystis* might be washed out rather than grow abundantly. Thus, the abundance of *Microcystis* decreased downstream in the river. In contrast, the abundance of Bacillariophyta increasing along the river might partly result from the resuspension of meroplanktonic species [47]. In spite of the significant relationships between lengths, TT and the total abundance, the lengths and TT only showed the phenomenon but not explained the spatial variations. It is the flow turbulence, represented as the velocity, explained the variations in phytoplankton abundance. Although the growth rate of *Microcystis* is slower in comparison to many green algae and diatoms [10,44,48], their buoyancy regulation can compensate for the low growth rate and help them float upwards to gain advantages in relatively stagnant lakes [49]. However, the buoyancy regulation of *Microcystis* rarely works in turbulent rivers [49] because the turbulent diffusivity enhances the vertical mixing of phytoplankton, which counter the formation of surface *Microcystis* blooms [50]. If the turbulent diffusivity is high enough, all the algae in rivers would be distributed uniformly in the vertical [15,49]. Under the circumstances, the fast-growth-rate algae are advantaged under the equal opportunity to acquire light [15,48] while the slow-growth-rate Cyanobacteria dominances are precluded [6]. Although some lakes, including the Dianchi Lake, are also turbulent [6,15], the turbulent diffusivity is much smaller compared with that in turbulent rivers. Thus, the buoyant *Microcystis* will still float upwards to have fully photosynthesis and in consequence impede their competitors in those lakes. Li *et al.* [16] found that flow velocities between 0 and 0.10 m/s would promote the rapid growth of phytoplankton. However, the higher flow velocity would inhibit phytoplankton growth. Although some researchers hold the opinion that flow velocities directly suppress the growth of algae [51], our analysis is not fully consistent with them. We suggest that the algae dominance is determined by two functions acting together: self characteristics of algae (buoyant or sinking) and flow turbulence. In the slow-flowing water, the influence of self characteristics outcompetes that of the flow turbulence. *Microcystis* float upwards while centric diatoms and chlorococcales colony green algae sink [29,52]. Accordingly, *Microcystis* undergo full photosynthesis to form dense surface blooms and then decrease underwater light intensity. However, other heavier and non-mobile algae are precluded. On the contrary, in the fast-flowing water, all the algae distribute uniformly due to high turbulent diffusivity. Hence, Chlorophyta and Bacillariophyta with faster growth rates dominate while Cyanobacteria with slower growth rates are precluded. We have developed a competition model of phytoplankton for light to explain the succession of phytoplankton in different aquatic environment, which will be discussed in detail in the next paper.

Turbidity is another important factor influencing the abundance of phytoplankton in the Tanglang River. It directly reflects light attenuation underwater as a water clarity metric and has a strong relationship with the total suspended solids [53]. Due to the interplay between the turbidity and the algae, the correlations shown in Table 3 were the comprehensive results of following two parts. On the one hand, suspended phytoplankton particles contribute to suspended solids as well as increasing the value of the turbidity. In the Tanglang River, besides a large amount of planktonic algae from the Dianchi Lake, the meroplanktonic diatoms resuspended from the bottom also contributed to the turbidity [52]. On the other hand, increased turbidity might increase light attenuation, which would affect light absorption by different phytoplankton [15]. Since most of the chlorococcales colony green algae are sensitive to low light, including *Pediastrum*, *Coelastrum*, and *Scenedesmus* [28], the low incident light might limit the growth of green algae and diatoms. Hence, the abundances of diatoms and green algae in the Tanglang River did not increase significantly like those in other nutrient-rich rivers [35,54,55]. When considering both the positive and the negative effects, the contributions to suspended solids outweighed the suppressed growth of green algae, which led to their weak positive correlation. Nevertheless, low light penetration suppressing the growth of diatoms outweighed their contributions to turbidity, resulting in the negative correlation. In addition, in view of the larger abundance of planktonic algae entering the Tanglang River from the Dianchi Lake upstream in September 2013, the higher self-shading of the phytoplankton might lead to less available penetrated light and, as a consequence, result in the slower increase in the abundances of Chlorophyta and Bacillariophyta in the Tanglang River compared to that in June. In that case, less Chlorophyta and Bacillariophyta would compete for resources with Cyanobacteria in the Tanglang River in September. Accordingly, the decrease rate of Cyanobacteria in September would be slower, which would also account for the less obvious spatial variations in phytoplankton abundances in September 2013.

5. Conclusions

(1) Based on environmental variables in Dataset1, the Pudu River was divided into three spatial groups: the Tanglang River which is directly influenced by the highly eutrophic Dianchi Lake upstream (C1), the lower Pudu River which is affected by the river reach upstream (the Tanglang River) and tributaries (C2), and three tributaries which have no eutrophic lakes upstream (C3). The study confirmed that the eutrophic Dianchi Lake upstream made the Tanglang River more heterogeneous in the space and distinguished it obviously from other reaches.

(2) The eutrophic Dianchi Lake upstream contributed the most phytoplankton to the Tanglang River downstream, especially a large amount of *Microcystis*. Therefore, unlike other rivers, the highest total abundance occurred at the origin of the Tanglang River. However, along the river, the impact of the

eutrophic Dianchi Lake gradually lessened and the *Microcystis* in the lower reaches of the Tanglang River were slowly replaced by fluvial algae including chlorococcales green algae and centric diatoms in June. The phytoplankton compositions in the lower reaches of the Tanglang River in June were similar to those found in other nutrient-rich rivers. Our results found that the toxic *Microcystis* were sustained for a long distance while not disappearing immediately after entering the Tanglang River.

(3) Intensified flow turbulence accounted for the longitudinal variations in the phytoplankton in the river downstream from the eutrophic lake when the nutrients were abundant and water temperature was suitable. In the Tanglang River, it was the lacustrine algae from the Dianchi Lake that comprised most of the algae. Hence, the influencing factors of the total abundance were consistent with those of the lacustrine algae. Strong negative correlations were found between the abundance of Cyanobacteria and the TT ($R = -0.617$, $P < 0.05$), and the reach lengths ($R = -0.696$, $P < 0.05$). The spatial variation in the velocity was the most obvious along the Tanglang River. Our results demonstrated that the hydrodynamic condition, represented by the velocity, could be the driving factor accounting for the spatial decrease of Cyanobacteria abundance and the spatial increase of Bacillariophyta and Chlorophyta abundances. Other variables, such as nutrient levels and WT, were not responsible for the spatial variations in phytoplankton abundances in the Tanglang River. Therefore, it is suggested that the flow turbulence could be used to suppress toxic *Microcystis* and stop them from further contaminating the river environment downstream when the nutrients level is high and WT is suitable.

Acknowledgments

We acknowledge the financial support of the National Natural Science Foundation of China (No. 51039002 and No. 51279078).

Author Contributions

Qian Yu and Zhaowei Liu conducted the field sampling, collected and analyzed the data. Yongcan Chen contributed to the data collection, and supervised the research. Nick van de Giesen and Dejun Zhu provided important advice on the structures of the manuscript and writing. The author and co-authors all contributed to the preparation of the manuscript.

Conflicts of Interest

The authors declare no conflict of interest.

References

1. Chen, Y.C.; Yu, Q.; Zhu, D.J.; Liu, Z.W. Possible influencing factors on phytoplankton growth and decay in rivers: Review and perspective. *J. Hydroelectr. Eng.* **2014**, *33*, 186–195. (In Chinese)
2. Hilton, J.; O'Hare, M.; Bowes, M.J.; Jones, J.I. How green is my river? A new paradigm of eutrophication in rivers. *Sci. Total Environ.* **2006**, *365*, 66–83.
3. Istvánovics, V.; Honti, M.; Vörös, L.; Kozma, Z. Phytoplankton dynamics in relation to connectivity, flow dynamics and resource availability—The case of a large, lowland river, the Hungarian Tisza. *Hydrobiologia* **2010**, *637*, 121–141.

4. Tavernini, S.; Pierobon, E.; Viaroli, P. Physical factors and dissolved reactive silica affect phytoplankton community structure and dynamics in a lowland eutrophic river (Po river, Italy). *Hydrobiologia* **2011**, *669*, 213–225.

5. Abonyi, A.; Leitão, M.; Lançon, A.M.; Padisák, J. Phytoplankton functional groups as indicators of human impacts along the River Loire (France). *Hydrobiologia* **2012**, *698*, 233–249.

6. Prygiel, J.; Leitão, M. Cyanophycean blooms in the reservoir of Val Joly (northern France) and their development in downstream rivers. *Hydrobiologia* **1994**, *289*, 85–96.

7. Paerl, H.W.; Xu, H.; McCarthy, M.J.; Zhu, G.; Qin, B.; Li, Y.; Gardner, W.S. Controlling harmful cyanobacterial blooms in a hyper-eutrophic lake (Lake Taihu, China): The need for a dual nutrient (N & P) management strategy. *Water Res.* **2011**, *45*, 1973–1983.

8. Qin, B.; Zhu, G.; Gao, G.; Zhang, Y.; Li, W.; Paerl, H.W.; Carmichael, W.W. A drinking water crisis in Lake Taihu, China: Linkage to climatic variability and lake management. *Environ. Manag.* **2010**, *45*, 105–112.

9. Bahnwart, M.; Hübener, T.; Schubert, H. Downstream changes in phytoplankton composition and biomass in a lowland river–lake system (Warnow River, Germany). *Hydrobiologia* **1998**, *391*, 99–111.

10. Reynolds, C.S. *Ecology of Phytoplankton*; Cambridge University Press: Cambridge, UK, 2006.

11. Allan, J.D.; Castillo, M.M. *Stream Ecology: Structure and Function of Running Waters*; Springer Science & Business Media: London, UK, 2007.

12. Hudon, C.; Paquet, S.; Jarry, V. Downstream variations of phytoplankton in the St. Lawrence River (Quebec, Canada). *Hydrobiologia* **1996**, *337*, 11–26.

13. Sullivan, B.E.; Prahl, F.G.; Small, L.F.; Covert, P.A. Seasonality of phytoplankton production in the Columbia River: A natural or anthropogenic pattern? *Geochim. Cosmochim. Acta* **2001**, *65*, 1125–1139.

14. Sherman, B.S.; Webster, I.T.; Jones, G.J.; Oliver, R.L. Transitions between Auhcoseira and Anabaena dominance in a turbid river weir pool. *Limnol. Oceanogr.* **1998**, *43*, 1902–1915.

15. Huisman, J.; Sharples, J.; Stroom, J.M.; Visser, P.M.; Kardinaal, W.E.A.; Verspagen, J.M.; Sommeijer, B. Changes in turbulent mixing shift competition for light between phytoplankton species. *Ecology* **2004**, *85*, 2960–2970.

16. Li, F.P.; Zhang, H.P.; Zhu, Y.P.; Xiao, Y.H.; Chen, L. Effect of flow velocity on phytoplankton biomass and composition in a freshwater lake. *Sci. Total Environ.* **2013**, *447*, 64–71.

17. Harris, G.P.; Baxter, G. Inter-annual variability in phytoplankton biomass and species composition in a subtropical reservoir. *Freshw. Biol.* **1996**, *35*, 545–560.

18. Yang, Y.H.; Zhou, F.; Guo, H.C.; Sheng, H.; Liu, H.; Dao, X.; He, C.J. Analysis of spatial and temporal water pollution patterns in Lake Dianchi using multivariate statistical methods. *Environ. Monit. Assess.* **2010**, *170*, 407–416.

19. Nanjing Institute of Geography and Limnology. *Environments and Sedimentation of Fault Lakes, Yunnan Province*; Science Press: Beijing, China, 1989. (In Chinese)

20. Wan, N.; Song, L.; Wang, R.; Liu, J. The spatio-temporal distribution of algal biomass in Dianchi Lake and its impact factors. *Acta Hydrobol. Sin.* **2008**, *32*, 184–188. (In Chinese)

21. Wei, Z.; Zheng, S.F.; Chu, Z.S.; Huang, G.Z.; Jin, X.C. Major growth control factors of *Microcystis* aeruginosa in Lake Dianchi. *Acta Sci. Circumst.* **2010**, *30*, 1472–1478. (In Chinese)

22. Qian, C.Y.; Deng, X.Y.; Xu, J.H.; Wang, R.N.; Zhao, J.C.; Zhu, Y.X. The investigation of the algae in the Tanglangchuan River. *J. Yunnan Univ.* **1985**, *7*, 123–137. (In Chinese)

23. Yu, Q.; Chen, Y.C.; Zhu, D.J.; Liu, Z.W. Pollutant source apportionment of the middle and lower reaches of the Pudu River in Southwest China. In Proceedings of the 35th IAHR World Congress, Chengdu, China, 8–13 September 2013; p. 165.

24. Jin, X.C.; Tu, Q.Y. *Criterion of Eutrophication Survey on Lakes*; Environmental Science: Beijing, China, 1990. (In Chinese).

25. Hu, H.J.; Wei, Y.X. *The Freshwater Algae in China: Systematics, Taxanomy and Ecology*; Science Press: Beijing, China, 2006. (In Chinese)

26. Simeonov, V.; Tsakovski, S.; Lavric, T.; Simeonova, P.; Puxbaum, H. Multivariate statistical assessment of air quality: A case study. *Microchim. Acta* **2004**, *148*, 293–298.

27. Singh, K.P.; Malik, A.; Sinha, S. Water quality assessment and apportionment of pollution sources of Gomti river (India) using multivariate statistical techniques—A case study. *Anal. Chim. Acta* **2005**, *538*, 355–374.

28. Reynolds, C.S.; Huszar, V.; Kruk, C.; Naselli-Flores, L.; Melo, S. Towards a functional classification of the freshwater phytoplankton. *J. Plankton. Res.* **2002**, *24*, 417–428.

29. Paerl, H.W.; Huisman, J. Blooms like it hot. *Science* **2008**, *320*, 57–58.

30. Ren, Y.; Pei, H.; Hu, W.; Tian, C.; Hao, D.; Wei, J.; Feng, Y. Spatiotemporal distribution pattern of cyanobacteria community and its relationship with the environmental factors in Hongze Lake, China. *Environ. Monit. Assess.* **2014**, *186*, 6919–6933.

31. Yu, Q.; Chen, Y.C.; Liu, Z.W.; Zhu, D.J.; Wang, H.R. Longitudinal succession of phytoplankton composition in lake-to-river system. In Proceedings of the 10th International Symposium on Ecohydraulics, Trondheim, Norway, 23–27 June 2014.

32. Becker, V.; Caputo, L.; Ordóñez, J.; Marcé, R.; Armengol, J.; Crossetti, L.O.; Huszar, V.L.M. Driving factors of the phytoplankton functional groups in a deep Mediterranean reservoir. *Water Res.* **2010**, *44*, 3345–3354.

33. Reynolds, C.S.; Descy, J.P. The production, biomass and structure of phytoplankton in large rivers. *Large Rivers* **1996**, *10*, 161–187.

34. Zhu, M.; Zhu, G.; Zhao, L.; Yao, X.; Zhang, Y.; Gao, G.; Qin, B.Q. Influence of algal bloom degradation on nutrient release at the sediment–water interface in Lake Taihu, China. *Environ. Sci. Pollut. Res.* **2013**, *20*, 1803–1811.

35. Bowes, M.J.; Gozzard, E.; Johnson, A.C.; Scarlett, P.M.; Roberts, C.; Read, D.S.; Armstrong, L.K. Spatial and temporal changes in chlorophyll-a concentrations in the River Thames basin, UK: Are phosphorus concentrations beginning to limit phytoplankton biomass? *Sci. Total Environ.* **2012**, *426*, 45–55.

36. Reynolds, C.S.; Glaister, M.S. Spatial and temporal changes in phytoplankton abundance in the upper and middle reaches of the River Severn. *Large Rivers* **1993**, *9*, 1–22.

37. Bum, B.K.; Pick, F.R. Factors regulating phytoplankton and zooplankton biomass in temperate rivers. *Limnol. Oceanogr.* **1996**, *7*, 1572–1577.

38. Lynam, C.P.; Cusack, C.; Stokes, D. A methodology for community-level hypothesis testing applied to detect trends in phytoplankton and fish communities in Irish waters. *Estuar. Coast. Shelf Sci.* **2010**, *87*, 451–462.

39. Chen, Y.C.; Zhang, D.; Tang, L. The spatial and temporal variations of phosphate concentrations and their relationships with algal growth in Lake Dianchi, China. *Ecol. Environ. Sci.* **2010**, *19*, 1363–1368. (In Chinese)

40. Devercelli, M.; O'Farrell, I. Factors affecting the structure and maintenance of phytoplankton functional groups in a nutrient rich lowland river. *Limnologica* **2013**, *43*, 67–78.

41. Baykal, T.; Açıkgöz, İ.; Udoh, A.U.; Yildiz, K. Seasonal variations in phytoplankton composition and biomass in a small lowland river-lake system (Melen River, Turkey). *Turk. J. Biol.* **2011**, *35*, 485–501.

42. Walks, D.J. Persistence of plankton in flowing water. *Can. J. Fish. Aquat. Sci.* **2007**, *64*, 1693–1702.

43. Mihaljević, M.; Stević, F. Cyanobacterial blooms in a temperate river-floodplain ecosystem: The importance of hydrological extremes. *Aquat. Ecol.* **2011**, *45*, 335–349.

44. Mitrovic, S.M.; Hardwick, L.; Dorani, F. Use of flow management to mitigate cyanobacterial blooms in the Lower Darling River, Australia. *J. Plankton Res.* **2011**, *33*, 229–241.

45. Grabowska, M. The role of a eutrophic lowland reservoir in shaping the composition of river phytoplankton. *Ecohydrol. Hydrobiol.* **2012**, *12*, 231–242.

46. Grabowska, M.; Mazur-Marzec, H. The effect of cyanobacterial blooms in the Siemianówka Dam Reservoir on the phytoplankton structure in the Narew River. *Oceabol. Hydrobiol. Stud.* **2011**, *40*, 19–26.

47. Istvánovics, V.; Honti, M. Efficiency of nutrient management in controlling eutrophication of running waters in the Middle Danube Basin. *Hydrobiologia* **2012**, *686*, 55–71.

48. Reynolds, C.S.; Descy, J.P.; Padisák, J. Are phytoplankton dynamics in rivers so different from those in shallow lakes? *Hydrobiologia* **1994**, *289*, 1–7.

49. Yu, Q.; Liu, Z.W.; Chen, Y.C.; Zhu, D.J. Modelling daily variation in the vertical distribution of *Microcystis. China Environ. Sci.* **2015**, in press. (In Chinese)

50. Paerl, H.W.; Hall, N.S.; Calandrino, E.S. Controlling harmful cyanobacterial blooms in a world experiencing anthropogenic and climatic-induced changes. *Sci. Total Environ.* **2011**, *409*, 1739–1745.

51. Long, T.Y.; Wu, L.; Meng, G.H.; Guo, W.H. Numerical simulation for impacts of hydrodynamic conditions on algae growth in Chongqing Section of Jialing River, China. *Ecol. Model.* **2011**, *222*, 112–119.

52. Huisman, J.; Arrayás, M.; Ebert, U.; Sommeijer, B. How do sinking phytoplankton species manage to persist? *Am. Nat.* **2002**, *3*, 245–254.

53. Bukaveckas, P.A.; MacDonald, A.; Aufdenkampe, A.; Chick, J.H.; Havel, J.E.; Schultz, R.; Angradi, T.R.; Bolgrien, D.W.; Jicha, T.M.; Taylor, D. Phytoplankton abundance and contributions to suspended particulate matter in the Ohio, Upper Mississippi and Missouri Rivers. *Aquat. Sci.* **2011**, *73*, 419–436.

54. Neal, C.; Hilton, J.; Wade, A.J.; Neal, M.; Wickham, H. Chlorophyll-a in the rivers of eastern England. *Sci. Total Environ.* **2006**, *365*, 84–104.

55. Istvànovic, V.; Honti, M. Phytoplankton growth in three rivers: The role of meroplankton and the benthic retention hypothesis. *Limnol. Oceanogr.* **2011**, *56*, 1439–1452.

9

A Simple Field Test to Evaluate the Maintenance Requirements of Permeable Interlocking Concrete Pavements

Terry Lucke [1,†,*]**, Richard White** [1,†]**, Peter Nichols** [1,†] **and Sönke Borgwardt** [2,†]

[1] Stormwater Research Group, University of the Sunshine Coast, Sippy Downs, Queensland 4558, Australia; E-Mails: rwhite@usc.edu.au (R.W.); pnichols@usc.edu.au (P.N.)

[2] Büro BWB Norderstedt, Kattendorf 24568, Germany; E-Mail: info@bwb-norderstedt.de

[†] These authors contributed equally to this work.

[*] Author to whom correspondence should be addressed; E-Mail: tlucke@usc.edu.au;

Academic Editor: Y. Jun Xu

Abstract: Adequate infiltration through Permeable Interlocking Concrete Pavements (PICPs) is critical to their hydraulic and stormwater treatment performance. Infiltration is affected by clogging caused by the trapping of fines in the PICP surface, which, over time, reduces treatment performance. Clogging can be reduced by periodic maintenance such as vacuum sweeping and/or pressure washing. Maintenance requirements can be indicated by measuring reduced infiltration rates. This paper compared infiltration results using the standard test (C1781M-14a) with the results of a new stormwater infiltration field test (SWIFT) developed in Australia to evaluate the maintenance requirements of PICPs. A strong correlation (Pearson's r = −0.714) was found between results using the two methods. This study found that the SWIFT was a reliable method for estimating the degree of clogging of PICPs while successfully overcoming some of the problems with the more technical existing test methodology such as horizontal water leakage (use of sealant), unrealistic pressure heads, speed of test, and portability. The SWIFT test is a simple, fast and inexpensive way for asset managers and local government employees to quickly assess the maintenance requirements of PICP installations in the field.

Keywords: permeable interlocking concrete pavements (PICP); stormwater pollution; clogging; infiltration testing

1. Introduction

Permeable pavement systems have been used globally for over two decades as a Water Sensitive Urban Design (WSUD) control measure to reduce both peak stormwater flows and pollution loads [1–4]. WSUD is of a similar design philosophy to Low Impact Development (LID) in the US, and Sustainable Urban Drainage Systems (SUDS) in Europe. Permeable pavements significantly reduce stormwater runoff volumes compared to conventionally constructed pavements. They also support increased evaporation which aids in further reducing runoff and peak stormwater flows. They also filter the stormwater within the pavement structure removing pollutants and improving water quality [1,3,5]. Stormwater treatment mainly takes place through the trapping of suspended solids during infiltration through the pavement structure [6–8].

The majority of modular permeable pavement systems are Permeable Interlocking Concrete Pavements (PICPs) which consist of concrete blocks or pavers with open joints which allow infiltration between the pavers (Figure 1). The stormwater infiltrates through the surface and bedding layers and then into the surrounding soil and groundwater, or discharged into conventional stormwater drainage.

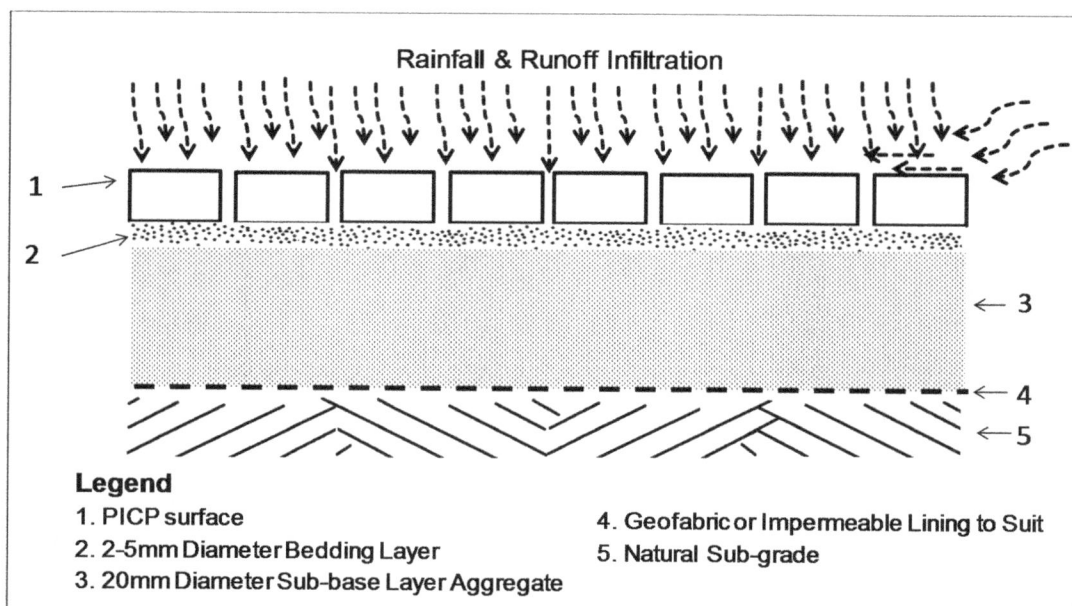

Figure 1. Typical Permeable Interlocking Concrete Pavement (PICP) structure.

The trapping of fine sediments is an important function of PICPs and the primary method of pollution removal. However, this process has also been shown to reduce the hydraulic performance of PICPs over time due to clogging [1,2,7]. Although infiltration rates of newly installed pavements have been shown to be high [9–11] these rates are known to diminish over time due to clogging, potentially leading to a decrease in useful lifespan, more maintenance and higher replacement costs [12,13]. Accelerated surface clogging can lead to more frequent maintenance intervals and increased costs. Previous research has demonstrated that clogging takes place in the spaces between and within the PICP structure, causing decreased infiltration capacity after several years [2,7,9,14]. Other reasons suggested for lower PICP infiltration rates over time include poor construction practices, lack of maintenance, and adverse environmental conditions such as dispersible soils or excessive tree litter at the PICP location [1,13,15,16].

Although some maintenance requirements have been recommended to reduce PICP clogging, such as vacuum sweeping or pressure washing [2,17–19], because they are a relatively new pavement technology, full maintenance requirements have yet to be specified across the full range of installation types and conditions. The lack of maintenance specifications for PICP systems has been highlighted as a possible factor limiting their wider use [1,15].

Monitoring changes in the PICP infiltration rate over time is the most common way to evaluate the long term performance of PICPs. Reduced PICP infiltration rates can indicate that clogging is occurring and pavement maintenance is required. However, measuring PICP infiltration rates has previously been difficult for a variety of reasons including the practical difficulties in applying existing testing methodologies, the amount of time taken to complete the existing methodology, the amount and type of specific equipment required to carry out the tests [11,20].

This paper presents results from a comparison between the standard methodology [21], and a new PICP field infiltration test methodology developed at the University of the Sunshine Coast in Australia. Arising from industry demand, the new test methodology, Stormwater Infiltration Field Test (SWIFT), has been developed as a rapid, low cost, and simple method to estimate PICP infiltration rates and maintenance requirements. A series of iterative design steps were undertaken to optimise the performance of the stormwater infiltration field test (SWIFT) prior to the full testing and comparisons reported in this article. This involved testing a range of elements relevant to the methodology including the volume of water released used during the tests (volumes from 4 litres to 10 litres were evaluated), the height of water release (60 mm, 240 mm and 300 mm heights were tested), and three different plug diameters were evaluated (20 mm, 40 mm and 50 mm diameters).

It is anticipated that tests performed over time at the same location using the SWIFT method may be used to quantify reductions in PICP surface infiltration rates, thereby identifying whether maintenance is required to restore infiltration rates to predefined levels. Classification of the degree of blocking (Table 1) was related to the categories suggested by Lucke and Beecham [1]. The minimum European PICP infiltration capacity of 97.2 mm/h [13,22] is also listed in Table 1 for reference purposes.

Table 1. Categories of pavement blockage and associated infiltration rates.

Average Infiltration Rate (mm/h)	Blockage Category
>2000	Unblocked
30–2000	Medium Blocked
<30	Fully Blocked
97.2	Minimum European PICP Infiltration rate

Approximate infiltration rates may be estimated from the number of bricks that are fully wetted by the SWIFT test. The SWIFT test has been developed to provide an estimate of PICP blockage directly at the point of measurement. However, because of the speed of the test, multiple measurements can be taken rapidly over large areas to determine an average blockage category rate of large pavement areas.

1.1. Description of the American Standard Testing and Materials (ASTM) C1781M-14a Single-Ring Infiltrometer Test (SRIT) Method [21]

The standard field test used to measure infiltration through PICPs uses a single-ring infiltrometer test (SRIT—Figure 2). As highlighted by Brown and Borst [23], the ASTM (C1781M-14a) test method does not specify which part of a pavement surface should be tested. Therefore prior to testing and placement of test equipment, consideration of upslope contributing drainage areas should be made to ensure test results contain areas of likely blockage. The SRIT involves the following steps:

i)   A 300 mm diameter ring is sealed to the PICP surface with plumbers' putty to prevent lateral water flow;

ii)   The surface to be tested should be pre-wet before any measurements are taken;

iii) Pre-wetting involves pouring 3.6 kg of water into the ring at a sufficient rate to maintain a constant head between two marked lines (between 10 and 15 mm from the base);

iv)   The time taken for the water to fully infiltrate through the surface (from the time the water hits the surface to the time it is no longer visible on the surface) is recorded;

v)   The elapsed time for pre-wetting test is recorded to the nearest 0.1 s;

vi)   The actual SRIT test is performed by repeating steps (ii) and (iii) above using 3.60 kg of water if the pre-wetting time t duration is ≥30 s, otherwise 18 kg of water is used;

vii) Record the appropriate mass of water (M) and the elapsed SRIT test time (T); and

viii) Calculate the infiltration rate using Equation (1).

$$I = \frac{KM}{D^2 T} \qquad (1)$$

where I = Infiltration rate (mm/h); M = mass of infiltrated water (kg); D = inner ring diameter (mm); T = time required for water to infiltrate the pavement surface (s); and K = constant value (4.58×10^9 in SI units).

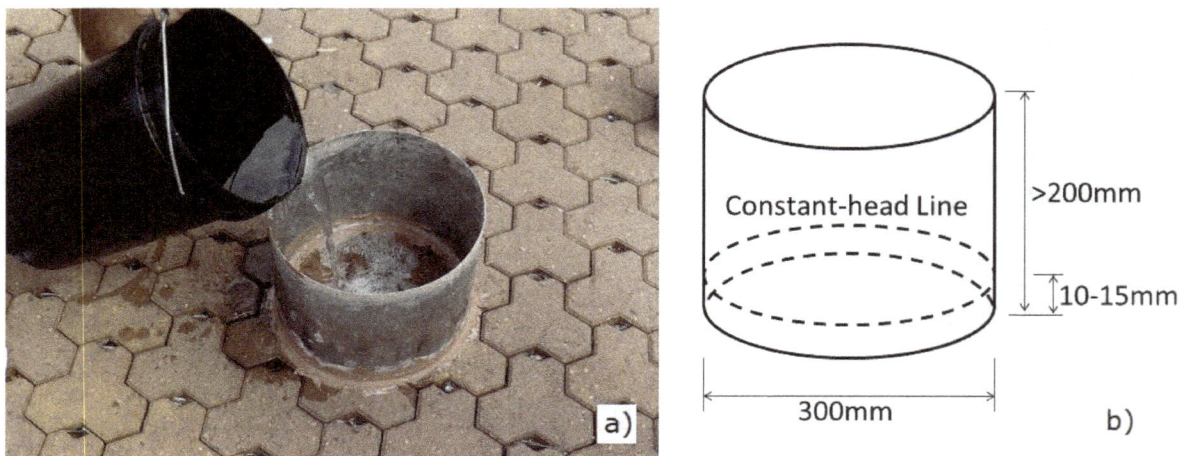

Figure 2. Single-Ring Infiltrometer Test (SRIT) used in this study. (**a**) SRIT during operation; (**b**) SRIT ring dimensions.

1.2. Description of the Stormwater Infiltration Field Test (SWIFT) Method

The SWIFT method uses a commonly available 20 L plastic bucket, with a 40 mm diameter hole cut into its base, to estimate PICP infiltration rates (Figure 3). As the SWIFT test relies on counting the number of fully wetted bricks, no pre-wetting of the surface is required. If a test is to be repeated at precisely the same location, a minimum antecedent dry period of 24 h is recommended. The SWIFT infiltration estimation method involves the following steps:

i)   Place bucket over one of the pavers of the surface to be tested so that the drainage hole is located directly above the centre of the paver;

ii)   Insert plug into bucket drain hole (making sure it fits snuggly to prevent leaks) and fill bucket with 6 L of water;

iii) Remove plug using attached chain or rope and allow all water to flow out of the bucket and onto the paving surface. Remove SWIFT device;

iv)   Count and record the number of individual bricks that are fully-wet across their entire surface (this step may be photographed for later analysis);

v)   Estimate the average infiltration rate, and categorise the state of pavement blockage and maintenance requirements using the information in Table 2 as a guide.

Figure 3. Stormwater infiltration field test (SWIFT) Infiltrometer used in this study. (**a**) SWIFT in use; (**b**) SWIFT dimensions.

2. Experimental Methodology

Field testing of both methodologies was carried out on a range of PICP pavement types, at three different PICP installations across the Sunshine Coast, Australia. The three installations were located at Sippy Downs (Site 1), Cotton Tree (Site 3) and Mary Cairncross Reserve, Maleny (Site 2), and incorporated a range of different construction techniques (Figures 4 and 5). The test pavement at Site 1 was comprised of six sub-sites (1a–1f, Figure 4) incorporating three different construction techniques, including varying aggregate sub-base depth layers, and different combinations of geotextile and impermeable layers. Geofabric liners were installed below the aggregate sub-base at Sites 1a and 4a and below the aggregate bedding layer at Sites 1–3. Site 1 was in effect six different pavements. Site 3 (Figure 5b) was a suburban roadside carpark area of approximately 750 m². The construction of Site 3 included a 50 mm deep sand bedding layer and a geofabric liner between the sand and the sub-base aggregate.

Figure 4. Six sub-site test locations at Site 1.

Figure 5. Site 2 (**a**) and Site 3 (**b**) PICP test locations.

The PICPs tested at Sites 1 and 3 were Ecotrihex® pavers (http://www.adbrimasonry.com.au/) with dimensions of 188 mm (L) × 92 mm (W) × 80 mm (H). These PICPs have apertures and spacing nibs that allow a close fit between pavers while maintaining a suitable joint width. The joints are typically filled with 2–5 mm bedding aggregate. The PICPs tested at Site 2 (Figure 5a) were Hydrapave™ pavers (http://www.boral.com.au/) with dimensions of 230 mm (L) × 115 mm (W) × 80 mm (H). A geofabric liner was installed between the 2–5 mm bedding aggregate layer and the 250 mm deep sub-base aggregate at Site 2.

The two replicates of each test method (C1781M-14a and SWIFT) were carried out at precisely the same location at each site (Figure 6). Replicate testing was undertaken on different days to ensure the PICP surface was dry to allow for the number of wet bricks to be counted.

Figure 6. Photograph showing SWIFT water stain (30 Fully-wetted bricks) over the precise location previously tested using the SRIT methodology.

Statistical Analysis

Replicate testing of pavements using both the standard SRIT and SWIFT methods provided scope for appropriate statistical analysis, and interpretation of any potential spatial or temporal variation. The potential variation in measurements between different field operators is also considered an important variable in the accuracy of any infiltrometer [20]. To overcome this potential source of error all measurements were undertaken by at least two operators in this study. The average of these measurements was used in the analysis of the results.

Correlation analysis (Pearson product-moment correlation coefficient) was used as a measure of the linear correlation (dependence) between the observed infiltration rate measurements from the two methodologies. This gives a value between +1 and −1 inclusive, where 1 is total positive correlation, 0 is no correlation, and −1 is total negative correlation. The variability of the results observed using the two methodologies was tested with the coefficient of variation (CV), using the ratio of the standard deviation to the mean.

3. Results and Discussion

Figure 7 shows the surface infiltration results observed using the SRIT test with the number of wet bricks observed after the SWIFT test for each of the test locations on Sites 1 and 2. The results of 47 different test locations are shown on Figure 7. An exponential trend line and its equation are also included on Figure 7. These show a correlation coefficient between the data points and the trend line of 0.854 which is quite high and demonstrates a good correlation. The general trend shown in Figure 7 is that as the surface infiltration rate decreases, the number of bricks wet by the SWIFT test increases. This result was anticipated and clearly demonstrates the feasibility of using the SWIFT test results to predict surface infiltration rates and potential maintenance requirements.

It was not possible to accurately measure the surface infiltration rates at Site 3 using the SRIT as the whole site was fully blocked. If the water level in the SRIT ring did not measurably decrease within the first 20 min of testing, the tests were discontinued and the sites were classified as fully blocked. It was thought that the reason Site 3 was fully blocked was because sand was used as the bedding layer instead of 2–5 mm aggregate. The small size of the sand between the PICP blocks appeared to become blocked very easily by sediment and organic particles. Using sand between PICPs is no longer recommended in any Australian PICP design guidelines.

The SWIFT test was still undertaken at the Site 3 test locations and the average number of wet bricks was 168 for these tests. Accordingly, Figure 7 does not contain any test data from Site 3. Inputting a value of 168 into the Figure 7 trendline equation produces an approximate infiltration rate of 5 mm/h (used in Figure 8) which is consistent with being effectively fully blocked.

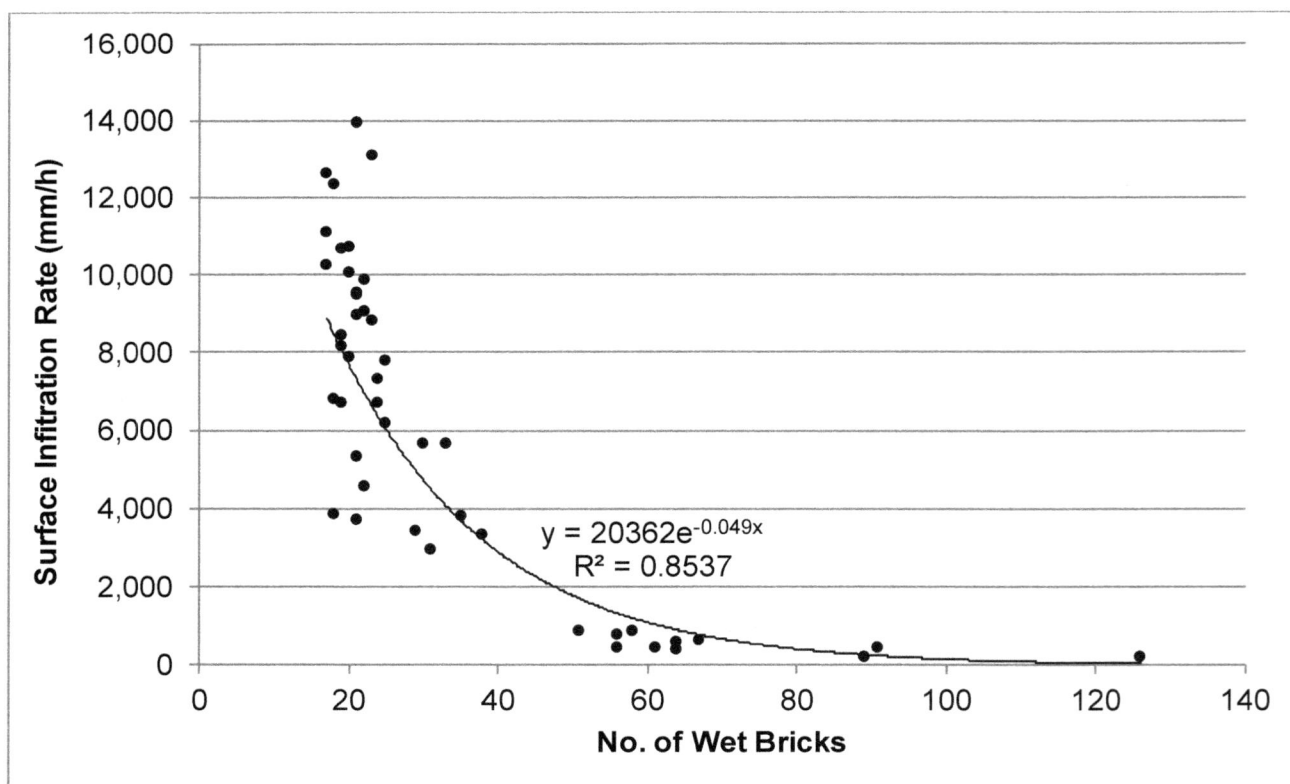

The chart shows data with the equation:
$$y = 20362e^{-0.049x}$$
$$R^2 = 0.8537$$

with axes "Surface Infiltration Rate (mm/h)" and "No. of Wet Bricks".

Figure 7. Comparison of PICP infiltration rates using SRIT and SWIFT methods.

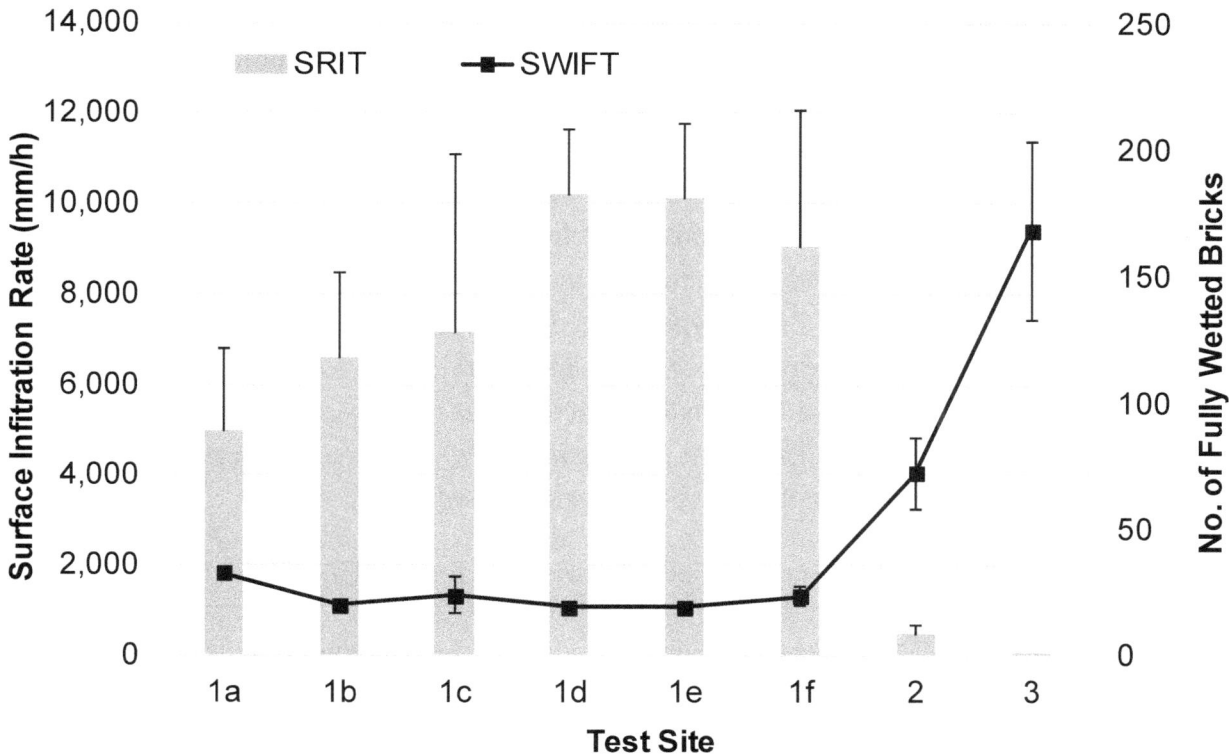

Figure 8. Comparison of PICP infiltration rates (± Standard Deviation) using SRIT and SWIFT methods.

A correlation analysis ($n = 58$) was also undertaken and the Pearson's Product-Moment Correlation Coefficient (Pearson's r) between the SRIT and SWIFT test results was found to be −0.714. This value is further evidence of the strong relationship that exists between the results.

Infiltration rates observed using both the SRIT and SWIFT methods varied between sites (Figure 8). Infiltration rates were substantially higher for all Site 1 tests compared with Sites 2 and 3. Average infiltration rates varied between 5 mm/h (calculated for Site 3) and 10,192 mm/h (Site 1d) using the SRIT. Figure 8 again demonstrates that the number of fully wetted bricks increased with decreasing infiltration rates. At the sites with lower measured infiltration rates during operation of the SWIFT method, counts of the number of wetted bricks increased six-fold (mean values ranged between 23 and 120). This shows a clear inverse relationship between infiltration rates and number of wet bricks, as expected.

The variability of the tests, calculated using the coefficient of variation (CV) differed substantially between methods (average SRIT CV = 30.1%; average SWIFT CV = 14.7%). The calculated CV was generally higher for the SRIT method compared to the results observed using the SWIFT method. This suggests that overall, the SWIFT had a lower degree of variability between tests when compared to the SRIT, providing more consistent results across all sites. However, the CV results did vary between sites with high infiltration compared to those with low infiltration. The calculated CV (30.7%) of the SRIT method used on sites with high infiltration was similar to those with low infiltration (28.1%). The calculated CV of the SWIFT method used on sites with high infiltration (12.9%) was calculated to be substantially lower than sites with lower infiltration (20.2%).

The low variability in SWIFT test results combined with the strong correlation between methods used, suggest that the new method may be used on PICPs to broadly categorise the degree of PICP clogging, and quickly and reliably determine PICP maintenance requirements.

Using the results shown in Figure 7, it was possible to relate the approximate number of wetted bricks expected from the SWIFT test to the infiltration rates and blockage categories listed in Table 1. Previous research by Borgwardt [7] investigated the reduction in PICP infiltration capacity over time. He found that the infiltration capacity of PICPs typically reduced exponentially to around 20% of the newly-installed infiltration rate after approximately 10 years in service. By combining Borgwardt's [7] results with those in Figure 7, suggested PICP maintenance requirements were estimated based on the SWIFT test results. These are shown in Table 2. It must be noted the maintenance intervals suggested in Table 2 are indicative only and should only be used as a general guideline.

Table 2. Suggested PICP maintenance requirements for SWIFT test results.

SWIFT—No. of Fully Wetted Bricks	Blockage Category	Maintenance Requirements
Less than 29	Unblocked	No maintenance required for foreseeable future
Between 29 and 133	Medium Blocked	Plan for maintenance within next 1–3 years
More than 133	Fully Blocked	Immediate maintenance required
European Comparison		
More than 109	Below Minimum European PICP Infiltration Rate	Immediate maintenance required

There are several fundamental differences between the SRIT and SWIFT methods compared in this study. The SRIT method calculates an actual surface infiltration rate (in mm/h) by dividing units of volume by area and time. By comparison, the SWIFT method does not allow the calculation of an actual pavement infiltration rate, rather it results in a specific number of fully-wetted pavement blocks. The blockage category and the corresponding suggested maintenance requirements for the pavement are then determined using the guidelines listed in Table 2. The SWIFT method was developed as a fast and convenient infiltration field test to assign broad infiltration categories to pavements in order to quickly determine pavement maintenance requirements. The SWIFT was not developed to determine PICP surface infiltration rates.

Secondly, the SRIT is conducted using the constant head (or sometimes the falling-head) method, and the SWIFT is conducted under little or no head (natural rainfall conditions). Infiltration rate tests involving artificial pressure heads are known to result in higher measured surface infiltration rates [20,24,25]. The effects of the head differences was not relevant in this comparative study because the SWIFT results are not intended to be used to calculate an actual infiltration rate, but only to rapidly identify pavement maintenance requirements. Future testing of the SWIFT on pavements with a wide range of blockage categories will enable a more precise conversion of SWIFT results into estimated infiltration rates. A more detailed investigation of the effects of different PICP block shapes and open spaces is also planned.

There were also numerous practical and time-saving advantages in using the SWIFT method. The main advantage was in not having to use a sealant to try to avoid water leakage from the rings. Setting up the SWIFT device was therefore not time-consuming like other test methods, and the testing was fast and efficient. In addition, no post-test clean-up was required as no sealant was used. Including set-up time, the duration of each SWIFT test was approximately two minutes. Furthermore, all of the equipment used for conducting the SWIFT test was inexpensive, and easy to obtain and assemble.

Where the SRIT has the advantage of being an accurate test of surface infiltration rates of permeable pavements, it also requires access to specialised equipment and has a significant setup time. The rapid nature of the SWIFT methodology, and low variability in results, particularly on sites with high infiltration, suggest it may also be suitable for carrying out multiple tests on large sites. This could significantly increase the speed of testing over larger areas and enable a fast classification of the overall state of pavement clogging across numerous and large sites. This would also allow for a more effective and efficient maintenance planning process.

The study results and strong correlations between tests suggest that the assignment of clogging categories and maintenance requirements using the SWIFT testing methodology (Table 2) may be an acceptable and reliable field test method for identifying the state of PICP clogging and for planning maintenance activities. The SWIFT test is a simple, fast and inexpensive way for asset managers and local government employees to quickly assess the maintenance requirements of PICP installations in the field. Although the SWIFT method still needs more research and evaluation to verify its suitability across a wide range of applications and *different* pavement types, it has the potential to be accepted on a global scale as the new benchmark for PICP infiltration testing. Further research is currently underway.

4. Conclusions

Measurement of infiltration rates of PICPs has previously been problematic for a variety of reasons, including the practical difficulties in applying existing test methodologies, access to the specialised equipment required for testing, and the time taken to undertake each test. This study examined and compared the performance of two PICP surface infiltration rate measurement methods to estimate the degree of PICP clogging: the Single-Ring Infiltrometer Test (ASTM C1781M-14a) method, and a newly-developed field test, the stormwater infiltration field test (SWIFT) method.

This study found that the SWIFT was a reliable method for estimating the degree of clogging of PICPs, and as indicated through correlation analysis, provides comparable results to the SRIT. A strong correlation (Pearson's r = −0.714) was found between results using the two methods. The new SWIFT methodology reliably categorised the degree of PICP clogging while successfully overcoming some of the problems with the more technical existing test methodology such as horizontal water leakage (use of sealant), unrealistic pressure heads, speed of test, and portability. The SWIFT also involves minimal setup costs, and is a device that is easily assembled from common items. The ease of conducting the test, and increased speed of testing when undertaking multiple tests, will reduce overall operator costs. The SWIFT test is a simple, fast and inexpensive way for asset managers and local government employees to quickly assess the maintenance requirements of PICP installations in the field.

Acknowledgments

The authors would like to thank the USC technical staff as well as Daniel Layton and Jayden Walter for assistance on this project. The project was jointly funded by the University of the Sunshine Coast, the Sunshine Coast Council and a grant (LP120200678) by the Australian Research Council.

Conflicts of Interest

The authors have declared no conflict of interest.

References

1. Lucke, T.; Beecham, S. Field Investigation of Clogging in a Permeable Pavement System. *J. Build. Res. Inf.* **2011**, *39*, 603–615.

2. Dierkes, C.; Kuhlmann, J.; Kandasamy, J.; Angelis, G. Pollution Retention Capability and Maintenance of Permeable Pavements. In Proceedings of the 9th International Conference on Urban Drainage, Portland, OR, USA, 8–13 September 2002.

3. Legret, M.; Nicollet, M.; Colandini, V.; Raimbault, G. Simulation of Heavy Metal Pollution from Stormwater Infiltration through a Porous Pavement with Reservoir Structure. *Water Sci. Technol.* **1999**, *39*, 119–125.

4. Pratt, C.J.; Newman, A.P.; Bond, P.C. Mineral Oil Bio-degradation within a permeable pavement: Long term observations. *Water Sci. Technol.* **1999**, *39*, 103–109.

5. Mullaney, J.; Lucke, T. Practical Review of Pervious Pavement Designs. *CLEAN Soil Air Water* **2014**, *42*, 111–124.

6. Beecham, S.; Pezzaniti, D.; Kandasamy, J. Stormwater treatment using permeable pavements. *Proc. ICE Water Manag.* **2012**, *165*, 161–170.

7. Borgwardt, S. Long Term *in-situ* Infiltration Performance of Permeable Concrete Block Pavement. In Proceedings of the 8th International Conference on Concrete Block Paving, San Francisco, CA, USA, 6–8 November 2006.

8. Nichols, P.W.B.; Lucke, T.; Dierkes, C. Comparing Two Methods of Determining Infiltration Rates of Permeable Interlocking Concrete Pavers. *Water* **2014**, *6*, 2353–2366.

9. Pratt, C.J. Permeable Pavements for Stormwater Quality Enhancement. In *Urban Stormwater Quality Enhancement—Source Control, Retrofitting and Combined Source Technology*; Torno, H.C., Ed.; American Society of Civil Engineers: Reston, VA, USA, 1990; pp. 131–155.

10. Brattebo, B.O.; Booth, D.B. Long term stormwater quantity and Quality Performance of Permeable Pavement Systems. *Water Res.* **2003**, *37*, 4369–4376.

11. Collins, K.A.; Hunt, W.F.; Hathaway, J.M. Hydrologic Comparison of Four Types of Permeable Pavement and Standard Asphalt in Eastern North Carolina. *J. Hydrol. Eng.* **2008**, *13*, 1146–1157.

12. Yong, C.; Deletic, A. Factors That Predict Clogging through Porous Pavements. In Proceedings of the 7th International Conference on Water Sensitive Urban Design, Melbourne, Australia, 21–23 February 2012.

13. Boogaard, F.; Lucke, T.; Beecham, S. Effect of Age of Permeable Pavements on Their Infiltration Function. *CLEAN Soil Air Water* **2014**, *42*, 146–152.

14. Pezzaniti, D.; Beecham, S.; Kandasamy, J. Influence of Clogging on the Effective Life of Permeable Pavements. *J. Water Manag.* **2009**, *162*, 76–87.

15. Bean, E.Z.; Hunt, W.F.; Bidelspach, D.A.; Smith, J.E. Evaluation of Four Permeable Pavement Sites in Eastern North Carolina for Runoff Reduction and Water Quality Impacts. *J. Irrig. Drain. Eng.* **2007**, *133*, 583–592.

16. Fassman, E.; Blackbourn, S. Urban Runoff Mitigation by a Permeable Pavement System over Impermeable Soils. *J. Hydrol. Eng.* **2010**, *15*, 475–485.

17. Baladès, J.D.; Legret, M.; Madiec, H. Permeable pavements: Pollution management tools. *Water Sci. Technol.* **1995**, *32*, 49–56.

18. Nolting, B.; Schönberger, O.; Harting, K.; Gabryl, P. Prüfung wasserdurchlässiger Flächenbeläge nach mehrjähriger Nutzungsdauer. Abschlussbericht eines Forschungsprojektes der FH Bochum, 2005. Available online: http://www.ikt.de/website/down/f0146langbericht.pdf (accessed on 21 May 2015). (In German)

19. Feldhaus, R.; Klein, N.; Röhrig, J.; Meier, G. Maßnahmen zur Niederschlagswasserbehandlung in kommunalen Trennsystemen am Beispiel des Regierungsbezirks Köln. Available online: http://www.lanuv.nrw.de/wasser/abwasser/forschung/pdf/Abschlussbericht.pdf (accessed on 21 May 2015). (In German)

20. Li, H.; Kayhanian, M.; Harvey, J.T. Comparative field permeability measurement of permeable pavements using ASTM C1701 and NCAT permeameter methods. *J. Environ. Manag.* **2013**, *118*, 144–152.

21. American Standard Testing and Materials (ASTM). *Standard Test Method for Surface Infiltration Rate of Permeable Unit Pavement Systems; C1781/C1781M-14a*; ASTM International: West Conshohocken, PA, USA, 2014.

22. Forschungsgesellschaft für Strassen und Verkehrswesen (FGSV). *Merkblatt für die Wasserdurchlässige Befestigung von Verkehrsflächen (Guidelines for the Laying of Trafficable Permeable Paving Systems)*; FGSV: Köln, Germany, 1998. (In German)

23. Brown, R.A.; Borst, M. Evaluation of Surface Infiltration Testing Procedures in Permeable Pavement Systems. *J. Environ. Eng.* **2014**, *140*, doi:10.1061/(ASCE)EE.1943-7870.0000808.

24. Borgwardt, S. Die Versickerung auf Pflasterflächen als Methode der Entwässerung von minderbelasteten Verkehrsflächen. Ph.D. Thesis, Universität Hannover, Hannover, Germany, 1995. (In German)

25. Dougherty, M.; Hein, M.; Martina, B.; Ferguson, B. Quick Surface Infiltration Test to Assess Maintenance Needs on Small Pervious Concrete Sites. *J. Irrig. Drain. Eng.* **2011**, *137*, 553–563.

10

Aquifer Recharge Estimation through Atmospheric Chloride Mass Balance at Las Cañadas Caldera, Tenerife, Canary Islands, Spain

Rayco Marrero-Diaz [1,2,†,*], **Francisco J. Alcalá** [3,4], **Nemesio M. Pérez** [1,2], **Dina L. López** [2,5], **Gladys V. Melián** [1,2], **Eleazar Padrón** [1,2] and **Germán D. Padilla** [1,2]

[1] Environmental Research Division, Institute of Technology and Renewable Energies (ITER), Granadilla de Abona, Tenerife, Canary Islands 38611, Spain; E-Mails: nperez@iter.es (N.M.P.); gladys@iter.es (G.V.M.); eleazar@iter.es (E.P.); german@iter.es (G.D.P.)

[2] Instituto Volcanológico de Canarias (INVOLCAN), Puerto de la Cruz, Tenerife, Canary Islands 38400, Spain; E-Mail: lopezd@ohio.edu

[3] Civil Engineering Research and Innovation for Sustainability (CERis), Instituto Superior Tecnico, Universtiy of Lisbon, Lisbon 1049-001, Portugal; E-Mail: francisco.alcala@ist.utl.pt

[4] Instituto de Ciencias Químicas Aplicadas, Facultad de Ingeniería, Universidad Autónoma de Chile, Santiago 7500138, Chile

[5] 316 Clippinger Laboratories, Department of Geological Sciences, Ohio University, Athens, OH 45701, USA

[†] Current address: Laboratorio Nacional de Energia e Geologia (LNEG), Alfragide, Lisbon 2610-999, Portugal

[*] Author to whom correspondence should be addressed; E-Mail: rayco.diaz@lneg.pt;

Academic Editor: María del Pino Palacios Díaz

Abstract: The atmospheric chloride mass balance (CMB) method was used to estimate net aquifer recharge in Las Cañadas Caldera, an endorheic summit aquifer area about 2000 m a.s.l. with negligible surface runoff, which hosts the largest freshwater reserve in Tenerife Island, Canary Islands, Spain. The wet hydrological year 2005–2006 was selected to compare yearly atmospheric chloride bulk deposition and average chloride content in recharge water just above the water table, both deduced from periodical sampling. The potential contribution of chloride to groundwater from endogenous HCl gas may invalidate the CMB method. The chloride-to-bromide molar ratio was an efficient tracer used to select recharge water samples having atmospheric origin of chloride. Yearly net aquifer recharge was 631 mm year^{-1}, *i.e.*, 69% of yearly precipitation. This result is in agreement with potential aquifer recharge estimated through an independent lumped-parameter rainfall-runoff model operated by the Insular Water Council of Tenerife. This paper illustrates basic procedures and routines to use the CMB method for aquifer recharge in active volcanic oceanic islands having sparse-data coverage and groundwater receiving contribution of endogenous halides.

Keywords: aquifer recharge; chloride mass balance; Cl/Br ratio; Las Cañadas; Tenerife

---

1. Introduction

In most oceanic volcanic islands, groundwater is the main natural freshwater resource because surface water is almost negligible due to the high permeability of the outcropping volcanic formations [1]. In Tenerife Island (Canary Islands, Spain), about 90% of total usable water resource has groundwater origin, *i.e.*, 200 Mm^3 of a total 226 Mm^3 according to 2004 official data [2]. This is a common circumstance to the other western Canary Islands [3]. The understanding of long-term balance between groundwater yield and aquifer recharge dynamics is key for the water management needed for sustainable groundwater-dependent economics in these isolated regions.

Oceanic volcanic islands are often characterized by a rugged topography that favours conditions for enhancing recharge in summit areas where precipitation and aquifer recharge rates are usually higher than in the lowlands and coastal plains [4–6]. However, the evaluation of aquifer recharge is a complex task subjected to different kinds of uncertainty [7–9] induced by the high spatial and temporal variability of precipitation [10], heterogeneities in soil and land use [11], few sampling and monitoring points [12], as well as the techniques used and the subsequent hydrological meaning and timing of estimates [13]. For instance, most aquifer recharge predictions are evaluated as difference in precipitation and actual evapotranspiration (E) deduced from non-global models [12]. Results may be biased when compared with experimental measures [14], regional evaluations [9,10], and calibrated numerical models [11]. The combination of several techniques is a way to identify the sources of uncertainty involved [14–16].

Tracer techniques are an alternative methodology to the most widely used physical and hydrodynamic techniques for aquifer recharge evaluations ([9,13,15], and references therein). The atmospheric chloride mass balance (CMB) method is of special interest for average net aquifer recharge at different spatiotemporal scales because it does not include E in the formulae thus reducing the overall uncertainty of estimates [8,9,14,17]. For average net aquifer recharge, the CMB method is based on the steady balance of chloride mass fluxes from (1) atmospheric bulk deposition [18]; (2) surface runoff leaving the area [19]; and (3) recharge water just arriving to the water-table [15,17]. In active volcanic islands having steep topography some additional processes must be considered, such as (1) the possibility of mixing chloride flow mass rates produced by recharge rates at different elevation [20]; (2) the variable rainfall-runoff partitioning and recharge mechanisms [14]; (3) the potential contribution of non-atmospheric chloride to groundwater [21]; and (4) the potential storage of chloride in the soil and vadose zones [8,17].

This paper widens the CMB method applicability for net aquifer recharge in active volcanic islands. For this objective, atmospheric bulk deposition, groundwater, and recharge water were periodically monitored in selected points in Las Cañadas Caldera (hereafter LCC) summit area in central-northern Tenerife Island (Figure 1) during the wet hydrological year 2005–2006. Additional research was conducted to identify the potential contribution of chloride to groundwater from endogenous HCl degassing from the volcanic-hydrothermal system because this external source of chloride may invalidate the widespread use of the CMB method for aquifer recharge evaluations in Tenerife Island, as reported in other active volcanic islands [5,6,21]. The chloride-to-bromide molar (hereafter Cl/Br) ratio was an efficient tracer used to select recharge water samples having atmospheric origin of chloride [21–23].

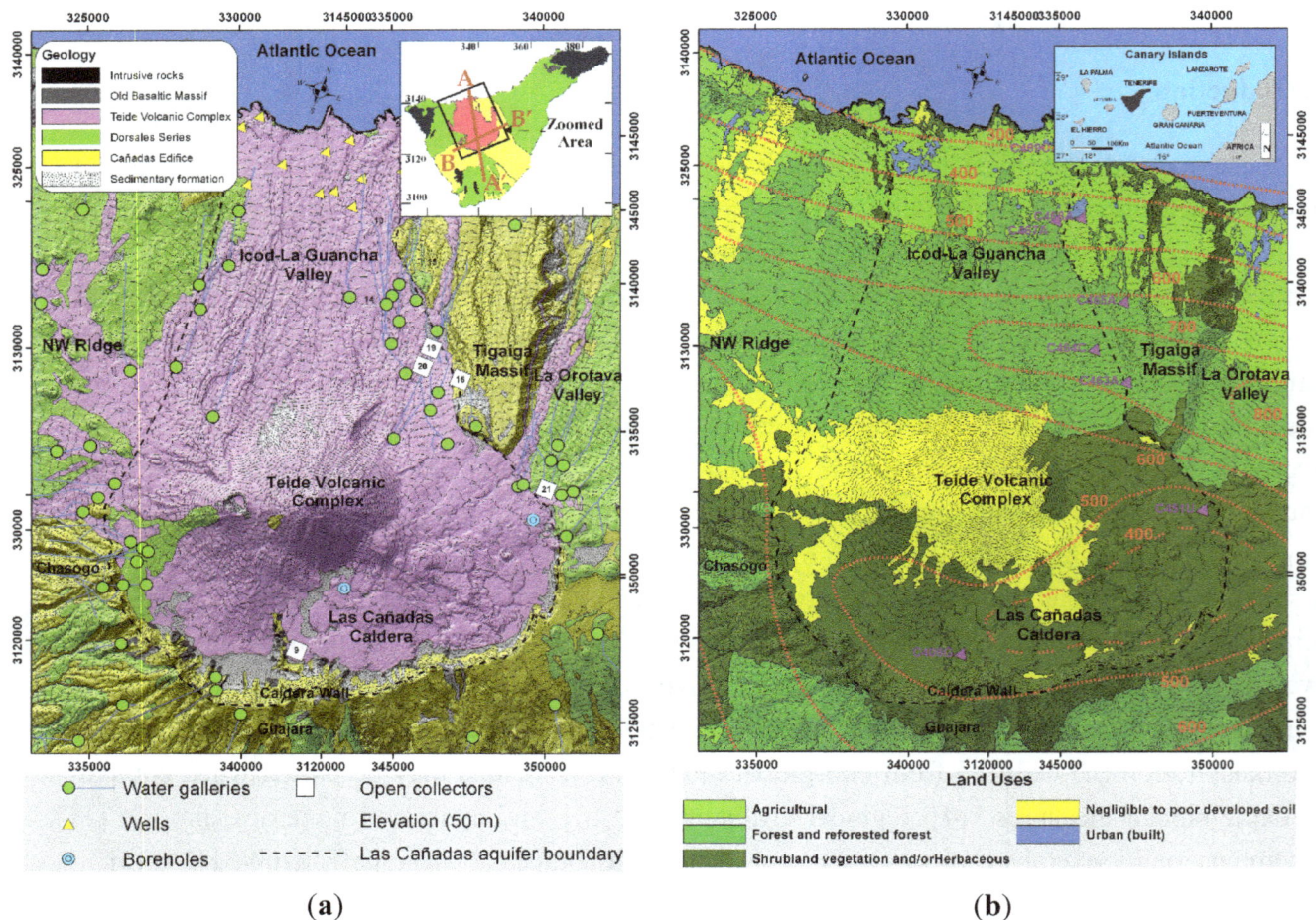

(a)   (b)

Figure 1. (a) Geological map of Las Cañadas aquifer system, showing the Teide Volcanic Complex materials partially filling both the Icod-La Guancha Valley and Las Cañadas Caldera; sites cited in the text, groundwater sampling points, and open collectors are included; **(b)** Land-use map of Las Cañadas aquifer system and mean annual precipitation contour map; weather stations (purple triangles) operated by the Spanish Agency for Meteorology (AEMET website: http://www.aemet.es/) are included. Data source: Insular Water Council of Tenerife (CIATFE) [27] and Cartographic Survey of Canary Islands (GRAFCAN).

The paper is organized as follows: Section 2 describes basic characteristics of the study area, Section 3 briefly describes the CMB method and interpretative basis for its application in the study area, and Section 4 presents the main results in LCC area. Section 5 discusses the hydrological meaning of results when compared with existing independent potential recharge estimates as well as the overall CMB method applicability in volcanic areas having contribution of endogenous halides to groundwater. Section 6 concludes with practical remarks for groundwater resources planning in Tenerife Island.

2. Study Area

Las Cañadas aquifer system is a 216-km^2 steep area on the northern side of Tenerife Island that ranges from the sea to the summit of the Teide Volcano. Las Cañadas aquifer is contained within the collapse-forming LCC, the Teide-Pico Viejo Volcanic Quaternary Complex (hereafter TPVC), and the landslide-formed Icod-La Guancha Valley (hereafter IGV) ([24], and references herein) (Figure 1a). The huge elevation from the coast to the Teide Volcano (peak elevation, 3718 m a.s.l.) induces remarkable climatic differences at the Las Cañadas aquifer. In spite of the geographical proximity to the Sahara Desert, air temperature in the coast and mid-slope of windward Tenerife is relatively mild during the whole year. The climate is dominated by humid trade winds from the northeast, which blow more than 90% in time in summer and less frequently in winter. In winter, low-pressure and oceanic storms also affect Tenerife Island weather ([25], and references therein). At the IGV, mean annual temperature varies in the 15–20 °C range while precipitation increases upslope from 300 mm·year^{-1} near the coast to 700 mm·year^{-1} on the mid-slope (Figure 1b) due to the intersection of the north-easterly trade winds. Above 1000–1500 m a.s.l., thermal inversion occurs and weather conditions becomes drier most of the year; precipitation occurs 43 days a year in average, 12 of them as snow [26]. Average annual precipitation in LCC ranges from 500 mm in windward areas to less of 300 mm in the center (Figure 1b), mainly from October to April; yearly precipitation may reach 1000 mm in wetter periods.

LCC is a 128-km^2 semi-elliptical, sparsely vegetated (Figure 1b) endorheic basin opened to the sea by the north side at the summit of the Las Cañadas aquifer system. LCC is considered the largest groundwater reserve in the island [27]. The TPVC Quaternary volcanic deposits lining LCC induces high fissuring and poorly developed soils of 0.2–0.3 m in thickness [28,29]. These conditions favour fast infiltration of rainfall and snow melt, null temporal storage of chloride in the soil and vadose zones, and negligible runoff [27,30].

The hydrogeological behaviour of the high-permeability TPVC deposits contrasts with the low-permeability Las Cañadas Edifice bedrock, the hydrothermal alteration core below TPVC, and the dyke-intrusion network across them [31,32] (Figure 2). Low-permeability materials act as barriers to the groundwater flow, thus allowing the groundwater storage in the high-permeability summit caldera above 1800 m a.s.l. [30,32,33] (Figure 2). The existence of a low-permeability clay-rich debris-avalanche deposit at the IGV bottom related to the landslide origin of this valley [24,31,34,35] enhances the relatively fast groundwater flow along this valley from TPVC and LCC to the northern coast (Figure 2). This conceptual hydrogeological functioning of Las Cañadas aquifer [33] is in accordance with the conceptual hydrogeological functioning proposed by [3,36] for this type of volcanic islands [5,6,37,38].

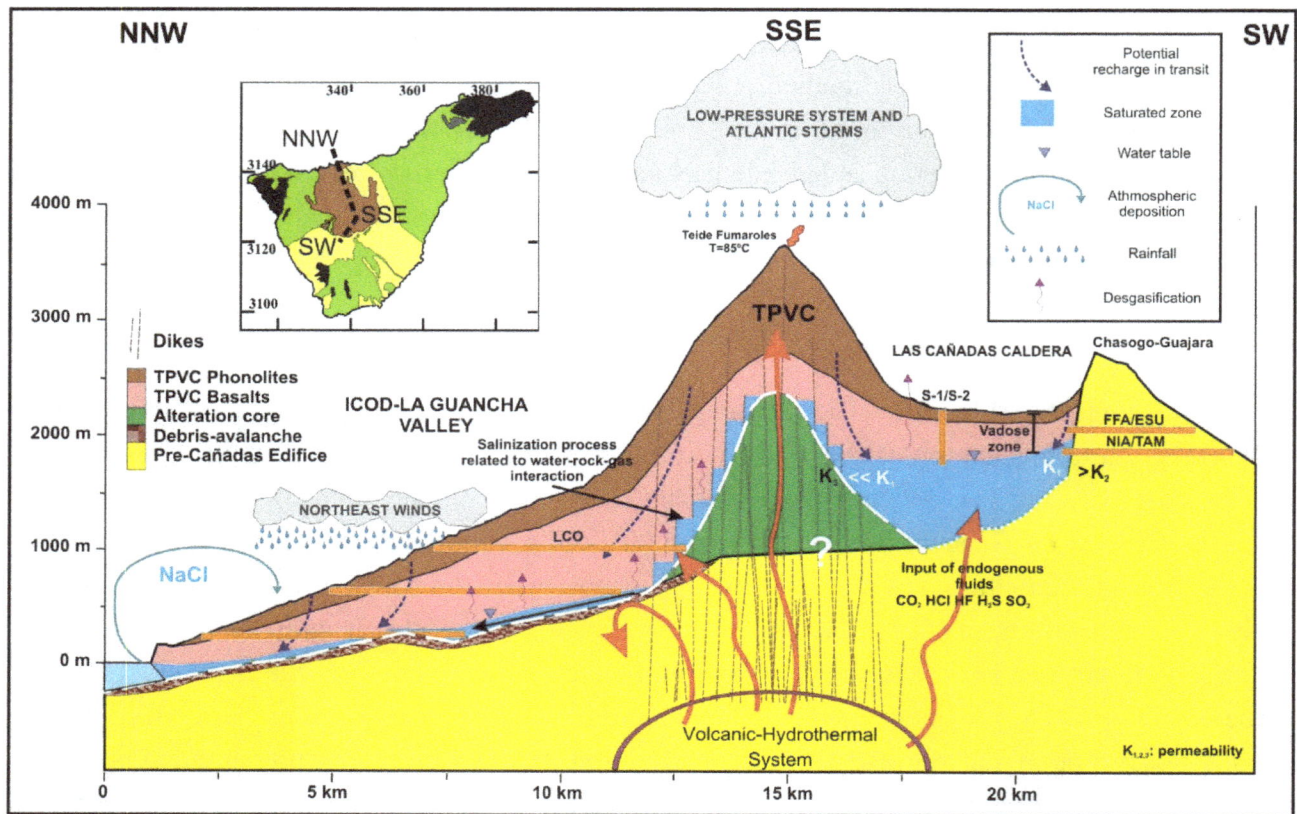

Figure 2. Conceptual hydrogeological model of Las Cañadas aquifer system, including Las Cañadas Caldera, Teide-Pico Viejo Volcanic Complex (TPVC), and Icod-La Guancha Valley areas, after [33]. Main hydrogeological processes affecting groundwater dynamics and hydrochemistry are showed.

The physical-chemical characteristics of groundwater discharged from galleries suggest that most of groundwater flowing from TPVC and LCC to surrounding areas occurs through the IGV [33]. Galleries or water mines are around 2 m × 2 m man-made sub-horizontal drains, usually with two or more branches and some-kilometers long (up to 6 km), depending on the distance needed to reach the aquifer saturated zone [1,2,33]. Groundwater hydrochemistry in LCC is sodium bicarbonate-rich with electrical conductivity above 1000 $\mu S\ cm^{-1}$ [33]. Hydrochemical processes are related to the emission of reactive deep-seated volcanic gases such as CO_2, H_2S, SO_2, and HCl from the volcanic-hydrothermal TPVC system [33,39–41]. Diffusion of these gases induce rock weathering, and enrichment in total dissolved inorganic carbon (TDIC) and in some chemical species such as Cl^-, HCO_3^-, Na^+, K^+, Mg^{2+}, and SiO_2 [42–44].

Water table depth varies in the 300–400-m range below the LCC surface (Figure 2), thus discarding direct evaporation from the saturated zone, as corroborated through similar signatures of ^{18}O and D stable isotopes in precipitation and recharge water [33]. The water-table geometry was inferred from self-potential and audiomagnetotelluric data [32]. Data showed two main eastern and central-western groundwater systems divided by the collapse caldera limit with different water-table trends, as recorded by CIATFE in scientific boreholes S-1 and S-2, respectively [30,45]. As plotted in Section 4.3, since 1994, the borehole S-1 records a drawdown oscillating trend of a few meters while the borehole S-2 records a drawdown linear trend around 0.12 m per month in average. These divergences are induced by a different basement geometry and testable influence of galleries draining groundwater nearby the borehole S-2 in the eastern system (Figure 1a) [30]. Tens of galleries drain the LCC aquifer at different depths and elevations (Figure 1a); no pumping wells yield the LCC aquifer.

Average transit time of infiltrated rainfall through the vadose zone is about 2 months, as deduced by [30] through water-table fluctuation techniques [46]. A good correlation between input precipitation events recorded at LCC and subsequent two-month delayed water-table rises in borehole S-1 was clearly observed in late autumn and winter rainy periods [30,45]. Groundwater response is a function of the saturation degree of the vadose zone as well as the rainfall distribution and intensity [30].

3. Methods and Interpretative Basis

3.1. Atmospheric Chloride Mass Balance (CMB) Method

The atmospheric chloride mass balance (CMB) is an appropriate, widely used method for net aquifer recharge in volcanic island environments [5,6,20,47,48]. Net aquifer recharge means the groundwater fraction that remains in the aquifer after discounting groundwater up-take by direct evaporation and deep-rooted vegetation from direct rainfall and runoff infiltration through the vadose zone after some delay, smoothing out the variability inherent to rainfall events [17,49,50]. In what follows, net aquifer recharge is expressed simply as recharge.

Chloride ion (hereafter Cl^- is expressed as Cl) is a conservative ideal tracer to perform mass balances because it does not undergo significant exchange with the environment, is chemically stable and highly soluble, has a known origin in most cases, and can be accurately measured by using common analytical methods [17,21]. The main atmospheric chloride sources are the marine aerosol dissolved in precipitation, salts contained in terrestrial dust, and natural and artificial volatile chloride compounds [51,52].

For a long enough period (e.g., one hydrological year) under steady-state conditions, in which the temporal storage of chloride in the soil and vadose zones can be assumed negligible, the CMB method is based on the steady balance of chloride mass fluxes from (1) atmospheric bulk deposition, which includes both chloride dissolved in precipitation and from atmospheric dust and marine aerosols [18]; (2) surface runoff leaving the area [19]; and (3) recharge water just arriving to the water-table [15,17]. The CMB method can be expressed as:

$$\sum \left(P_i \cdot C_{P_i} \right) = \sum \left(R_i \cdot C_{R_i} \right) + \sum \left(S_i \cdot C_{S_i} \right) \pm \sum F_i \qquad (1)$$

where P, R, and S are precipitation, recharge, and surface runoff for the successive i events sampled, respectively; C_P, C_R, and C_S are the average Cl concentration in the corresponding P, R, and S i water samples, respectively; and F is other non-atmospheric chloride mass fluxes adding to the groundwater such as endogenous halides in volcanic areas receiving deep warm gases. For a yearly balance P, R, and S are expressed in consistent units as $m \cdot year^{-1}$; C_P, C_R, and C_S as $g \cdot m^{-3} \equiv mg \cdot L^{-1}$; and the Cl mass fluxes A ($=P \cdot C_P$), $R \cdot C_R$, $S \cdot C_S$, and F as $g \cdot m^{-2} \cdot year^{-1}$. Actual evapotranspiration (E) is not involved into CMB equations because it is chloride-free water vapour [8].

3.2. The Chloride Mass Balance (CMB) Method Application in the Study Area

Because the CMB method may present limitations in some disadvantageous hydrogeological contexts [53,54], some additional processes must be considered for recharge evaluations in Las Cañadas aquifer system, such as (1) the possibility of mixing chloride flow mass rates produced by recharge infiltrated at different elevations [20]; (2) the variable rainfall-runoff partitioning and recharge mechanisms, land use, and soil-vegetation conditions with elevation [14]; (3) the potential contribution of non-atmospheric chloride to groundwater from endogenous halides and human activities [21]; and (4) the potential storage of chloride in the soil and vadose zones [8,17].

In low and mid-slope areas, along the IGV, most of the groundwater sampled in deep penetrating galleries integrates the variable chloride mass flux produced by recharge infiltrated at different elevations (Figure 2). Groundwater is highly mineralized as result of a significant rock-water interaction enhanced by long residence times and subsequent input of warm and reactive volcanic gases contributing chloride to groundwater; intensive agriculture may be an additional source of chloride to groundwater. Tritium and radiocarbon data suggest long residence time in the order of hundreds to thousands years for some of that water [55,56]. The limited number of sampling points having a traceable atmospheric origin of chloride prevents the usage of this water for recharge evaluations.

In summit areas, LCC nevertheless meets conditions to use the CMB method for recharge evaluations. Groundwater from recharge infiltrated in that site has a traceable atmospheric origin of chloride, as corroborated by similar signatures of ^{18}O and D stable isotopes in recharge water and precipitation [33]. Recharge in transit through the vadose zone just above the water-table can be sampled in some galleries draining the vadose zone to avoid the diffusion of endogenous halides to groundwater that occurs in the saturated continuous media. For practical purposes, this recharge in transit can be assumed a reliable proxy of net aquifer recharge having a less smoothed and delayed signal of atmospheric chloride [14,17], and negligible contribution of endogenous halides [33]. Diffuse and variable combinations of diffuse and preferential recharge mechanisms are intersected and drained by galleries, thus being these water points reliable to collect bulk chloride mass flow of different recharge mechanisms before arriving to the saturated zone.

In LCC, some specific hydrological boundary conditions described above were of assistance in evaluating recharge, such as (1) direct evaporation from the saturated zone is null; (2) surface runoff leaving the area is null, thus $S \cdot C_S = 0$; (3) fast infiltration induces negligible temporal storage of chloride in the soil and vadose zones; and (4) long-term F due to direct diffusion of halides to groundwater can be assumed negligible. So, Equation (1) can be expressed as:

$$\sum \left(P_i \cdot C_{P_i} \right) = \sum \left(R_i \cdot C_{R_i} \right) \tag{2}$$

For the preparation of yearly A, successive monthly $A_i = P_i \cdot C_{P_i}$ values were added. Sampling frequency was programmed to cover the average percolation time of about two months reported by [30], as in [14,16]. Some gaps in data sets were filled by proportional correction [18]; for a full n-day period having an analysed P' fraction of total P it was assumed that $A^C = \left(365P/nP' \right) A^M$ where superscripts M and C denote measured A in the n-day period and calculated yearly A, respectively. This correction is easy but if gaps tend to correspond to a given period of the year or omit exceptional events, yearly results may be biased [9,50]. The C_R data from successive monthly and bimonthly i

samples in galleries were averaged by the corresponding discharge flow to deduce average-weighted yearly values [14].

The CMB method application in the study area was based on (1) short percolation time through the vadose zone [30]; (2) steady conditions in sampling points with no-forced trends induced by pumping and climate in flow and C_R [33]; and (3) traceable similar atmospheric origin of the average-weighted C_P and C_R thought the use of the Cl/Br ratio [21–23].

The C_P and C_R concentrations in filtered unacidified rainwater and groundwater samples were analysed by the Environmental Research Division Laboratory (ITER) with a DIONEX DX-500® high-performance liquid (anionic) chromatograph. Regarding to the precision of the analyses, relative standard deviations for the peak areas were below 2%. Analytical detection limit was 0.01 mg L^{-1} for Cl and Br.

4. Results

4.1. Hydrological Regime of the Study Area

The hydrological regime in the LCC area was deduced from the 23-year C406G and 35-year C451U AEMET weather stations time series (Figure 1b). Average yearly precipitation (P) was 402.5 mm and 390.9 mm in C406G and C451U stations, respectively (Figure 3b,c). Cumulative deviation from average yearly P (Figure 3d) showed five-year long positive and negative phases which coarsely follow the decadal North Atlantic Oscillation cycles (Figure 3a). Because only daily records of P and temperature were available from 1986 onward, the non-global Hargreaves model was used for E, as proposed by FAO [57]; average yearly evapotranspiration (E) was 845.6 mm in C406G station (Figure 3e). Samplings and field surveys programmed to evaluate recharge were carried out in the wet hydrological year 2005–2006 (Figure 3). In this year, yearly P recorded in C406G and C451U stations was 797.8 mm and 1031.6 mm, respectively. In the year 2005–2006, an arithmetic mean of 914.7 mm year^{-1} was the yearly P established in the LCC recharge area with a yearly E-to-P (hereafter E/P) ratio around 1.08.

4.2. Atmospheric Chloride Bulk Deposition

Yearly atmospheric chloride bulk deposition in the LCC area was estimated by adding successive monthly precipitation records and the corresponding average chloride concentration measured in five non-permanent open collectors installed from 1316 m a.s.l. to 2159 m a.s.l. in the LCC recharge area (Figure 1a, Table 1). A full description of monthly values, sampling procedures, and statistical treatment of data is included in [33].

Precipitation measured in open collectors 9 and 21 was 729.2 and 680.7 mm·year^{-1}, respectively. These figures are slightly lower than precipitation recorded in nearby C406G and C451U weather stations (Figure 3). The difference is attributable to failures in collecting rainwater samples in the open collectors. So arithmetic mean P = 914.7 mm·year^{-1} from C406G and C451U weather stations and average-weighed C_P = 5.37 mg·L^{-1} from collectors 9 and 21 (Table 1) were used as input data to calculate recharge in the year 2005–2006. Average-weighed Cl/Br = 255 is probably a biased value because Br was only measured in three rainwater samples in collectors 9 and 21.

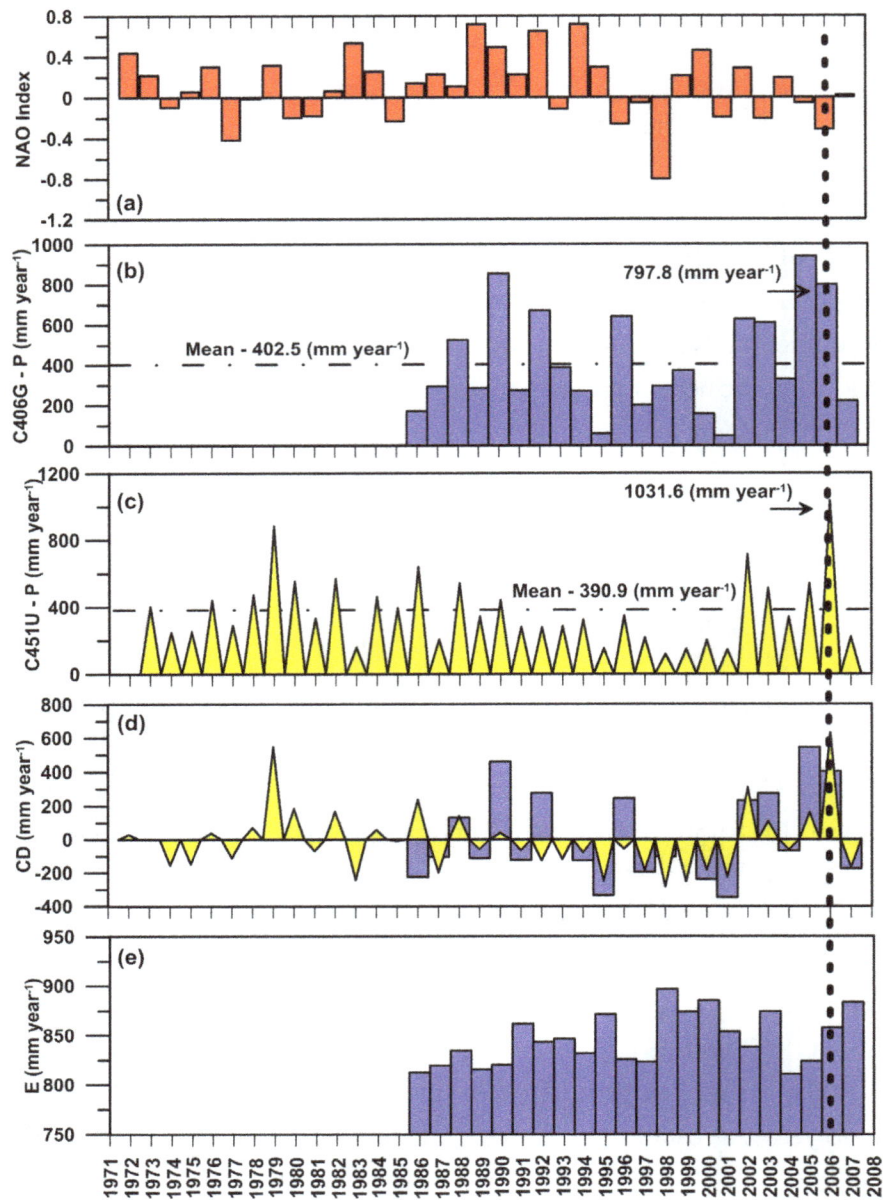

Figure 3. Hydrological regime in the LCC area from hydrological years 1971–1972 to 2007–2008. (**a**) Normalized North Atlantic Oscillation (NAO) index (NAO website: http://www.ncdc.noaa.gov/teleconnections/nao/); (**b**) yearly precipitation (P, mm·year^{-1}) in C406G weather station at 2160 m a.s.l.; (**c**) yearly precipitation (P, mm·year^{-1}) in C451U weather station at 2118 m a.s.l.; (**d**) cumulative deviation from average yearly precipitation (CD, mm·year^{-1}) in C406G and C451U stations; and (**e**) yearly evapotranspiration (E, mm·year^{-1}) in C406G station. Horizontal dashed lines correspond to average yearly precipitation in C406G and C451U stations. Vertical dashed line corresponds to yearly precipitation recorded in the year 2005–2006.

4.3. Average Chloride Concentration in Recharge Water

Selective sampling were programmed in the year 2005–2006 to deduce average-weighted yearly C_R in recharge water and groundwater having different residence times, water-rock interactions, and contribution of endogenous chloride. Data from the literature were also used to enlarge data sets, in particular from the CIATFE hydrochemical database [33]. A total of 16 sampling points were selected in the LCC area (Figure 4) attending to the groundwater flow type and the origin of chloride deduced in previous research [21,22,33] by means of C_R and Cl/Br values, as defined in Section 3.2 (Table 2).

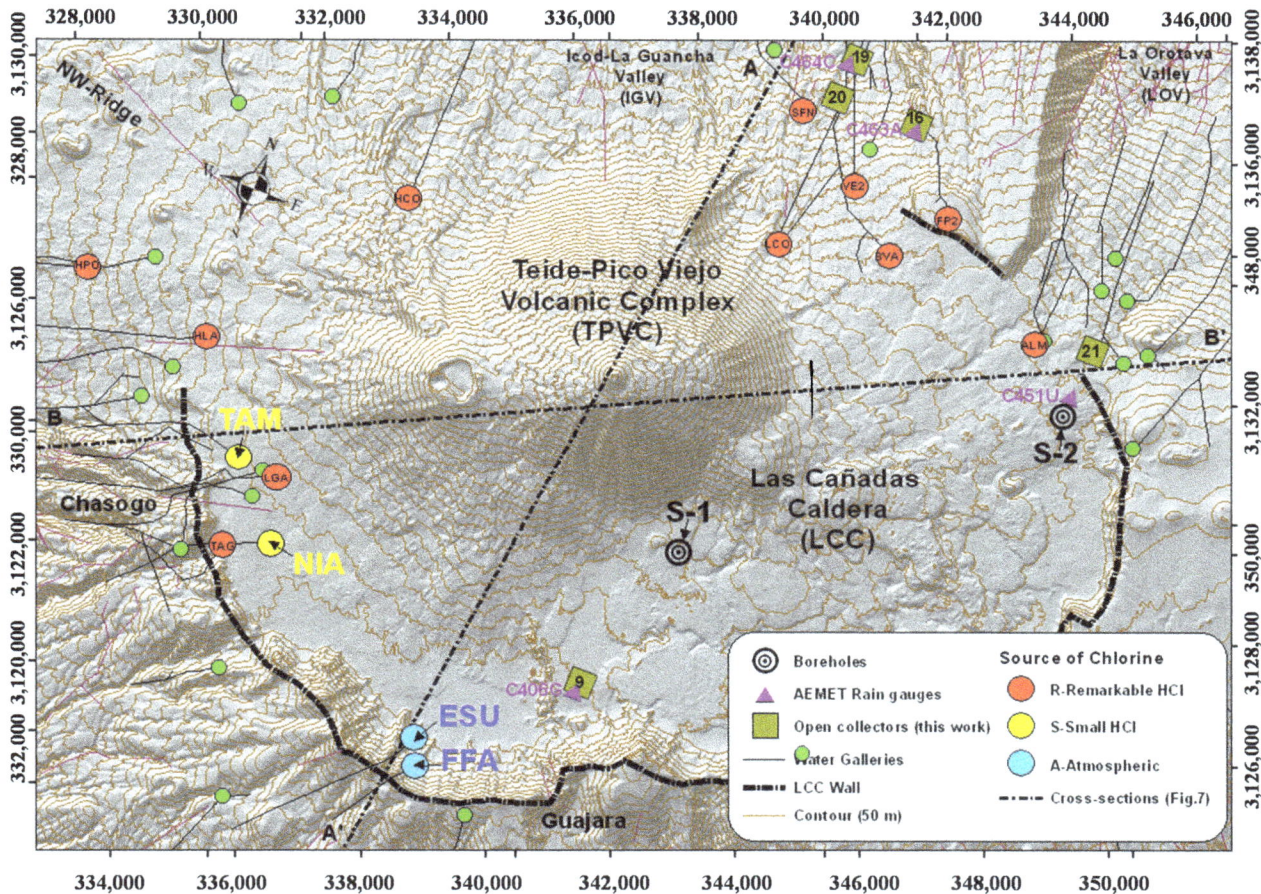

Figure 4. Selected groundwater sampling points in the Las Cañadas Caldera area, clustered by groundwater flow type and origin of chloride by means of C_R and Cl/Br values as in Table 2. Open collectors for atmospheric bulk deposition and AEMET weather stations are showed.

The galleries ESU and FFA are located just above the water table. They drain potential recharge in transit bypassing a 150-m thick vadose zone [30,33] as combination of diffuse infiltration in porous media and preferential percolation thought fissures [33,58], giving average-weighted yearly C_R = 7.73 mg·L^{-1} and Cl/Br = 144 (Table 2). Tritium contents in ESU (0.4 ± 1.1 Tritium Units) and FFA (3.0 ± 1.1 Tritium Units) determined by [55] in the year 1984 corroborates a predominant contribution of modern recharge.

Table 1. Yearly atmospheric chloride bulk deposition data from 5 non-permanent open collectors in the Las Cañadas Caldera area.

Station	Coordinates		Elevation m a.s.l.	Sampling Period		n days	P mm	P' mm	Cl mg L^{-1}	Br mg·L^{-1}	Cl/Br	A^M g·m^{-2}	A^C g m^{-2}·year^{-1}
				From	To								
9	-16.3735	28.1328	2149	28 September 2005	17 October 2006	384	729.2	653.3	5.62	0.16	145	3.67	3.90
16	-16.3640	28.1940	1711	28 September 2005	04 October 2006	369	636.0	590.0	5.50	0.02	890	3.25	3.46
19	-16.3731	28.2010	1316	22 September 2005	04 October 2006	377	552.9	552.6	5.91	0.04	300	3.28	3.19
20	-16.3632	28.1937	1475	22 September 2005	04 October 2006	377	676.9	673.9	3.50	0.03	726	2.34	2.28
21	-16.3359	28.1817	2059	28 September 2005	17 October 2006	383	680.9	636.7	5.18	0.01	368	3.30	3.36

Notes: P—Total precipitation recorded in the study period; P'—Fraction of analysed P; A^M—Measured atmospheric chloride bulk deposition in the n-period monitoring; A^C—Calculated yearly atmospheric chloride bulk deposition as $A^C = (365P/nP')A^M$.

Table 2. Chloride and bromide content, and Cl/Br ratio in groundwater in the Las Cañadas Caldera area. Sampling points were clustered into 3 groups attending to groundwater flow type and the origin of chloride.

Groups	n	Flow (L·s^{-1})	Cl			Cl/Br			Br			Recharge Area Elevation (m a.s.l.)	Outlet Discharge Elevation (m a.s.l.)	Source of Data	Source of Chloride
			M	±1σ	CV	M	±1σ	CV	M	±1σ	CV				
1-Recharge	9	0.1–29.3	7.73	4.58	0.59	144	87	0.61	0.04	0.16	0.25	2074–2184	1895–2040	[1,2]	A
2-Shallow	16	2.0–47.0	20.95	11.20	0.53	338	144	0.43	0.09	0.01	0.11	2077–2133	1525–1779	[1,2]	S
3-Deep	11	4.0–237.2	25.55	11.20	0.44	740	207	0.28	0.08	0.03	0.38	1533–2115	790–1780	[1]	R

Notes: Recharge water (Group 1) includes ESU and FFA galleries; Shallow groundwater (Group 2) includes ALM, BVA, FP2, HLA, HCO, HPO, LCO, LGA, TAG, SFN, and VE2 galleries, after [33]. See location of sampling points in Figure 4. M—Average-weighed yearly value; ±1σ—Standard deviation of M; CV—Coefficient of variation of M. For Cl, M and ±1σ in mg·L^{-1}; for the rest, dimensionless. Source of data: 1—[33]; 2 the CIATFE hydrochemical database, in [33]. Source of chloride: A—Atmospheric only; S—Small contribution of volcanogenic halides; R—Remarkable contribution of volcanogenic halides.

The NIA and TAM galleries drain the shallow water-table [33], giving average-weighted yearly C_R = 14.96 mg·L^{-1} and Cl/Br = 413. These figures are induced by a mixture of recharge in transit in low-water-table periods and short-turnover-time groundwater receiving contribution of endogenous chloride in high-water-table periods (Table 2).

Peripheral galleries penetrating from outside to the saturated zone at different elevation and depth provided average-weighed yearly C_R = 25.55 mg·L^{-1} and Cl/Br = 740. These are long-turnover-time groundwater samples showing a remarkable contribution of endogenous chloride (Table 2).

Average-weighed C_R and Cl/Br values increased from recharge to long-turnover-time groundwater (Table 2) due to (1) the combination of an increasing smoothing effect as the percolation time of recharge in transit increases in Group 1; and (2) the progressive contribution of endogenous chloride as the average turnover time of groundwater increases in Groups 2 and 3 (Table 2). The overall effect is shown by lower coefficients of variation of C_R and Cl/Br in Group 3 (Table 2). C_R and Cl/Br values are in agreement with findings reported in Tenerife Island by [21]. Group 1 was selected to calculate yearly recharge in the LCC recharge area by using C_R data from galleries ESU and FFA. Note that no-forced trends in C_R and flow dynamics were observed in galleries ESU and FFA (Figure 5). This is a prerequisite of steady conditions for recharge evaluations [9,50].

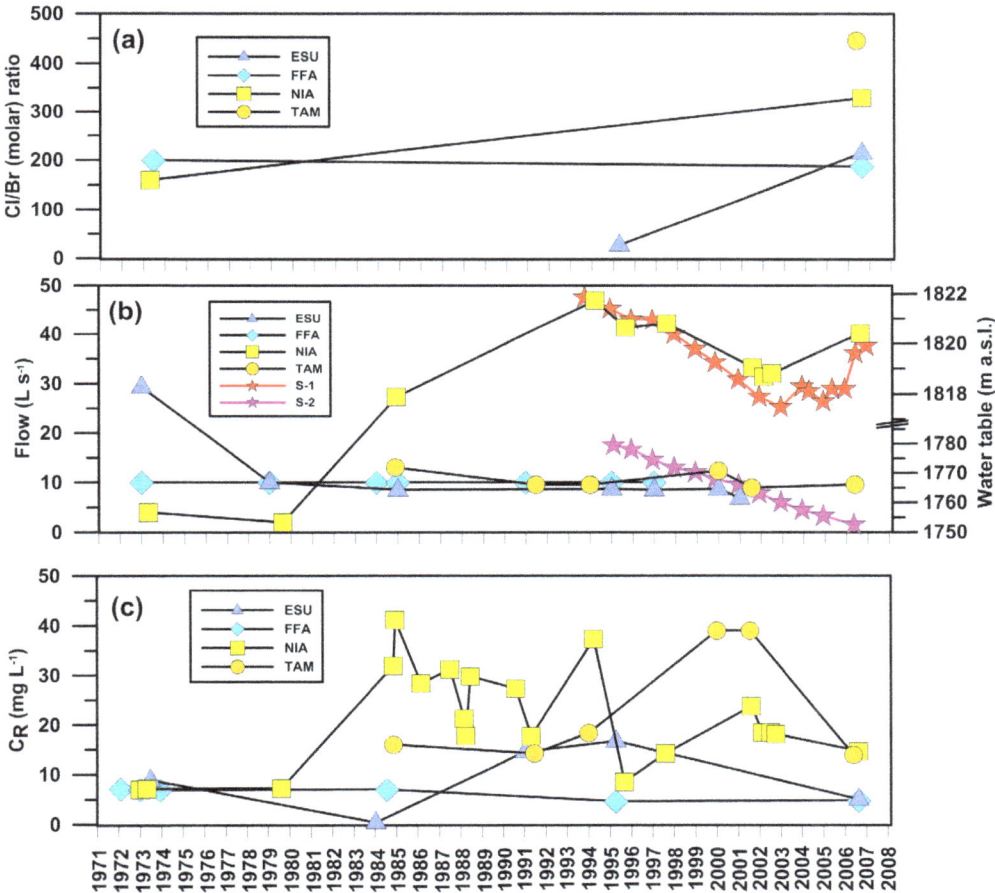

Figure 5. Temporal evolution of data in selected sampling points. (**a**) Cl/Br molar ratio; (**b**) Discharge flow (L·s^{-1}) in galleries and piezometric level (m a.s.l.) in S-1 and S-2 boreholes; and (**c**) C_R (mg·L^{-1}) in galleries and boreholes S-1 and S-2. Data source: [30,33], and CIATFE hydrochemical database in [33].

4.4. Recharge Evaluation

In the LCC recharge area the Cl/Br ratio was 255 in atmospheric bulk deposition (Table 1) and 144 in recharge water (Table 2). As commented in Section 4.2, Br was only measured in three rainwater samples in collectors 9 and 21. Similar few Br data exist for recharge water [33]. This low statistical significance may be the reason of the negative bias of Cl/Br values, when compared with Cl/Br data ranging from 250 to 300 in high-elevation recharge waters in Tenerife Island reported by [21] (Figure 6). However, the small negative bias in Cl/Br values does not prevent the recharge evaluation based on the premise of an atmospheric origin of C_R for a proper CMB application.

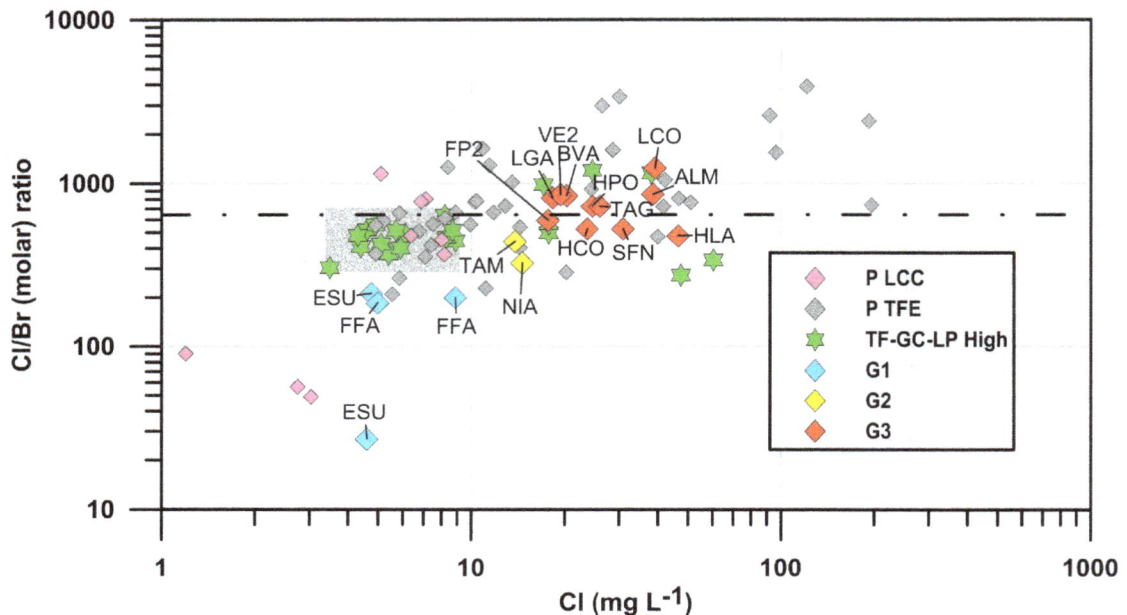

Figure 6. Plot of Cl (mg·L^{-1}) *vs.* Cl/Br molar ratio from different groups of samples studied. G1, recharge water; G2, shallow groundwater; and G3, deep groundwater as in Table 2 [33]. Data from atmospheric bulk deposition samples in the LCC area (P LCC) and in other areas of Tenerife Island (P TFE) after [33]. Groundwater samples from high and mid-slopes areas (over 1000 m a.s.l.) in Tenerife (TF), Gran Canaria (GC), and La Palma (LP) Islands (TF-GC-LP High) after [21]. Cl/Br = 655 ± 4 for global seawater (dashed line) and Cl/Br = 484 ± 93 (20 samples) for recharge waters in high and mid-slope areas in TF, GC, and LP Islands (dashed grey square) were taken from [21].

Using average-weighted yearly C_P = 5.37 mg·L^{-1} (Table 1), C_R = 7.73 mg·L^{-1} (Table 2) and P = 914.7 mm·year^{-1} from AEMET weather stations into Equation (2), calculated yearly R in the LCC recharge area was 631.1 mm·year^{-1} in year 2005–2006, *i.e.*, 69% of yearly P. Note that the recharge area elevation for selected galleries (Table 2) was similar to the elevation of open collectors for sampling bulk deposition (Table 1) and AEMET weather stations used for yearly P, thus allowing reliable comparisons of R and P.

5. Discussion

Las Cañadas Caldera (LCC) is a high-permeability endorheic basin favouring high recharge rates. The CMB method was used to estimate net aquifer recharge (R) in the wet hydrological year 2005–2006. This year provides information on the maximum threshold of renewable groundwater resources expected at the area. Yearly R was 631.1 mm·year^{-1}. The recharge-to-precipitation (hereafter R/P) ratio was 0.69; note that the E/P ratio was 1.08 in this year (Figure 3e). This figure is in good agreement with the R/P ratio about 0.65 obtained for similar wetter years (Figure 3) through an independent lumped-parameter rainfall-runoff model operated by CIATFE [2,59] from hydrological years 1944–1945 to 2003–2004 [30]. The positive relative difference was about 0.04.

About the CIATFE model, it was programmed to evaluate water resources throughout Tenerife Island [2,59]. For this purpose (1) the island was discretized into an 1 km × 1 km grid; (2) in each grid node precipitation was measured or interpolated from existing 230 weather stations, evaporation and transpiration coefficients were calculated, and overland runoff was evaluated; (3) in each grid node, potential aquifer recharge is deduced as difference in precipitation, evapotranspiration, and runoff, taking the cumulative water surplus from high to lower grid-nodes into account; (4) piezometric records were used for calibration; and (5) daily estimates were added to get yearly values. In the LCC area, the CIATFE model provided an average R/P ratio about 0.68 for the 60-year modeling period, as deduced from average yearly precipitation of 489 mm·year^{-1} and potential recharge of 334 mm·year^{-1} [30].

A low divergence among potential recharge estimates obtained through accurate physical (e.g., soil water balances combined with hydrodynamic methods) and tracer (e.g., the CMB method) techniques was also documented in other geographical domains having similar high R/P ratios and variable contribution of diffuse and preferential recharge mechanisms [10,14,16]. The low relative difference between the CMB method and the CIATFE model is attributable to (1) homogeneity of the system response to recharge inputs; (2) negligible water losses by surface and interflow runoff induced by the particular LCC aquifer boundary; (3) accuracy of actual evapotranspiration estimated by the CIATFE model; (4) fast infiltration induced by the sparsely-vegetated condition of the area and the high permeability of the vadose zone, thus favoring reliable comparisons of average-weighted yearly C_P and C_R; and (5) high representativeness of C_R data from galleries ESU and FFA. Galleries are proposed as optimal sampling points for potential recharge in transit through the vadose zone, a desirable condition in active volcanic areas in order to avoid the contribution of volcanogenic halides to groundwater. However, low divergence does not necessarily provide an additional confidence in both recharge estimates [12,60]; this is a necessary but not sufficient condition for confident results.

6. Conclusions

In view of the results, four conclusions arise: (1) potential recharge estimates deduced from sampling C_R in the vadose zone just above the water-table prove to be a reliable proxy of net aquifer recharge; (2) the apparent linearity in R/P ratios among the CMB method and the CIATFE rainfall-runoff model estimates allows for comparisons; (3) results provide groundwater resources around 80.8 Mm^3 in the wet 2005–2006 hydrological year and 42.8 Mm^3 in an average-rainfall period [30] taking the 128 km^2 recharge area of LCC into account, thus contributing between 12% and

23% to the total 358 Mm³ renewable groundwater resources estimated in Tenerife Island (2037 km²) through the CIATFE model [2]; and (4) longer data series of C_P and C_R are desirable to use the CMB method for long-term recharge evaluations and calibrations. This recharge evaluation allows for planning groundwater sustainability. In the LCC aquifer, average yearly discharge of 37.6 Mm³ was recorded in galleries draining the LCC aquifer in the period 2002–2006 [33,44], thus the discharge-to-recharge ratio is 0.88 for an average hydrological period. The 0.12 difference is attributed to diffuse discharge to other aquifers.

In large areas of Tenerife Island, as well as in other active oceanic volcanic islands, the contribution of endogenous halides from the volcanic-hydrothermal system [39,41–43] to groundwater is the largest source of uncertainty to use the CMB method when groundwater samples are used for net aquifer recharge, irrespective of the recharge-to-precipitation ratio induced by local climate and soil properties. Permanent precipitation of halides in the vadose zone may occur by combination of two successive high-order controls (1) a testable degassing of endogenous halides through the vadose zone, as reported by [43] in Tenerife Island; and (2) low recharge-to-precipitation ratios such as reported in drylands [61], thus favoring temporary retentions. The second control is negligible in the LCC area. So the combined effect results in a negligible precipitation of endogenous halides in the vadose zone. Unfortunately, there is no vertical Cl/Br for assessing potential soil halide dissolution occurring.

In spite this constraint, LCC is a rare, suitable area to use the CMB method. The suitability of the CMB method is supported by two basic steady conditions (1) no-forced trends induced by pumping and climate were identified in C_R and flow dynamics in recharge sampling points (Figure 5); and (2) similar low Cl/Br ratios were found in atmospheric bulk deposition and recharge water (Figure 6) as prerequisite for asserting the atmospheric origin of C_R and discarding a relevant precipitation of endogenous halides in the vadose zone.

In Tenerife Island, the Cl/Br ratio in high and mid-slope recharge water vary in the 150–400 range, whereas the progressive HCl degassing in volcanic-hydrothermal systems can raise Cl/Br up to 1300 (Figures 6 and 7) as pointed by [21]. The Cl/Br values reported in this work are in good agreement with existing information in other active volcanic-hydrothermal systems [62,63]. This available Cl/Br database was of interest in the spatial identification and zoning of disadvantageous areas having preferential contribution of endogenous halides, thus limiting the CMB application.

This paper evaluates recharge dynamics in sparse-data areas in high-elevation volcanic oceanic islands. The paper contributes to enhance the applicability of the CMB method in hydrogeological systems where reported contributions of endogenous halides to groundwater are often a main cause to invalidate the application. The mapping of the Cl/Br ratio in groundwater in Tenerife Island (Figure 7) is a contribution of this paper with application for planning aquifer recharge studies and volcanic degassing monitoring in the future. Understanding recharge dynamics is key for groundwater resources management focussed in sustainable socioeconomic development in these isolated regions.

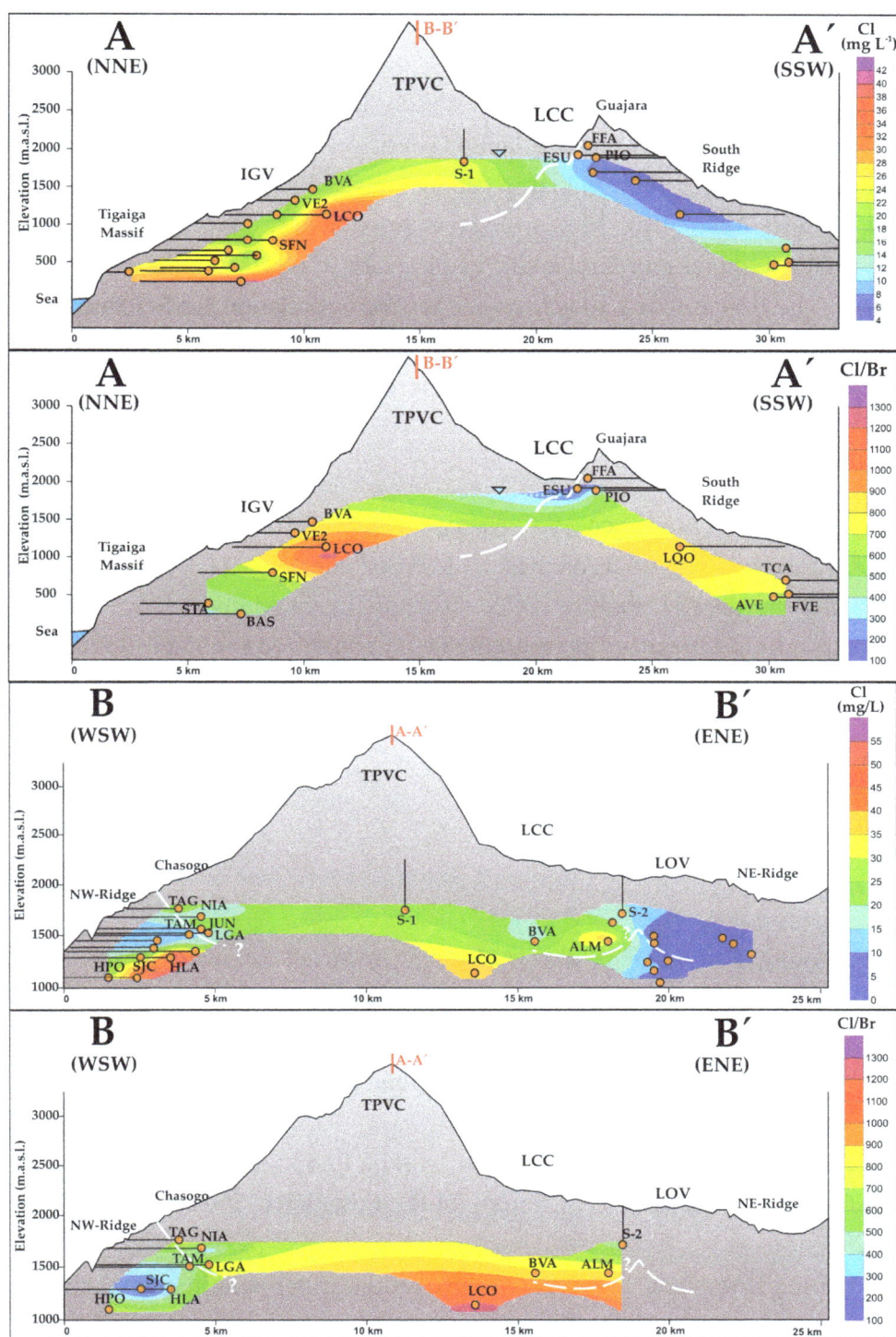

Figure 7. Distribution maps of (**a**) chloride content $(mg \cdot L^{-1})$ and (**b**) Cl/Br ratio in groundwater in Tenerife Island (see location of cross-sections A-A´ and B-B´ in Figure 1a and Figure 4). Horizontal and vertical lines mean projected locations of galleries described in Table 2 and scientific boreholes S-1 and S-2, respectively (see location in Figure 4). Dashed blue line in cross-section A-A´ is the water-table position. White line in cross-sections A-A´ and B-B´ represents the upper boundary of the basement, as inferred by [33].

Acknowledgments

This research was partially supported by the Cabildo Insular de Tenerife (Island Council of Tenerife) and Gobierno de Canarias (Canary Islands Government). Authors are grateful to different Water Communities of Tenerife for their valuable support in field surveys, the Insular Water Council of Tenerife and the Spanish Agency for Meteorology to provide data, and the undergraduate student group of University of La Laguna and University of Granada in Spain for their support in the field surveys and lab works. The first author also acknowledges the Foundation for Science and Technology of Portugal for the post-doctoral scholarship SFRH/BPD/76404/2011. Valuable comments and suggestions by two anonymous reviewers were greatly appreciated.

Author Contributions

Rayco Marrero-Diaz and Francisco J. Alcalá conceived the subject of the article; Rayco Marrero-Diaz, Francisco J. Alcalá, Nemesio M. Pérez and Dina L. López designed the work; Rayco Marrero-Diaz and Germán D. Padilla collected data; Rayco Marrero-Diaz, Gladys V. Melián and Eleazar Padrón carried out laboratory measurements and analyzed the data; Rayco Marrero-Diaz, Francisco J. Alcalá and Dina L. López wrote the manuscript.

Conflicts of Interest

The authors declare no conflict of interest.

References and Notes

1. Falkland, A.; Custodio, E. *Guide on the Hydrology of Small Islands*; Studies and Reports in Hydrology, UNESCO: Paris, France, 1991; Volume 49, pp. 51–130.
2. Braojos, J.J.; Farrujia, I.; Fernández, J.D. Los recursos hídricos en Tenerife frente al cambio climático. In Proceedings of the III Congreso de Ingeniería Civil, Territorio y Medio Ambiente, Zaragoza, Spain, 25–27 October 2006; p. 16.
3. Custodio, E.; Cabrera, M.C. The Canary Islands. In *Water, Agriculture Environment Spain, Can We Square the Circle?* de Stefano, L., Llamas, M.R., Eds.; CRC Press: Boca Raton, FL, USA, 2013; pp. 281–290.
4. Join, J.L.; Folio, J.L.; Robineau, B. Aquifers and groundwater within active shield volcanoes. Evolution of conceptual models in the Piton de la Fournaise volcano. *J. Volcanol. Geotherm. Res.* **2005**, *147*, 187–201.
5. Herrera, C.; Custodio, E. Conceptual hydrogeological model of volcanic Easter Island (Chile) after chemical and isotopic surveys. *Hydrogeol. J.* **2008**, *16*, 1329–1348.
6. Herrera, C.; Custodio, E. Groundwater flow in a relatively old oceanic volcanic island: The Betancuria area, Fuerteventura Island, Canary Islands, Spain. *Sci. Total Environ.* **2014**, *496*, 531–550.
7. Milly, P.C.D.; Eagleson, P.S. Effects of spatial variability on annual average water balance. *Water Resour. Res.* **1987**, *23*, 2135–2143.

8. Scanlon, B.R. Uncertainties in estimating water fluxes and residence times using environmental tracers in an arid unsaturated zone. *Water Resour. Res.* **2000**, *36*, 395–409.

9. Alcalá, F.J.; Custodio, E. Natural uncertainty of spatial average aquifer recharge through atmospheric chloride mass balance in continental Spain. *J. Hydrol.* **2015**, *524*, 642–661.

10. Contreras, S.; Boer, M.; Alcalá, F.J.; Domingo, F.; García, M.; Pulido-Bosch, A.; Puigdefábregas, J. An ecohydrological modelling approach for assessing long-term recharge rates in semiarid karstic landscapes. *J. Hydrol.* **2008**, *351*, 42–57.

11. Hugman, R.; Stigter, T.Y.; Monteiro, J.P.; Nunes, L.M. Influence of aquifer properties and the spatial and temporal distribution of recharge and abstraction on sustainable yields in semi-arid regions. *Hydrol. Process.* **2012**, *26*, 2791–2801.

12. España, S.; Alcalá, F.J.; Vallejos, A.; Pulido-Bosch, A. A GIS tool for modelling annual diffuse infiltration on a plot scale. *Comput. Geosci.* **2013**, *54*, 318–325.

13. Scanlon, B.R.; Keese, K.E.; Flint, A.L.; Flint, L.E.; Gaye, C.B.; Edmunds, W.M.; Simmers, I. Global synthesis of groundwater recharge in semiarid and arid regions. *Hydrol. Process.* **2006**, *20*, 3335–3370.

14. Alcalá, F.J.; Cantón, Y.; Contreras, S.; Were, A.; Serrano-Ortiz, P.; Puigdefábregas, J.; Solé-Benet, A.; Custodio, E.; Domingo, F. Diffuse and concentrated recharge evaluation using physical and tracer techniques: Results from a semiarid carbonate massif aquifer in southeastern Spain. *Environ. Earth Sci.* **2011**, *63*, 541–557.

15. Coes, A.L.; Spruill, T.B.; Thomasson, M.J. Multiple-method estimation of recharge rates at diverse locations in the North Caroline Coastal Plains, USA. *Hydrog. J.* **2007**, *15*, 773–788.

16. Andreu, J.M.; Alcalá, F.J.; Vallejos, Á.; Pulido-Bosch, A. Recharge to aquifers in SE Spain: Different approaches and new challenges. *J. Arid Environ.* **2011**, *75*, 1262–1270.

17. Custodio, E.; Llamas, M.R.; Samper, J. *Aquifers Recharge Evaluation in Water Planning*; IAH-Spanish Chapter and ITGE: Madrid, Spain, 1997; p. 455.

18. Alcalá, F.J.; Custodio, E. Atmospheric chloride deposition in continental Spain. *Hydrol. Process.* **2008**, *22*, 3636–3650.

19. Maréchal, J.C.; Varma, M.R.R.; Riotte, J.; Vouillamoz, J.M.; Kumar, M.S.M.; Ruiz, L.; Sekhar, M.; Braun, J.J. Indirect and direct recharges in a tropical forested watershed: Mule Hole, India. *J. Hydrol.* **2009**, *364*, 272–284.

20. Custodio, E. Estimation of aquifer recharge by means of atmospheric chloride deposition balance in the soil. *Contrib. Sci.* **2010**, *6*, 81–97.

21. Alcalá, F.J.; Custodio, E. Using the Cl/Br ratio as a tracer to identify the origin of salinity in aquifers in Spain and Portugal. *J. Hydrol.* **2008**, *359*, 189–207.

22. Custodio, E.; Herrera, C. Utilización de la relación Cl/Br como trazador hidrogeoquímico en hidrología subterránea. *Boletín Geológico y Minero* **2000**, *111*, 49–68.

23. Cruz-Fuentes, T.; Cabrera, M.C.; Heredia, J.; Custodio, E. Groundwater salinity and hydrochemical processes in the volcano-sedimentary aquifer of La Aldea, Gran Canaria, Canary Islands, Spain. *Sci. Total Environ.* **2014**, *484*, 154–166.

24. Carracedo, J.C.; Rodríguez Badiola, E.; Guillou, H.; Paterne, M.; Scaillet, S.; Pérez Torrado, F.J.; Paris, R.; Fra-Paleo, U.; Hansen, A. Eruptive and structural history of Teide Volcano and rift zones of Tenerife, Canary Islands. *Geol. Soc. Am.* **2007**, *119*, 1027–1051.

25. Garcia Herrera, R.; Gallego, D.; Hernández, E. Influence of the North Atlantic Oscillation on the Canary Islands Precipitation. *J. Clim.* **2001**, *14*, 3889–3904.

26. Bustos, J.J.; Delgado, F.S. *Climatología del Parque Nacional de Las Cañadas del Teide*; Sección de Estudios y Desarrollo del C.M.T.; de Canarias Occidental, Instituto Nacional de Meteorología: Canarias, Spain, 2000.

27. Water Planning of Tenerife Island (PHI). *DECRETO 319/1996, de 23 de diciembre*; Boletín Oficial de Canarias: Canarias, Spain, 1977.

28. Fernández, E.; Tejedor, M.L.; Hernández, J. Andosoles Canarios (VII). *An. Edafol. Agrobiol.* **1975**, *24*, 359–369.

29. Fernández, E.; Tejedor, M.L.; Hernández, J. Andosoles Canarios (IX). *An. Edafol. Agrobiol.* **1975**, *24*, 383–394.

30. Farrujia, I.; Braojos, J.J.; Fernández, J.D. Evolución cuantitativa del sistema acuífero de Tenerife. In Proceedings of the III Congreso de Ingeniería Civil, Territorio y Medio Ambiente, Zaragoza, Spain, 25–27 October 2006; p. 12.

31. Márquez, A.; López, I.; Herrera, R.; Martín-González, F.; Izquierdo, T.; Carreño, F. Spreading and potential instability of Teide volcano, Tenerife, Canary Islands. *Geophys. Res. Lett.* **2008**, *35*, 1–5.

32. Villasante-Marcos, V.; Finizola, A.; Abella, R.; Barde-Cabusson, S.; Blanco, M.J.; Brenes, B.; Cabrera, V.; Casas, B.; de Agustín, P.; Di Gangi, F.; *et al.* Hydrothermal system of Central Tenerife Volcanic Complex, Canary Islands (Spain), inferred from self-potential measurements. *J. Volcanol. Geotherm. Res.* **2014**, *272*, 59–77.

33. Marrero, R. Modelo Hidrogeoquímico del Acuífero de Las Cañadas del Teide (Tenerife, Islas Canarias). Doctoral Thesis, Universidad Politécnica de Cataluña, Barcelona, Spain, 2010.

34. Navarro, J.M.; Coello, J. Depressions originated by landslide processes in Tenerife. In Proceedings of the ESF Meeting on Canarian Volcanism, Lanzarote, Spain, 30 November–3 December 1989; pp. 150–152.

35. Ablay, G.J.; Hürlimann, M. Evolution of the north flank of Tenerife by recurrent giant landslides. *J. Volcanol. Geotherm. Res.* **2000**, *103*, 135–169.

36. Custodio, E. Groundwater in volcanic hard rocks. In *Groundwater in Fractured Rocks*; Krásný, J., Sharp, J.M., Jr., Eds.; Selected Papers (Prague Conference), No. 9; IAH-AIH, Taylor & Francis: London, UK, 2007; pp. 95–108.

37. Ingebritsen, S.E.; Scholl, M.A. The Hydrogeology of Kilauea volcano. *Geothermics* **1993**, *22*, 255–270.

38. Cabrera, M.C.; Custodio, E. Groundwater flow in a volcanic–sedimentary coastal aquifer: Telde area, Gran Canaria, Canary Islands, Spain. *Hydrogeol. J.* **2004**, *12*, 305–320.

39. Pérez, N.M.; Sturchio, N.C.; Williams, S.N.; Carracedo, J.C.; Coello, J. Geochemical characteristics of the volcanic-hydrothermal gases in Teide, Timanfaya, Taburiente, and Teneguía volcanoes, Canary Islands, Spain. In Proceedings of the III Congreso Geológico España and VIII Congreso Latinoamericano de Geologia, Salamanca, Spain, 12–16 January 1992; pp. 463–467.

40. Soler, V.; Castro-Almazán, J.A.; Viñas, R.T.; Eff-Darwich, A.; Sánchez-Moral, S.; Hillaire-Marcel, C.; Farrujia, I.; Coello, J. High CO_2 levels in boreholes at El Teide Volcano

Complex (Tenerife, Canary Islands): Implications for volcanic activity monitoring. *Pure Appl. Geophys.* **2004**, *161*, 1519–1532.

41. Melián, G.V.; Tassi, F.; Pérez, N.M.; Hernández, P.; Sortino, F.; Vaselli, O.; Padrón, E.; Nolasco, D.; Barrancos, J.; Padilla, G.; *et al*. A magmatic source for fumaroles and diffuse degassing from the summit crater of Teide Volcano (Tenerife, Canary Islands): A geochemical evidence for the 2004–2005 seismic–volcanic crisis. *Bull. Volcanol.* **2012**, *74*, 1465–1483.

42. Albert-Beltran, J.F.; Araña, V.; Diez, J.L.; Valentin, A. Physical-chemical conditions of the Teide volcanic system (Tenerife, Canary Islands). *J. Volcanol. Geotherm. Res.* **1990**, *43*, 321–332.

43. Valentin, A.; Albert-Beltrán, J.F.; Diez, J.L. Geochemical and geothermal constraints on magma bodies associated with historic activity, Tenerife (Canary Islands). *J. Volcanol. Geotherm. Res.* **1990**, *44*, 251–264.

44. Marrero, R.; López, D.L.; Hernández, P.A.; Pérez, N.M. Carbon dioxide discharged through the Las Cañadas Aquifer, Tenerife, Canary Islands. *Pure Appl. Geophys.* **2008**, *165*, 147–172.

45. Farrujia, I.; Braojos, J.J.; Fernández, J. Ejecución de dos sondeos profundos en Las Cañadas del Teide. In Proceedings of the VII Simposio de Hidrogeología, Murcia, Spain, 28 May–1 June 2001; pp. 661–672.

46. Healy, R.W.; Cook, P.G. Using groundwater levels to estimate recharge. *Hydrogeol. J.* **2002**, *10*, 91–109.

47. Hagedorn, B.; El-Kadi, A.I.; Mair, A.; Whittier, R.B.; Ha, K. Estimating recharge in fractured aquifers of a temperate humid to semiarid volcanic island (Jeju, Korea) from water table fluctuations, and Cl, CFC-12 and ^{3}H chemistry. *J. Hydrol.* **2011**, *409*, 650–662.

48. Mair, A.; Hagedorn, B.; Tillery, S.; El-Kadi, A.I.; Westenbroek, S.M.; Ha, K.; Koh, G.-W. Temporal and spatial variability of groundwater recharge on Jeju Island, Korea. *J. Hydrol.* **2013**, *501*, 213–226.

49. Lerner, D.N.; Issar, A.S.; Simmers, I. *Groundwater Recharge: A Guide to Understanding and Estimating Natural Recharge*; IAH International Contributions to Hydrogeology: Heise, Hannover, 1990; p. 345.

50. Alcalá, F.J.; Custodio, E. Spatial average aquifer recharge through atmospheric chloride mass balance and its uncertainty in continental Spain. *Hydrol. Process.* **2014**, *28*, 218–236.

51. Erickson, D.J.; Merrill, J.T.; Duce, R.A. Seasonal estimates of global atmospheric sea-salt distributions. *J. Geophys. Res.* **1986**, *91*, 1067–1072.

52. Edmunds, W.M.; Shand, M.P. *Natural Groundwater Quality*; Blackwell Publishing, Ltd.: Oxford, UK, 2009; p. 469.

53. Somaratne, N. Pitfalls in application of the conventional chloride mass balance (CMB) method in karst aquifers and use of the generalized CMB method. *Environ. Earth Sci.* **2015**, *2015*, doi:10.1007/s12665-015-4038-y.

54. Subyani, A.; Sen, Z. Refined chloride mass balance method and its application in Saudi Arabia. *Hydrol. Process.* **2006**, *20*, 4373–4380.

55. Custodio, E.; Hoppe, J.; Hoyos-Limón, A. Aportaciones al conocimiento hidrogeológico de Tenerife utilizando isótopos ambientales. In *Simp. Canar. Agua 2000*; Puerto de la Cruz: Tenerife, Spain, 1987; p. 25.

56. Skupien, E.; Poncela, A. Trabajos realizados para la caracterización isotópica de las aguas subterráneas de Tenerife en el ámbito del proyecto AQUAMAC II para cumplimiento de la Directiva Marco de Aguas. In Proceedings of the Las aguas subterráneas en la Directiva Marco de Aguas, Santa Cruz de Tenerife, Spain, 26–27 September 2007; p. 14.

57. Allen, R.G.; Pereira, L.S.; Raes, D.; Smith, M. Crop evapotranspiration: Guidelines for computing crop water requirements. In *Irrigation and Drainage*; Paper 56 UN-FAO: Rome, Italy, 1998; p. 465.

58. Navarro, J.M. *Geología e Hidrogeología del Parque Nacional del Teide*; Ministerio de Agricultura, Pesca y Alimentación, Subdirección General de Espacios Naturales, Parque Nacional del Teide: Tenerife, Spain, 1995; p. 103.

59. Braojos, J.J. Definición de la recarga a través del balance hídrico en las Islas Canarias occidentales. Modelación. Proceedings of the La Evaluación de la Recarga a los Acuíferos en la Planificación Hidrológica, Las Palmas de Gran Canaria, Spain, 20–24, January 1997; Custodio, E., Llamas, M.R., Samper, J., Eds.; pp. 267–277.

60. Somaratne, N.; Smettem, K.; Frizenschaf, J. Three criteria reliability analyses for groundwater recharge estimations. *Environ. Earth Sci.* **2014**, *72*, 2141–2151.

61. Cartwright, I.; Weaver, T.R.; Fulton, S.; Nichol, C.; Reid, M.; Cheng, X. Hydrogeochemical and isotopic constraints on the origins of dryland salinity, Murray Basin, Victoria, Australia. *Appl. Geochem.* **2004**, *19*, 1233–1254.

62. Delmelle, P.; Bernard, A. Geochemistry, mineralogy, and chemical modelling of the acid water lake of Kawah Ijen volcano, Indonesia. *Geochim. Cosmochim. Acta* **1994**, *58*, 2445–2460.

63. Bureau, H.E.; Métrich, N. An experimental study of bromine behaviour in water-saturated silicic melts. *Geochim. Cosmochim. Acta* **2003**, *67*, 1689–1697.

11

Sensitivity Analysis in a Complex Marine Ecological Model

Marcos D. Mateus * and Guilherme Franz

MARETEC, Instituto Superior Técnico, Universidade de Lisboa, Av. Rovisco Pais, Lisboa 1049-001, Portugal; E-Mail: guilherme.franz@tecnico.ulisboa.pt

* Author to whom correspondence should be addressed; E-Mail: marcos.mateus@tecnico.ulisboa.pt;

Academic Editor: Y. Jun Xu

Abstract: Sensitivity analysis (SA) has long been recognized as part of best practices to assess if any particular model can be suitable to inform decisions, despite its uncertainties. SA is a commonly used approach for identifying important parameters that dominate model behavior. As such, SA address two elementary questions in the modeling exercise, namely, how sensitive is the model to changes in individual parameter values, and which parameters or associated processes have more influence on the results. In this paper we report on a local SA performed on a complex marine biogeochemical model that simulates oxygen, organic matter and nutrient cycles (N, P and Si) in the water column, and well as the dynamics of biological groups such as producers, consumers and decomposers. SA was performed using a "one at a time" parameter perturbation method, and a color-code matrix was developed for result visualization. The outcome of this study was the identification of key parameters influencing model performance, a particularly helpful insight for the subsequent calibration exercise. Also, the color-code matrix methodology proved to be effective for a clear identification of the parameters with most impact on selected variables of the model.

Keywords: biogeochemical modeling; complex models; sensitivity analysis; parameter perturbation

1. Introduction

Sensitivity analysis (SA) can be basically described as the process to evaluate the contribution of input parameters to model behavior. Several parameters in ecosystem models represent specific process coefficients that are only measured with difficulty (if possible at all). As a consequence, there are uncertainties related with the parameterization and the nonlinearity of interactions within the model. This raises two basic questions: (a) how sensitive is the model to changes in individual parameter values; and (b) which parameters or associated processes have more influence on specific output variables? Answering the first question may reveal which parameters have more influence on the model and need more attention. The answer to the second question will help to understand the simulated system.

Performing a SA is an effective way to answer these questions and assess model performance [1–5], thus justifying its widespread use to quantitatively assess the influence of parameters in model performance, and identify those having the most significant impact on the results [6–12]. In complex models with a significant number of parameters, the SA may help in the selection of the most relevant parameters for the calibration process [4,7,13,14].

Sometimes the large number of input values in the habitual "one at a time" SA perturbation method requires excessive computation time [14–16]. Also, it leads to serious difficulties in visualizing the results in a comprehensive manner, especially in complex ecosystem models with dozens of parameters. To avoid a sensitivity matrix with numerous columns or rows, some proposed methodologies do not analyze single parameters, but rather clusters of related parameters in group-collecting sensitivity analysis, reducing the total amount of required variations by at least a factor 1/9 [17]. Other approaches, such as metamodeling, link model inputs to the output through a known relationship, emulating the original model but with less computational demands [10,18]. Despite all the advantages of such SA tests, one of the drawbacks is that it can mask the importance of individual parameters in model performance, providing instead a black-box response for the effect of individual parameters [19].

The aim of this paper is to perform a SA on a complex marine biogeochemical model that reproduces the dynamics of living groups (producers, consumers and decomposers), main nutrients (N, P and Si), oxygen and several pools of organic matter. We have used a schematic mesocosm application to prevent any influence of physical transport processes on the outcome of the model. To simplify the analysis, we have developed a color-code matrix to visualize the results of the SA. With this study we intend to identify key parameters influencing model performance, to focus on their variation in the subsequent calibration process.

2. Model and Methods

A process-oriented zero-dimensional ecological model (MOHID-Life [20]) is applied to simulate biogeochemical constituents in a virtual mesocosm. The choice of a zero-dimensional model application, instead of a real case application, is justified because it avoids a substantial increase in model run time. Also, it guarantees that only ecological processes are responsible for the change in the state-variable over time, by eliminating the effect of transport on the state-variables. The only physical process considered in the simulations is the sinking of particulate organic matter. MOHID-Life consists of a series of coupled first-order differential equations representing the major biogeochemical processes influencing the water quality, as well as the dynamic of several groups of primary and secondary producers, the microbial loop, nutrient recycling and oxygen dynamics. MOHID-Life is a complex model that accounts for the cycles of carbon, nitrogen, phosphorus, silica and oxygen. A detailed description of the model options and philosophy, as well as its parameterization, has been previously described and the values of the parameters tested here have been derived from these studies [20,21].

The flexible parameterization of the model allows configuration of different degrees of complexity, by reducing or increasing the number of phytoplankton and zooplankton groups. So, to reduce the volume of results generated by the runs, and aiming at simplicity in the analysis, the reference simulation for this study is simplified by considering only one producer (diatoms) and one consumer (microzooplankton). Because the model philosophy is built on the concept of a Generic Type Model [20], this simplification in the application still allows assessing the importance of parameters, since the model uses the same code for all groups of producers and for consumer.

A reference run is used to provide the benchmark results for the parameter sensitivity analysis (see Section 2.3). The results of the sensitivity analysis are classified by distinguishing model parameters with a qualitatively different effect on the outcome of the simulation. The effect of single parameter variation on model performance is analyzed by singly and sequentially altering the standard value of each parameter with up- and down-variations of 10% in a series of separate runs, while holding all other terms constant. A raised and lowered 10% parameter perturbation is frequently adopted [15,22,23], hence its use in the present study.

2.1. Sensitivity Index

The choice of a sensitive index varies significantly in the literature [1,10,24,25]. From all the indexes available to quantify parameter sensitivity, we chose one able to quantify normalized sensitivity, but with the capacity to reveal up (positive) and down (negative) variation. The choice of index was based on its simplicity, and directly related to the objective of knowing the influence of any given parameter on the outcome of the simulation, *i.e.*, the impact of a parameter and the nature of variation in the results.

Normalized sensitivity, $S_{(p)}$, is defined as the relative change in model output divided by the relative change in the parameter value. It is calculated as:

$$S_{(p)} = \frac{\left(V_{(p)} - V_s\right)/V_s}{\left(p - p_s\right)/p_s} \tag{1}$$

where all variables with an S in the lower index represent standard case values (V_s the value of a given variable for the standard case with parameter p_s), and $V_{(p)}$ is the value for the case when the parameter is given the value p. As an example, to evaluate the impact of a negative perturbation (-10%) of the parameter *bio_si_diss* on silicate concentration, we would have $p_s = 0.01$ and $p = 0.009$, and the respective model results for each parameter value, $V_s = 3.07$ and $V_{(p)} = 2.85$, respectively. In this particular case, $S_{(p)} = 0.071$.

This method is proposed by Fasham *et al.* [26] and adopted in other studies (e.g., [23,27]). According to this sensitivity index, a negative parameter perturbation (10% below) with a negative index result means a positive perturbation in the end result of a given property (meaning a higher end value compared with the reference run value), whereas a positive index result means a negative perturbation in the result (a lower end value compared with the reference run). Conversely, a positive parameter perturbation (10% above) will give a negative index result if a negative end result is achieved, and a positive index result with a positive end result.

The degree of model sensitivity towards any given parameters can be defined as sensitive ($S > 0.1$, meaning a change of more than 1% in the result when compared with the reference value), highly sensitive ($S > 1$, meaning a change of more than 10%), and extremely sensitive ($S > 10$, meaning a change of more than 100%). Whenever $S < 0.1$, it can be said that the model is not sensitive to that parameter. However, it must be kept in mind that the bias achieved by omitting some variations is a frequent or potential error that might occur in a systematic SA. In addition, it is difficult to define the magnitude of parameter perturbation avoiding non-realistic values, but at the same time covering the actual range of the parameter [26].

2.2. Variables of Interest

Not all state-variables were selected to monitor the sensitivity of parameters. This choice relied on the assumption that an excessive number of state-variables would compromise the comprehensibility of the analysis. The biomass of functional groups, concentration of nutrients and labile-DOC (DOMl), and chlorophyll content in phytoplankton were selected as sensitivity indicators (Table 1), mostly because they correspond to properties for which field data is more frequently available. The phytoplankton carbon content to bacterioplankton carbon content ratio (P_c:B_c) was set to evaluate the varying microbial community composition in response to different parameters values.

Table 1. Selected variables to assess model sensitivity.

Symbol	Description	Initial Value
P_c	Producers biomass	$1 \ mg \cdot C \cdot m^{-3}$
Z_c	Consumers biomass	$0.5 \ mg \cdot C \cdot m^{-3}$
B_c	Decomposer biomass	$1 \ mg \cdot C \cdot m^{-3}$
NH_4	Ammonium concentration	$4 \ mmol \cdot N \cdot m^{-3}$
NO_3	Nitrate concentration	$10 \ mmol \cdot N \cdot m^{-3}$
PO_4	Phosphate concentration	$1 \ mmol \cdot P \cdot m^{-3}$
Si	Silicate concentration	$6 \ mmol \cdot Si \cdot m^{-3}$
Chla	Chlorophyll concentration	$0.01 \ mg \cdot C \cdot m^{-3}$
DOMl	Labile DOM concentration	$1 \ mg \cdot C \cdot m^{-3}$
P_c:B_c	Producers-decomposer ratio	1

2.3. Simulations Runs

The model was applied to a virtual mesocosm corresponding to a schematic reservoir with $3 \times 3 \times 5$ square cells with 1 m each. A spin-up period of 16 months was used to stabilize the model, after which the SA simulations run for 3 months (May to July). The long spin-up period aimed at stabilizing the model after the "chaotic" oscillations characteristic from the first year run. The standard model parameterization used in the simulations is presented in Table 2 to Table 5, and have been taken from previous studies [20,21]. The initial values for the properties addressed in the SA are provided in Table 1. Environmental conditions used in the model forcing are shown in Figure 1. A schematic water temperature time-series was used and solar radiation corresponds to conditions representative for mid-latitudes in the northern hemisphere. The term "variable" is loosely used throughout the sensitivity analysis discussion because, with the exception of P_c:B_c, the remaining correspond to state-variables in model.

Table 2. List of parameters for Producers (diatoms).

Parameter	Description	Reference Value	Units
min_nc_ratio	Minimum N:C ratio	0.006284	mmol·N·(mg C)$^{-1}$
max_nc_ratio	Maximum N:C ratio	0.0251	mmol·N·(mg C)$^{-1}$
min_pc_ratio	Minimum P:C ratio	0.000393	mmol·P·(mg C)$^{-1}$
max_pc_ratio	Maximum P:C ratio	0.001572	mmol·P·(mg C)$^{-1}$
max_chln_ratio	Maximum Chl:N ratio	3	mg·Chl·(mmol N)$^{-1}$
chl_degrad_rate	Chlorophyll degradation rate	0	day^{-1}
alpha_chl	Initial slope of the photosynthesis-light curve	3.0025	mg·C·m^2 (mg Chl·W·d)$^{-1}$
affinity_nh4	Affinity for NH$_4$ (uptake rate)	0.0025	(mg C)$^{-1}$·m^{-3}·day^{-1}
affinity_no3	Affinity for NO$_3$ (uptake rate)	0.0025	(mg C)$^{-1}$·m^{-3}·day^{-1}
affinity_po4	Affinity for PO$_4$ (uptake rate)	0.0025	(mg C)$^{-1}$·m^{-3}·day^{-1}
exu_nut_stress	Exudation under nutrient stress	0.05	Dimensionless
exc_dom_sl_frac	Excreted DOM fraction diverted to semi-labile pool	0.4	Dimensionless
max_assimil	Maximum assimilation rate	2.5	day^{-1}
max_store_fill	Maximum rate of storage filling	1	day^{-1}
min_lysis	Minimum lysis rate	0.05	day^{-1}
mort_dom_sl_frac	DOM fraction diverted to semi-labile pool	0.4	Dimensionless
ref_temp	Reference temperature	30	°C
rel_excess_si	Release rate of excess silicate	1	day^{-1}
resp_basal	Basal respiration rate	0.15	day^{-1}
resp_frac_prod	Respired fraction of production	0.1	Dimensionless
sedim_min	Minimum sedimentation rate	0	m·day^{-1}
sedim_nut_stress	Nutrient stress sedimentation rate	5	m·day^{-1}
si_uptake_ks	Silicate uptake Michaelis constant	0.3	mmol·Si·m^{-3}
nut_stress_thresh	Exudation under nutrient stress	0.7	Dimensionless

Table 3. List of parameters for consumers (microzooplankton).

Parameter	Description	Reference Value	Units
max_nc_ratio	Maximum N:C ratio	0.0167	mmol·N·(mg C)$^{-1}$
max_pc_ratio	Maximum P:C ratio	0.00185	mmol·P·(mg C)$^{-1}$
min_nc_ratio	Minimum N:C ratio	0.015	mmol·N·(mg C)$^{-1}$

Table 3. *Cont.*

Parameter	Description	Reference Value	Units
min_pc_ratio	Minimum N:C ratio	0.0017	$mmol \cdot P \cdot (mg \cdot C)^{-1}$
excre_up_frac	Excreted fraction of uptake	0.5	Dimensionless
mort_o2_dep	Oxygen-dependent mortality rate	0.25	day^{-1}
mort_pom_frac	Fraction of excretion to POM	0.5	Dimensionless
mort_rate	Temperature-independent mortality rate	0.05	day^{-1}
o2_ks	Oxygen half saturation constant	0.25	$mg \cdot O_2 \cdot l^{-1}$
ref_temp	Reference temperature	20	°C
rest_resp_@10c	Rest respiration	0.02	day^{-1}
assimil_effic	Assimilica efficiency	0.5	Dimensionless
max_spec_up_@10c	Maximum specific uptake	1.2	day^{-1}
graz_up_ks	Half saturation value for uptake	40	$mg \cdot C \cdot m^{-3}$
p_graz_avail	Availability of prey	0.9	Dimensionless

Table 4. List of parameters for decomposer (bacterioplankton).

Parameter	Description	Reference Value	Units
min_nc_ratio	Minimum N:C ratio	0.016652	$mmol \cdot N \cdot (mg \cdot C)^{-1}$
max_nc_ratio	Maximum N:C ratio	0.01972	$mmol \cdot N \cdot (mg \cdot C)^{-1}$
min_pc_ratio	Minimum P:C ratio	0.001096	$mmol \cdot P \cdot (mg \cdot C)^{-1}$
max_pc_ratio	Maximum P:C ratio	0.001665	$mmol \cdot P \cdot (mg \cdot C)^{-1}$
ass_effic	Assimilation efficiency	0.3	Dimensionless
ass_effic_low_o2	Assimilation efficiency at low oxygen	0.2	Dimensionless
dens_dep_mort	Density-dependent mortality rate	0.5	day^{-1}
lys_ref_con	Mortality density dependent concentration	100	$mg \cdot C \cdot m^{-3}$
max_spec_up_@10c	Maximum specific uptake at reference temperature	5	day^{-1}
dom_up_ks	Half saturation constant for DOM uptake	10.6	$(mg \cdot C)^{-1} \cdot m^{-3}$
mort_dom_sl_frac	Fraction of DOM to semi-labile pool	0.2	Dimensionless
mort_pom_frac	Fraction of mortality products to particulate organic matter	0.4	Dimensionless
mort_rate	Density-independent mortality rate	0.05	day^{-1}
o2_ks	Oxygen half saturation constant	0.01	$mmol \cdot O_2 \cdot m^{-3}$
o2_low_ass_efic	Assimilation efficiency at low oxygen	1.6	$mmol \cdot O_2 \cdot m^{-3}$
rest_resp_@10c	Rest respiration at reference temperature	0.01	day^{-1}
ref_temp	Reference temperature	30	°C
po4_ks	Affinity for PO_4 (uptake rate)	0.0025	$(mg \cdot C)^{-1} \cdot m^{-3} \cdot day^{-1}$
nh4_ks	Affinity for NH_4 (uptake rate)	0.0025	$(mg \cdot C)^{-1} \cdot m^{-3} \cdot day^{-1}$
no3_ks	Affinity for NO_3 (uptake rate)	0.0025	$(mg \cdot C)^{-1} \cdot m^{-3} \cdot day^{-1}$

Table 5. General parameters (not necessarily related with any functional group).

Parameter	Description	Reference Value	Units
bio_si_diss	Biogenic silica dissolution rate	0.01	day^{-1}
pom_bac_ks	POM hydrolysis half saturation constant	32	$mg \cdot C \cdot m^{-3}$
pom_bac_vmax	Maximum rate for POM hydrolysis	0.5	day^{-1}
domsl_bac_ks	DOMsl hydrolysis half saturation constant	200	$mg \cdot C \cdot m^{-3}$
domsl_bac_vmax	Maximum rate for DOMsl hydrolysis	1	day^{-1}
nitrifradlim	Light intensity threshold for nitrification	4	$W \cdot m^{-2}$
nitrifrate	Nitrification rate	0.04	day^{-1}
nit_in_coef	First-order nitrification inhibition coefficient	0.6	$1\ mg^{-1}$

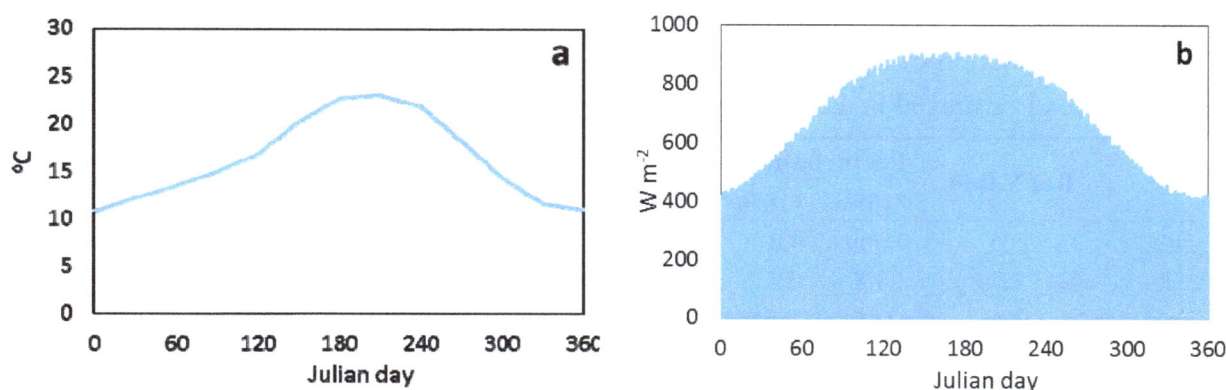

Figure 1. Environmental conditions used to force the model: (**a**) Surface temperature and (**b**) Daily maximum solar radiation.

2.4. Interpretation of the Results

A simple Visual Basic (VBA) macro was developed in Excel environment to process the huge amount of data produced by the model outputs for all the simulations. The macro loads the model results into a spreadsheet and then calculates the sensitivity index of the selected variables to each parameter. In the next step the matrix is interpreted to convert the quantitative index into a qualitative descriptor (the degree of model sensitivity), attributing the + and − symbols to indicate a positive or negative perturbation, respectively. The degree of sensitivity is then given by + and − for sensitive ($S > 0.1$), ++ or −− for highly sensitive ($S > 1$), and + + + or − − − extremely sensitive ($S > 10$). Finally, to allow a better visualization of the results, a color code is used to identify the deferent degrees of sensitivity, positive or negative. No symbol or color is assigned when the model is not sensitive to a given parameter ($S < 0.1$). A step-wise description of this process is presented in Tables 6–8.

Table 6. Representation of the processing and interpretation of the SA result, Step 1: the sensitivity index is calculated from the results of the simulations. Only a fraction of the results are shown for illustrative purposes. The description of parameters and variables can be found in Tables 1–5.

Ammonia	Nitrate	Inorganic Phosphorus	Silicate Acid	Biogenic Silica	Oxygen
2.790	10.600	0.945	2.850	3.180	8.560
2.820	10.600	0.948	3.070	2.960	8.560
2.790	10.600	0.941	3.070	2.960	8.560
2.800	10.600	0.945	3.070	2.960	8.560
2.800	10.600	0.945	3.070	2.960	8.560
2.810	10.600	0.945	3.070	2.960	8.560
3.080	10.400	0.945	3.070	2.960	8.560
2.810	10.600	0.945	3.070	2.960	8.560

Table 7. Representation of the processing and interpretation of the SA result, Step 2: the sensitivity index is arranged in a matrix that relates input parameters and output variables.

Parameter	Ref Value	Perturbation		Pc		Zc		Bc	
		+10%	+10%	+10%	+10%	−10%	+10%	−10%	+10%
bio_si_diss	0.01	0.009	0.011	2.892	2.795	0.097	0.000	0.000	0.000
pom_bac_ks	32	28.8	35.2	5.073	10.032	0.097	0.097	0.204	0.204
pom_bac_vmax	0.5	0.45	0.55	−26.672	−6.268	−0.194	−0.291	−0.476	−0.340
domsl_bac_ks	200	180	220	0.000	0.000	0.000	0.000	0.000	0.000
domsl_bac_vmax	1	0.9	1.1	0.000	0.000	0.000	0.000	0.000	0.000
nitrifradlim	4	3.6	4.4	0.000	0.000	0.000	0.000	0.000	0.000
nitrifrate	0.04	0.036	0.044	0.016	0.016	0.000	0.000	0.000	0.000
nit_in_coef	0.6	0.54	0.66	0.000	0.000	0.000	0.000	0.000	0.000

Table 8. Representation of the processing and interpretation of the SA result, Step 3: for a straightforward interpretation, the sensitivity value is converted to a qualitative descriptor, using signs (+ and −) and a color scale to indicate positive or negative perturbation, and the degree of sensitivity of a variable to a given parameter.

Parameter	Pc		Zc		Bc	
	+10%	+10%	−10%	+10%	−10%	+10%
bio_si_diss	++	++				
pom_bac_ks	++	++			−	−
pom_bac_vmax	− − −	− −	−	−	−	−
domsl_bac_ks						
domsl_bac_vmax						
nitrifradlim						
nitrifrate						
nit_in_coef						

Notes: ▮ sensitive; ▮ highly sensitive; ▮ extremely sensitive.

3. Results

The effect of 68 parameters was tested for their positive and negative variation influence on 11 variables, resulting in a matrix of results with 1496 values. For simplicity, only the most relevant parameter influences on model results, *i.e.*, denoting moderate or high model sensitivity toward that parameter, are discussed. The results are presented in tables for each functional group parameters and for the general parameters, where the impact of each parameter perturbation is expressed in a qualitative way. The tables provide a condensed view of the complexity of the interrelations between parameters and variables within the model.

The simulated period was chosen to have temporal variability, namely the formation and destruction of spring bloom of phytoplankton, thus avoiding a period of model stasis. Results of the reference run are illustrated in Figure 2.

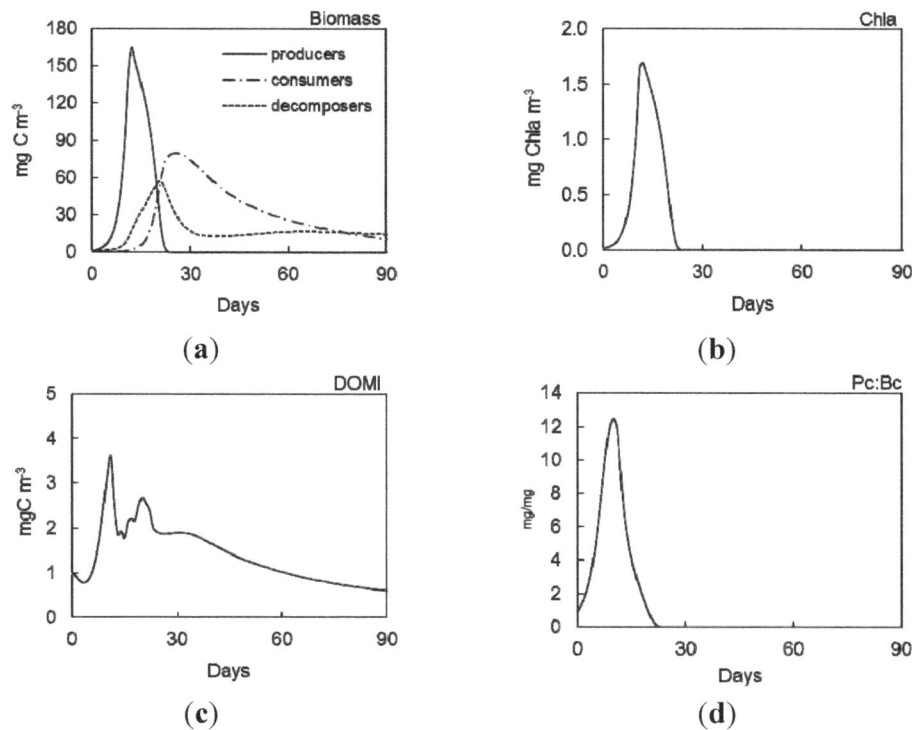

Figure 2. Results for (**a**) biomass of producers, consumers and decomposers; (**b**) chlorophyll *a* (Chla) concentration; (**c**) labile dissolved organic carbon; and (**d**) producers:decomposers ratio.

3.1. General Parameters

Variables sensitivity to general parameters is represented in Table 9, where it is possible to see that only three parameters (*bio_si_diss, pom_bac_ks* and *pom_bac_vmax*) have a significant impact on the results (>1%). Biogenic silica dissolution rate (*bio_si_diss*) perturbation induces a moderate change in results, especially in producers (also reflected in chlorophyll) and consequently on the $P_c:B_c$ ratio.

Pom_bac_ks and *pom_bac_vmax* induce small changes in nutrient variables, but have a high impact on producers (and all related variables), especially *pom_bac_vmax* ($S > 10\%$). A negative perturbation of this parameter leads to the increase in the biomass of producers, and a positive perturbation has the contrary effect. This parameter controls the rate at which hydrolysis converts POM to DOM, a substrate for decomposers. This means the higher *pom_bac_vmax* is, more DOM is made available for decomposers; these will compete with producers for nutrients thus impacting their growth.

3.2. Producers' Parameters

The SA of producers' parameters (Table 10), shows a minor impact of their perturbation in the result of consumers and decomposers, the exception being for *max_assimil* and *ref_temp* with a sensitivity of $S > 0.1$ in consumers. Overall, nutrient affinity to both forms of nitrogen were the parameters with less effect on any of the variables. The model is extremely sensitive ($S > 10$) only to three parameters, namely, exudation under nutrient stress (*exu_nut_stress*), maximum assimilation rate (*max_assimil*), and reference temperature (*ref_temp*). The effect of these parameters is particularly strong in all phytoplankton related variables (biomass and Chla content), a fact also observed in other studies [15]. In this particular case, the simulation ends in a situation of apparent nutrient shortage, and as nutrient limitation increases, the influence of parameters governing growth and response to nutrient stress increases. This is observed in the decrease of *exu_nut_stress*, meaning less exudation under nutrient stress, resulting in a positive perturbation on the biomass of producers.

The highest observed sensitivity is for the *ref_temp* upper perturbation in Chla. Overall, *ref_temp* is the parameter to which producers are more sensitive, an expected occurrence given the control of temperature on several physiological processes.

3.3. Consumers' Parameters

Globally, the model is more sensitive to consumers' parameters. SA results in Table 11 show that only some variables are sensitive ($S > 0.1$) to minimum and maximum nutrient:carbon ratios, and that the model is highly or extremely sensitive to all other parameter, especially the variables related to phytoplankton (biomass, Chla, P_c:B_c). Consumers parameterization is important to the control of both producers and consumers groups, and not so relevant for the decomposers and organic matter dynamics.

The highest sensitivity values observed are for *assimil_effic* and *p_graz_avail* by Chla and P_c. The sensitive index observed in Chla for *assimil_effic* has the highest value in the SA. Figure 3 illustrates the influence of these parameters on the temporal evolution of some variables, and it is possible to notice that, while Z_c and B_c are not as sensitive as P_c, their dynamics is affected by this parameter. When compared to the parameter sets of other groups, consumer parameters have the strongest effect on model behavior, most probably due to the grazing control that consumers have on producers.

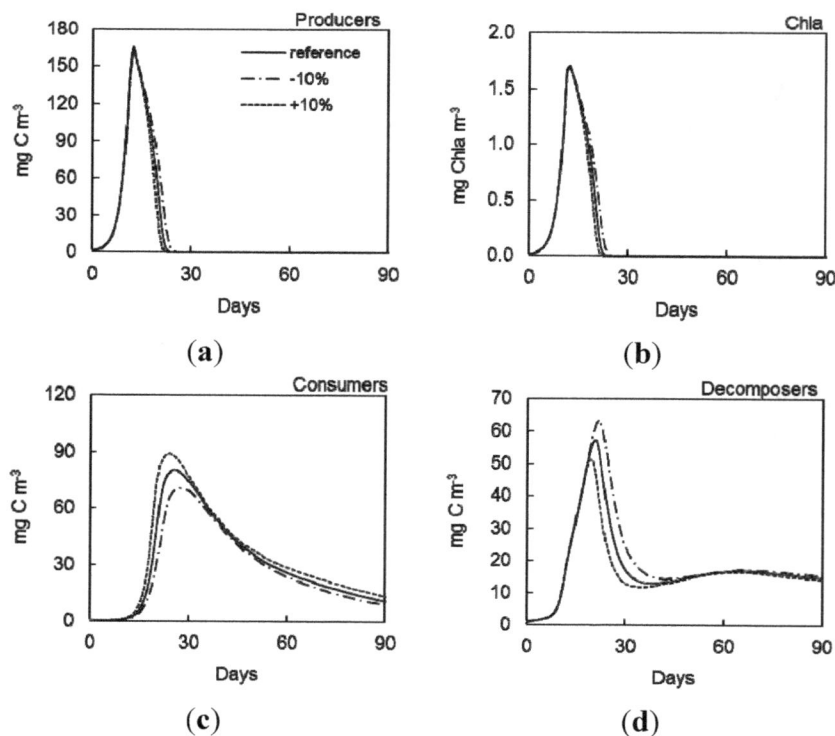

Figure 3. Temporal evolution of the response of (**a**) producers biomass; (**b**) chlorophyll *a* (Chla); (**c**) consumers biomass; and (**d**) decomposers biomass to the perturbation of *assimil_effic*.

Table 9. SA result matrix for the general parameters impact on the analyzed variables. Color code: blue for sensitive ($S > 0.1$), orange for highly sensitive ($S > 1$) and red for extremely sensitive ($S > 10$). The signs + and − stand for positive and negative perturbation, respectively. A description of the variables and parameters is found in Tables 1–5.

Parameter	Pc		Zc		Bc		NH4		NO3		PO4		Si		Chla		DOMl		Pc:Bc	
	−10%	+10%	−10%	+10%	−10%	+10%	−10%	+10%	−10%	+10%	−10%	+10%	−10%	+10%	−10%	+10%	−10%	+10%	−10%	+10%
bio_si_diss	++	++											−		++	++	+		++	++
pom_bac_ks	++	++			−	−									++	++			++	++
pom_bac_vmax	−−	−−			−	−									−−−	−−−		−	−−−	−−−
domsl_bac_ks																				
domsl_bac_vmax																				
nitrifradlim																				
nitrifrate							−		−											
nit_in_coef																				

Table 10. SA result matrix for producers' parameters impact on the analyzed variables. Color code: blue for sensitive ($S > 0.1$), orange for highly sensitive ($S > 1$) and red for extremely sensitive ($S > 10$). The signs + and − stand for positive and negative perturbation, respectively. A description of the variables and parameters is found in Tables 1 and 2, respectively.

Parameter	Pc		Zc		Bc		NH4		NO3		PO4		Si		Chla		DOMl		Pc:Bc	
	−10%	+10%	−10%	+10%	−10%	+10%	−10%	+10%	−10%	+10%	−10%	+10%	−10%	+10%	−10%	+10%	−10%	+10%	−10%	+10%
min_nc_ratio	−														−				−	
max_nc_ratio	++	++					−		−						++	++	+	−	++	++
min_pc_ratio	−																		−	
max_pc_ratio		+														+	+			+
max_chln_ratio	++	++													++	++			++	++
chl_degrad_rate																				
alpha_chl	++	++													++	++			++	++
affinity_nh4															−		+			

Table 10. *Cont.*

Parameter	Pc −10%	Pc +10%	Zc −10%	Zc +10%	Bc −10%	Bc +10%	NH4 −10%	NH4 +10%	NO3 −10%	NO3 +10%	PO4 −10%	PO4 +10%	Si −10%	Si +10%	Chla −10%	Chla +10%	DOMl −10%	DOMl +10%	Pc:Bc −10%	Pc:Bc +10%
affinity_no3																				
affinity_po4	−							−							−				−	
exu_nut_stress	−−−	−−−													−−−	−−−	+		−−−	−
exc_dom_sl_frac	−	−	−	−											−	−		−	−	−
max_assimil	++	+++		−											++	+++			++	+++
max_store_fill																				
min_lysis	−−	−−													−−	−−	+		−	−
mort_dom_sl_frac	−	−	−	−											−	−			−	−
ref_temp	−−−	−−−	−	−				−							−−−	−−−	+		−−−	
rel_excess_si	−−	−−													−−	−−			−	−
resp_basal	−−	−−													−−	−−	+		−	−
resp_frac_prod	−−	−−													−−	−−	+		−	−
sedim_min																				
sedim_nut_stress								−												
si_uptake_ks	−−	−−													−−	−−	+		−−	−
nut_stress_thresh																				

Table 11. SA result matrix for consumers' parameters impact on the analyzed variables. Color code: blue for sensitive ($S > 0.1$), orange for highly sensitive ($S > 1$) and red for extremely sensitive ($S > 10$). The signs $+$ and $-$ stand for positive and negative perturbation, respectively. A description of the variables and parameters is found in Tables 1 and 3, respectively.

Parameter	Pc −10%	Pc +10%	Zc −10%	Zc +10%	Bc −10%	Bc +10%	NH4 −10%	NH4 +10%	NO3 −10%	NO3 +10%	PO4 −10%	PO4 +10%	Si −10%	Si +10%	Chla −10%	Chla +10%	DOMl −10%	DOMl +10%	Pc:Bc −10%	Pc:Bc +10%
max_nc_ratio	−																			
max_pc_ratio							−										+		−	
min_nc_ratio																				
min_pc_ratio		+					−	−				−						+		+
excre_up_frac	−−−	−−	+	++	+	+		−							−−	−	+	+	−−−	−
mort_o2_dep	++	++	+	+	+	+		−							++	++			++	++
mort_pom_frac	++	++			+	+									++	++			++	++
mort_rate	++	+++			+	+									++	+++	+	−	++	+++
o2_ks	++	++			+	+									++	++		−	++	++
ref_temp	++	+++	++	++	+	+	+	+						−	++	+++	+	+	++	+++
rest_resp_@10c	++	+++	−−	−−	+	+									++	+++	−	−	++	+++
assimil_effic	−−−	−−	−	−			−	−			−		−	−	−−	−	+	+	−−−	−
max_spec_up_@10c	−−−	−−			+	+									++	−			++	−
graz_up_ks	++	+++	−	−				+							++	+++		−	++	+++
p_graz_avail	−−−	−−					−	−							−−	−−		−	−	−
d_graz_avail	−	−	+	+	−	−									−	−	+	+		−

Table 12. SA result matrix for decomposers parameters impact on the analyzed variables. Color code: blue for sensitive ($S > 0.1$), orange for highly sensitive ($S > 1$) and red for extremely sensitive ($S > 10$). The signs + and − stand for positive and negative perturbation, respectively. A description of the variables and parameters is found in Tables 1 and 4, respectively.

Parameter	Pc −10%	Pc +10%	Zc −10%	Zc +10%	Bc −10%	Bc +10%	NH4 −10%	NH4 +10%	NO3 −10%	NO3 +10%	PO4 −10%	PO4 +10%	Si −10%	Si +10%	Chla −10%	Chla +10%	DOMl −10%	DOMl +10%	Pc:Bc −10%	Pc:Bc +10%
min_nc_ratio		−														−		−		−
max_nc_ratio	+						+	+							+				+	
min_pc_ratio																	+			
max_pc_ratio																	+	−		
ass_effic																				
ass_effic_low_o2																				
dens_dep_mort	++	++													++	++	−	+	++	++
lys_ref_con	− −	− −													− −	− −	−	−	− −	− −
max_spec_up_@10c	−	−													−	−	− −	−		−
dom_up_ks	+	+													+	+	++		++	++
mort_dom_sl_frac	++	++													++	++			++	− −
mort_pom_frac	++	+		+	+	+									++	+			++	+
mort_rate	++														++	+	+	+	++	
o2_ks	+	+													+	+	+		+	+
o2_low_ass_efic																				
rest_resp_@10c	++	++		−	+	+									++	++	−	−	++	++
ref_temp	++	+++	+	+	+	+									++	+++	++	++	++	+++
po4_ks																	+			
nh4_ks																				
no3_ks																				

3.4. Decomposers' Parameters

The results for the SA analysis on decomposers parameters are illustrated in Table 12. The model only shows to be extremely sensitive ($S > 10$) to the positive perturbation of *ref_temp*, and the effect of the perturbation in the outcome of the model can be seen in Figure 4. Despite the oscillation seen in the time series, the simulations converge to a similar result. As in producers, the strong control of temperature on physiologic functions makes the model extremely sensitivity to temperature reference values.

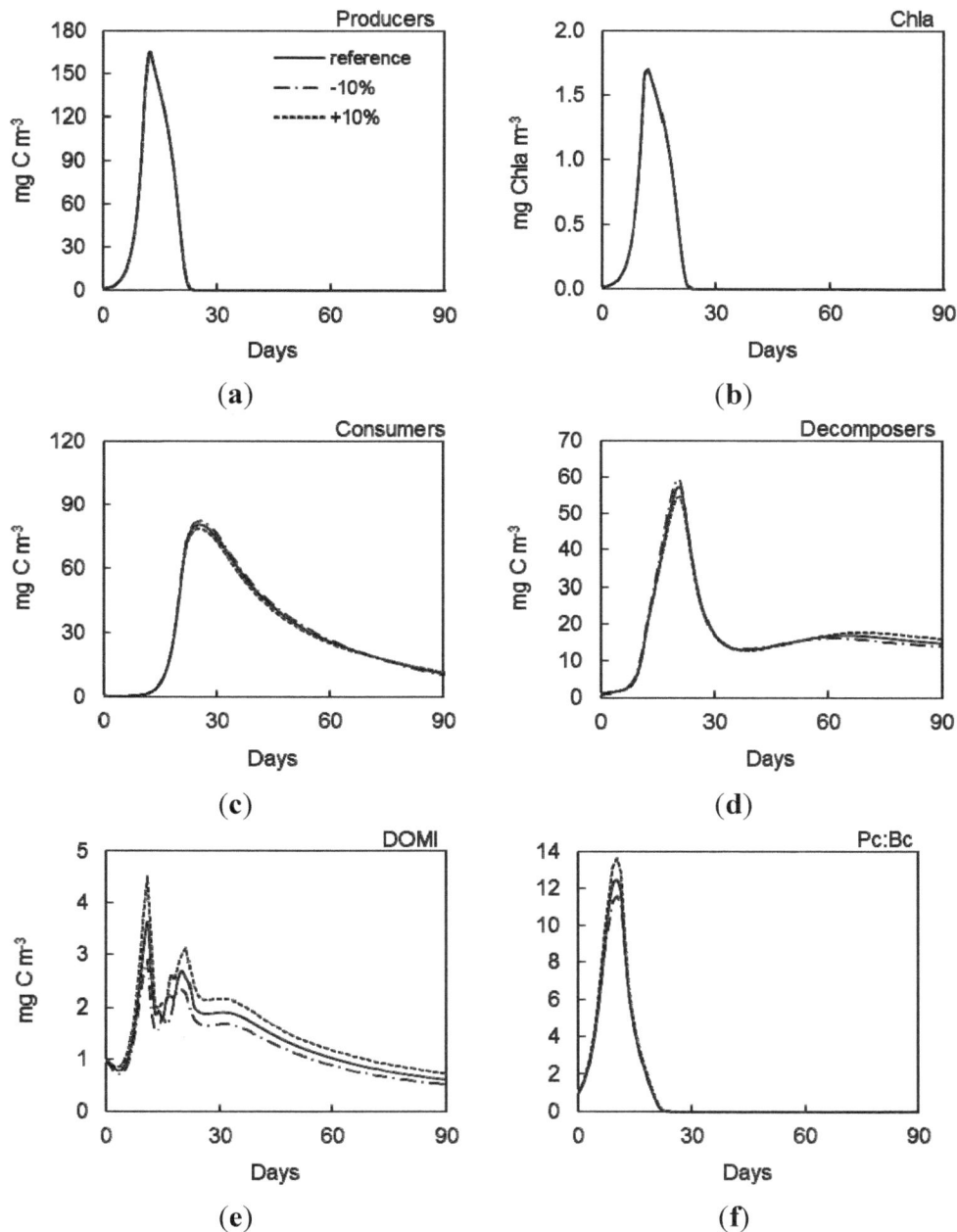

Figure 4. Temporal evolution of the response of (**a**) producers biomass, (**b**) chlorophyll *a* (Chla), (**c**) consumers biomass, (**d**) decomposers biomass, (**e**) labile dissolved organic carbon, and (**f**) producers:decomposers ratio to the perturbation of *ref_temp* (decomposers).

Producers and related variables are the most sensitive to decomposers parameter perturbation. This is an expected outcome, since the microbial loop control much of the available nutrients (via respiration of organic matter), affecting directly the dynamics of nutrient uptake by producers. This effect, however, is not noticed in the nutrient variables, meaning that they are being consumed as they become available. DOMl also shows sensitivity to the perturbation of the parameters that control its mineralization by bacteria, as seen in Figure 4. The affinity for nutrient does not show any relevant disturbance on model response, in a similar way as observed for producers.

4. Discussion

Generally, the model seems to be sensitive to most parameters. The observed sensitivity, however, is not detected in all model compartments, implying that a single parameter may have an impact on one or several state-variables, but not on the overall outcome of the model. Similar observations have been reported in similar studies [3,4,25,28], and can be explained by the complexity of the model in which all state-variables are dependent on processes controlled by multiple parameters.

The model is extremely sensitive to only three parameters in producers: Exudation under nutrient stress (*nut_stress_thresh*), maximum assimilation rate (*max_assimil*), and reference temperature. Overall, reference temperature (*ref_temp*) is the parameter to which producers are more sensitive, an expected occurrence given the control of temperature on several physiological processes [28,29]. Results show that consumer's parameterization is particularly relevant to the control of producers, and not so important for decomposers. This is observed in the number of parameters in consumers for which P_c and Chla show high or extremely high sensitivity (Table 11). These state-variables are particularly sensitive to parameters related with the feeding and assimilation in consumers (*assimil_effic*, *max_spec_up_@10c*, *graz_up_ks* and *p_graz_avail*).

Consumers' parameters have the strongest effect on the behavior of the model, probably through grazing control on producers. Some studies have shown a similar control of consumers (zooplankton) parameters on model performance [2,15,30,31], denoting the importance of this group in ecological models for pelagic systems. Decomposers have the lowest number of parameters for which variables are extremely sensitive. This is only observed for the reference temperature (*ref_temp*) for this functional group (Figure 4).

Of all variables studied in this SA, nutrients were the least affected in the number of parameters to which they are sensitive. Nutrients sensitivity to parameter perturbation never exceeded 10% in the final result. From this, it is possible to conclude that nutrient variables are the most robust. It can be hypothesized, however, that model sensitivity to parameters related with nutrient dynamics may increase under nutrient limitation.

Silica is the variable affected by the lower number of different parameters, because it is the only variable depending only on producers. Remarkably, no producer parameter disruption causes any significant change in the final outcome of silica. Biogenic silica dissolution rate (*bio_si_diss*) perturbation produced a minor impact (<10%) on silica, implying that the model is not very sensitive to it. Nevertheless, this parameter showed to be highly sensitive to other state-variables, namely P_c and Chla. This impact is probably due to limitation by silica in producers. Small variations in these parameters will affect the amount of dissolved silica to diatoms, because it mediates the conversion of biogenic silica to dissolved silica. Also, the other parameters to which the model silica compartment is sensitive are related to consumers grazing activity (*ref_temp*, *assimil_effic* and *max_spec_up_@10c*). The grazing pressure influences silica abundance, because this element is not assimilated by consumers, but instead diverted to the biogenic silica pool.

Phosphate only reflects the variation of two parameters, the upper perturbation of *max_pc_ratio* and the lower perturbation of *min_nc_ratio* (Table 11), because they disturb the immobilization and mineralization of phosphorus. Hence, their perturbation has a direct impact on phosphate, even though not a significant impact (<10%). Nitrate is not a product of mineralization, so it is not affected by any consumers or decomposer parameters, and is only affected by the nitrification rate perturbation (both positive and negative) and by *max_nc_ratio*. Of all groups, consumers have more parameters affecting ammonium dynamics, probably because of its important role in nutrient mineralization through grazing and subsequent nutrient recycling. Most of the parameters related to grazing activity have an effect on ammonium.

Ultimately, one of the main outcomes of SA is to identify where a relatively minor change in a factor's value leads to a major change in model output [32,33]. This is an essential procedure in conventional model development, illustrated by the volume of published works on the topic and stressed in official guidelines regarding the modeling of environmental systems (see [34,35]). In our study we opted for a local SA, where only one parameter varies at a time, while all the others remain fixed at predetermined values, instead of a global analysis in which all parameters vary [7,36]. In global SA, or total SA (*sensu* [37]), all factors are changed together across the full multidimensional input space. This means that the main effect of parameters and the interactions terms involving them are both included, enabling to determine if parameter influence is mostly due to interactions terms rather than to linear effects. Several studies on complex ecosystem models have shown that global SA are more useful because they address parameter interactions and potential nonlinear relationships between parameters and model outputs [12,38–41].

With the simple parameter perturbation approach used in this study it is only possible to suppose potential interactions between parameters controlling model behavior. Taking the most sensitive parameter in consumers as an example, *ref_temp* and *max_spec_up_@10c*, it is possible to infer a potential interaction in the control of consumers, producers and nutrients, because both parameters are used to determine the grazing rate. The same could be inferred for two of the most sensitive parameters in producers, *ref_temp* and *max_assimil*, since both mediate the maximum rate of photosynthesis [20].

Global SA may provide useful information on the most uncertain parameters, and it certainly shows that model parameterization should be done with carefully determined empirical values because of the interaction effects [29]. Nonetheless, a drawback in this approach relies on the extensive computation needed to address all interaction terms and total effects for each input parameter [41,42]. Because global SA aims to characterize models performance and uncertainty [42], it may represent an excessive effort when the aim is to identify key parameters to address in the calibration of the model. Similarly, it may not be as effective as local SA on illustrating the direct relationship between the perturbations of a single parameter and model output, given the volume of results produced. While not providing information on nonlinear responses and parameter interaction (or their combined effect), local SA still provides fundamental information to identify the most sensitive parameters governing model behavior.

This is significant information because the factors the model is most sensitive to usually have stronger main rather than interaction effects [29].

Graphical representation of SA results, especially global SA, has become a challenge of its own given the continuously increase in complexity of ecological models. A way to reduce the volume of results, and to achieve simplification, relies on grouping parameters and assessing model sensitivity to these assemblages [12,43–45]. While these representations have its own benefits, they also have the obvious disadvantage of covering-up the impact of single parameters. This is, in fact, one of the aims of this methodology [46–49]. Since the objective of the work presented here was to identify the key parameters controlling the model behavior, it relied on an approach that explicitly addresses the impact of single parameters. This is a standard and essential practice in process-oriented water-quality and ecological models, because it precedes the calibration exercise [4,5,50–55]. Despite the argumentation on the constraints of the one-factor-at-a-time SA [4,6,7,28], we have simplified the complexity in the interpretation of results by using a color-code matrix, thus minimizing some of its drawbacks. With this approaches it becomes easier to relate parameters (model input) and state-variables (model output) and have instantaneous perception on the magnitude of this relationship. When limited data sets are available (for a few state-variables, for instance), the color-code tables may well be an efficient way to calibrate a model adjusting only a few parameters.

5. Conclusions

Most parameters exert some influence on the selected outputs of the model, but the study shows that the proposed degree of variation in the standard parameterization does not yield a much different scenario, or leads to any significant change in the model results. In the proposed methodology, the result tables provide a condensed view on the sensitivity of state-variables to individual parameters. These tables simplify the interpretation of SA methods that produce significant amounts of results, and can be used as look-up tables to identify sensitive responses in real cases applications.

The methodology presented here is based on local SA. While not providing the same volume of information on the model behavior as a global SA, it yields fundamental information to identify the most sensitive parameters governing model behavior. As such, our approach proved to be a simple and objective way to assist in the calibration exercise of complex models.

Acknowledgments

The authors would like to thank the reviewers for providing valuable comments that helped us to improve the manuscript.

Author Contributions

Marcos D. Mateus performed the computer simulations and sensitivity tests; Guilherme Franz performed further analysis and has interpreted the results. Both authors developed the color-code matrix and worked equally on the preparation of the manuscript.

Conflicts of Interest

The authors declare no conflict of interest.

References

1. Gan, Y.J.; Duan, Q.Y.; Gong, W.; Tong, C.; Sun, Y.W.; Chu, W.; Ye, A.Z.; Miao, C.Y.; Di, Z.H. A comprehensive evaluation of various sensitivity analysis methods: A case study with a hydrological model. *Environ. Modell. Softw.* **2014**, *51*, 269–285.

2. Yoshie, N.; Yamanaka, Y.; Rose, K.A.; Eslinger, D.L.; Ware, D.M.; Kishi, M.J. Parameter sensitivity study of the nemuro lower trophic level marine ecosystem model. *Ecol. Model.* **2007**, *202*, 26–37.

3. Klepper, O. Multivariate aspects of model uncertainty analysis: Tools for sensitivity analysis and calibration. *Ecol. Model.* **1997**, *101*, 1–13.

4. Arhonditsis, G.B.; Brett, M.T. Eutrophication model for Lake Washington (USA): Part I. Model description and sensitivity analysis. *Ecol. Model.* **2005**, *187*, 140–178.

5. Wade, A.J.; Hornberger, G.M.; Whitehead, P.G.; Jarvie, H.P.; Flynn, N. On modeling the mechanisms that control in-stream phosphorus, macrophyte, and epiphyte dynamics: An assessment of a new model using general sensitivity analysis. *Water Resour. Res.* **2001**, *37*, 2777–2792.

6. Saltelli, A.; Annoni, P. How to avoid a perfunctory sensitivity analysis. *Environ. Modell. Softw.* **2010**, *25*, 1508–1517.

7. Cariboni, J.; Gatelli, D.; Liska, R.; Saltelli, A. The role of sensitivity analysis in ecological modelling. *Ecol. Modell.* **2007**, *203*, 167–182.

8. Campolongo, F.; Cariboni, J.; Saltelli, A. An effective screening design for sensitivity analysis of large models. *Environ. Model. Softw.* **2007**, *22*, 1509–1518.

9. Ratto, M.; Tarantola, S.; Saltelli, A. Sensitivity analysis in model calibration: Gsa-glue approach. *Comput. Phys. Commun.* **2001**, *136*, 212–224.

10. Borgonovo, E.; Castaings, W.; Tarantola, S. Model emulation and moment-independent sensitivity analysis: An application to environmental modelling. *Environ. Model. Softw.* **2012**, *34*, 105–115.

11. Van Griensven, A.; Meixner, T.; Grunwald, S.; Bishop, T.; Diluzio, A.; Srinivasan, R. A global sensitivity analysis tool for the parameters of multi-variable catchment models. *J. Hydrol.* **2006**, *324*, 10–23.

12. Harper, E.B.; Stella, J.C.; Fremier, A.K. Global sensitivity analysis for complex ecological models: A case study of riparian cottonwood population dynamics. *Ecol. Appl.* **2011**, *21*, 1225–1240.

13. Rigosi, A.; Marce, R.; Escot, C.; Rueda, F.J. A calibration strategy for dynamic succession models including several phytoplankton groups. *Environ. Model. Softw.* **2011**, *26*, 697–710.

14. Ratto, M.; Castelletti, A.; Pagano, A. Emulation techniques for the reduction and sensitivity analysis of complex environmental models. *Environ. Model. Softw.* **2012**, *34*, 1–4.

15. Rodrigues, M.; Oliveira, A.; Costa, M.; Fortunato, A.B.; Zhang, Y. Sensitivity analysis of an ecological model applied to the ria de aveiro. *J. Coast. Res.* **2009**, *56*, 448–452.

16. Makler-Pick, V.; Gal, G.; Gorfine, M.; Hipsey, M.R.; Carmel, Y. Sensitivity analysis for complex ecological models—A new approach. *Environ. Modell. Softw.* **2011**, *26*, 124–134.

17. Kohler, P.; Wirtz, K.W. Linear understanding of a huge aquatic ecosystem model using a group-collecting sensitivity analysis. *Environ. Modell. Softw.* **2002**, *17*, 613–625.

18. Bayarri, M.J.; Berger, J.; Steinberg, D.M. Special issue on computer modeling. *Technometrics* **2009**, *51*, 353–353.

19. Oakley, J.E.; O'Hagan, A. Probabilistic sensitivity analysis of complex models: A bayesian approach. *J. R. Stat. Soc. B* **2004**, *66*, 751–769.

20. Mateus, M. A process-oriented model of pelagic biogeochemistry for marine systems. Part I: Model description. *J. Marine Syst.* **2012**, *94*, S78–S89.

21. Mateus, M.; Leitão, P.C.; de Pablo, H.; Neves, R. Is it relevant to explicitly parameterize chlorophyll synthesis in marine ecological models? *J. Marine Syst.* **2012**, *94*, S23–S33.

22. Cochrane, K.L.; James, A.G.; Mitchellinnes, B.A.; Pitcher, G.C.; Verheye, H.M.; Walker, D.R. Short-term variability during an anchor station study in the southern benguela upwelling system—A simulation-model. *Prog. Oceanogr.* **1991**, *28*, 121–152.

23. Anderson, T.R. Modeling the influence of food cn ratio, and respiration on growth and nitrogen-excretion in marine zooplankton and bacteria. *J. Plankton Res.* **1992**, *14*, 1645–1671.

24. Risbey, J.; van der Sluijs, J.; Kloprogge, P.; Ravetz, J.; Funtowicz, S.; Quintana, S.C. Application of a checklist for quality assistance in environmental modelling to an energy model. *Environ. Model. Assess.* **2005**, *10*, 63–79.

25. Estrada, V.; Diaz, M.S. Global sensitivity analysis in the development of first principle-based eutrophication models. *Environ. Model. Softw.* **2010**, *25*, 1539–1551.

26. Fasham, M.J.R.; Ducklow, H.W.; Mckelvie, S.M. A nitrogen-based model of plankton dynamics in the oceanic mixed layer. *J. Marine Res.* **1990**, *48*, 591–639.

27. Blumberg, A.F.; Georgas, N. Quantifying uncertainty in estuarine and coastal ocean circulation modeling. *J. Hydraul. Eng.* **2008**, *134*, 403–415.

28. Morris, D.J.; Speirs, D.C.; Cameron, A.I.; Heath, M.R. Global sensitivity analysis of an end-to-end marine ecosystem model of the north sea: Factors affecting the biomass of fish and benthos. *Ecol. Model.* **2014**, *279*, 114–117.

29. Wang, F.; Mladenoff, D.J.; Forrester, J.A.; Keough, C.; Parton, W.J. Global sensitivity analysis of a modified century model for simulating impacts of harvesting fine woody biomass for bioenergy. *Ecol. Model.* **2013**, *259*, 16–23.

30. Steele, J.H.; Henderson, E.W. A simple-model for plankton patchiness. *J. Plankton Res.* **1992**, *14*, 1397–1403.

31. Steele, J.H.; Henderson, E.W. The role of predation in plankton models. *J. Plankton Res.* **1992**, *14*, 157–172.

32. Link, J.S.; Ihde, T.F.; Harvey, C.J.; Gaichas, S.K.; Field, J.C.; Brodziak, J.K.T.; Townsend, H.M.; Peterman, R.M. Dealing with uncertainty in ecosystem models: The paradox of use for living marine resource management. *Prog. Oceanogr.* **2012**, *102*, 102–114.

33. Arhonditsis, G.B.; Brett, M.T. Evaluation of the current state of mechanistic aquatic biogeochemical modeling. *Mar. Ecol. Prog. Ser.* **2004**, *271*, 13–26.

34. European Commission. *Impact Assessment Guidelines sec(2009)92*. Available online: http://ec.europa.eu/smart-regulation/impact/commission_guidelines/docs/iag_2009_en.pdf (accessed on 23 April 2015).

35. U.S. Environmental Protection Agency. *Guidance on the Development, Evaluation and Application of Regulatory Environmental Models*; Office of the Federal Register, National Archives and Records Administration : Washington, DC, USA, 2009.

36. Saltelli, A.; Ratto, M.; Andres, T.; Campolongo, F.; Cariboni, J.; Gatelli, D.; Saisana, M.; Tarantola, S. *Global Sensitivity Analysis: The Primer*; Wiley-Interscience: Chichester, West Sussex, UK, 2008.

37. Saltelli, A.; Tarantola, S.; Chan, K.P.S. A quantitative model-independent method for global sensitivity analysis of model output. *Technometrics* **1999**, *41*, 39–56.

38. Chu-Agor, M.L.; Muñoz-Carpena, R.; Kiker, G.; Emanuelsson, A.; Linkov, I. Exploring vulnerability of coastal habitats to sea level rise through global sensitivity and uncertainty analyses. *Environ. Model. Softw.* **2011**, *26*, 593–604.

39. Miao, Z.; Lathrop, R.G., Jr.; Xu, M.; la Puma, I.P.; Clark, K.L.; Hom, J.; Skowronski, N.; van Tuyl, S. Simulation and sensitivity analysis of carbon storage and fluxes in the New Jersey pinelands. *Environ. Model. Softw.* **2011**, *26*, 1112–1122.

40. Zador, J.; Zsely, I.G.; Turanyi, T. Local and global uncertainty analysis of complex chemical kinetic systems. *Reliab. Eng. Syst. Saf.* **2006**, *91*, 1232–1240.

41. Francos, A.; Elorza, F.J.; Bouraoui, F.; Bidoglio, G.; Galbiati, L. Sensitivity analysis of distributed environmental simulation models: Understanding the model behaviour in hydrological studies at the catchment scale. *Reliab. Eng. Syst. Saf.* **2003**, *79*, 205–218.

42. Bennett, N.D.; Croke, B.F.W.; Guariso, G.; Guillaume, J.H.A.; Hamilton, S.H.; Jakeman, A.J.; Marsili-Libelli, S.; Newham, L.T.H.; Norton, J.P.; Perrin, C.; *et al.* Characterising performance of environmental models. *Environ. Model. Softw.* **2013**, *40*, 1–20.

43. Knights, A.M.; Piet, G.J.; Jongbloed, R.H.; Tamis, J.E.; White, L.; Akoglu, E.; Boicenco, L.; Churilova, T.; Kryvenko, O.; Fleming-Lehtinen, V.; *et al.* An exposure-effect approach for evaluating ecosystem-wide risks from human activities. *ICES J. Marine Sci.* **2015**, *72*, 1105–1115.

44. Alvarez, M.C.; Franco, A.; Perez-Dominguez, R.; Elliott, M. Sensitivity analysis to explore responsiveness and dynamic range of multi-metric fish-based indices for assessing the ecological status of estuaries and lagoons. *Hydrobiologia* **2013**, *704*, 347–362.

45. Brun, R.; Reichert, P.; Kunsch, H.R. Practical identifiability analysis of large environmental simulation models. *Water Resour. Res.* **2001**, *37*, 1015–1030.

46. Hornberger, G.M.; Spear, R.C. Eutrophication in peel inlet—I. The problem-defining behavior and a mathematical model for the phosphorus scenario. *Water Res.* **1980**, *14*, 29–42.

47. Spear, R.C.; Hornberger, G.M. Eutrophication in peel inlet—II. Identification of critical uncertainties via generalized sensitivity analysis. *Water Res.* **1980**, *14*, 43–49.

48. Beven, K.; Binley, A. The future of distributed models—Model calibration and uncertainty prediction. *Hydrol. Process.* **1992**, *6*, 279–298.

49. Bastidas, L.A.; Gupta, H.V.; Sorooshian, S.; Shuttleworth, W.J.; Yang, Z.L. Sensitivity analysis of a land surface scheme using multicriteria methods. *J. Geophys. Res.-Atmos.* **1999**, *104*, 19481–19490.

50. Blackwell, A.L.; Blackwell, C.C., Jr. Minimum sensitivity parameter estimation for dynamic ecosystem models1. In *Developments in Environmental Modelling*; William, K., Lauenroth, G.V.S., Marshall, F., Eds.; Elsevier: Amsterdam, The Netherlands, 1983; Volume 5, pp. 189–194.

51. Loehle, C. A hypothesis testing framework for evaluating ecosystem model performance. *Ecol. Modell.* **1997**, *97*, 153–165.

52. Omlin, M.; Brun, R.; Reichert, P. Biogeochemical model of lake zurich: Sensitivity, identifiability and uncertainty analysis. *Ecol. Model.* **2001**, *141*, 105–123.

53. Meixner, T.; Gupta, H.V.; Bastidas, L.A.; Bales, R.C. Sensitivity analysis using mass flux and concentration. *Hydrol. Process.* **1999**, *13*, 2233–2244.

54. McIntyre, N.R.; Wagener, T.; Wheater, H.S.; Chapra, S.C. Risk-based modelling of surface water quality: A case study of the charles river, massachusetts. *J. Hydrol.* **2003**, *274*, 225–247.

55. Reckhow, K.H.; Chapra, S.C. Modeling excessive nutrient loading in the environment. *Environ. Pollut.* **1999**, *100*, 197–207.

Permissions

The contributors of this book come from diverse backgrounds, making this book a truly international effort. This book will bring forth new frontiers with its revolutionizing research information and detailed analysis of the nascent developments around the world.

We would like to thank all the contributing authors for lending their expertise to make the book truly unique. They have played a crucial role in the development of this book. Without their invaluable contributions this book wouldn't have been possible. They have made vital efforts to compile up to date information on the varied aspects of this subject to make this book a valuable addition to the collection of many professionals and students.

This book was conceptualized with the vision of imparting up-to-date information and advanced data in this field. To ensure the same, a matchless editorial board was set up. Every individual on the board went through rigorous rounds of assessment to prove their worth. After which they invested a large part of their time researching and compiling the most relevant data for our readers.

The editorial board has been involved in producing this book since its inception. They have spent rigorous hours researching and exploring the diverse topics which have resulted in the successful publishing of this book. They have passed on their knowledge of decades through this book. To expedite this challenging task, the publisher supported the team at every step. A small team of assistant editors was also appointed to further simplify the editing procedure and attain best results for the readers.

Apart from the editorial board, the designing team has also invested a significant amount of their time in understanding the subject and creating the most relevant covers. They scrutinized every image to scout for the most suitable representation of the subject and create an appropriate cover for the book.

The publishing team has been an ardent support to the editorial, designing and production team. Their endless efforts to recruit the best for this project, has resulted in the accomplishment of this book. They are a veteran in the field of academics and their pool of knowledge is as vast as their experience in printing. Their expertise and guidance has proved useful at every step. Their uncompromising quality standards have made this book an exceptional effort. Their encouragement from time to time has been an inspiration for everyone.

The publisher and the editorial board hope that this book will prove to be a valuable piece of knowledge for researchers, students, practitioners and scholars across the globe.

List of Contributors

Mikołaj Piniewski
Department of Hydraulic Engineering, Warsaw University of Life Sciences (WULS-SGGW), Nowoursynowska str. 159, Warszawa 02-776, Poland
Potsdam Institute for Climate Impact Research (PIK), P.O. Box 60 12 03, Potsdam 14412, Germany

Paweł Marcinkowski
Department of Hydraulic Engineering, Warsaw University of Life Sciences (WULS-SGGW), Nowoursynowska str. 159, Warszawa 02-776, Poland

Ignacy Kardel
Department of Hydraulic Engineering, Warsaw University of Life Sciences (WULS-SGGW), Nowoursynowska str. 159, Warszawa 02-776, Poland

Marek Giełczewski
Department of Hydraulic Engineering, Warsaw University of Life Sciences (WULS-SGGW), Nowoursynowska str. 159, Warszawa 02-776, Poland

Katarzyna Izydorczyk
European Regional Centre for Ecohydrology of the Polish Academy of Sciences, Tylna str. 3, Łódź 90-364, Poland

Wojciech Frątczak
European Regional Centre for Ecohydrology of the Polish Academy of Sciences, Tylna str. 3, Łódź 90-364, Poland
Regional Water Management Authority in Warsaw, 13B Zarzecze, Warszawa 03-194, Poland

Sean Curneen
Department of Civil, Structural and Environmental Engineering, Trinity College Dublin, Dublin 2, Ireland

Laurence Gill
Department of Civil, Structural and Environmental Engineering, Trinity College Dublin, Dublin 2, Ireland

D. James Morré
MorNuCo, Inc., 1201 Cumberland Avenue, Suite B, Purdue Research Park, West Lafayette, IN 47906, USA

Dorothy M. Morré
MorNuCo, Inc., 1201 Cumberland Avenue, Suite B, Purdue Research Park, West Lafayette, IN 47906, USA

Clemens Kittinger
Institute of Hygiene, Microbiology and Environmental Medicine, Medical University Graz, Graz 8010, Austria

Rita Baumert
Institute of Hygiene, Microbiology and Environmental Medicine, Medical University Graz, Graz 8010, Austria

Bettina Folli
Institute of Hygiene, Microbiology and Environmental Medicine, Medical University Graz, Graz 8010, Austria

Michaela Lipp
Institute of Hygiene, Microbiology and Environmental Medicine, Medical University Graz, Graz 8010, Austria

Astrid Liebmann
Institute of Hygiene, Microbiology and Environmental Medicine, Medical University Graz, Graz 8010, Austria

Alexander Kirschner
Division Water Hygiene, Institute for Hygiene and Applied Immunology, Medical University of Vienna, Vienna 1090, Austria
Interuniversity Cooperation Centre for Water and Health, Vienna, Austria

Andreas H. Farnleitner
Interuniversity Cooperation Centre for Water and Health, Vienna, Austria
Institute of Chemical Engineering, Research Group Environmental Microbiology and Molecular Ecology, Vienna University of Technology, Vienna 1090, Austria

Andrea J. Grisold
Institute of Hygiene, Microbiology and Environmental Medicine, Medical University Graz, Graz 8010, Austria

Gernot E. Zarfel
Institute of Hygiene, Microbiology and Environmental Medicine, Medical University Graz, Graz 8010, Austria

Mattia Callegari
EURAC Research, European Academy of Bozen/Bolzano, Institute for Applied Remote Sensing, viale Druso, Bolzano 1-39100, Italy
Department of Earth and Environmental Science, University of Pavia, via Ferrata 1-27100 Pavia, Italy

Paolo Mazzoli
Research and development (R&D) Unit Suedtirol, Geographic Environmental COnsulting (GECO) Sistema— srl, via Maso della Pieve 60, Bolzano 39100, Italy

Ludovica de Gregorio
EURAC Research, European Academy of Bozen/Bolzano, Institute for Applied Remote Sensing, viale Druso, Bolzano 1-39100, Italy

Claudia Notarnicola
EURAC Research, European Academy of Bozen/Bolzano, Institute for Applied Remote Sensing, viale Druso, Bolzano 1-39100, Italy

Luca Pasolli
Informatica Trentina Spa, via G. Gilli 2, Trento 38121, Italy

Marcello Petitta
EURAC Research, European Academy of Bozen/Bolzano, Institute for Applied Remote Sensing, viale Druso, Bolzano 1-39100, Italy

Alberto Pistocchi
European Commission, Directorate-General Joint Research Centre (DG JRC), via E.Fermi 2749, Ispra (VA) 21027, Italy

Huijie Li
Key Laboratory of Plant Nutrition and the Agri-Environment in Northwest China, Ministry of Agriculture, Northwest Agriculture and Forestry University, Yangling 712100, China

Jun Yi
Key Laboratory of Plant Nutrition and the Agri-Environment in Northwest China, Ministry of Agriculture, Northwest Agriculture and Forestry University, Yangling 712100, China

Jianguo Zhang
Key Laboratory of Plant Nutrition and the Agri-Environment in Northwest China, Ministry of Agriculture, Northwest Agriculture and Forestry University, Yangling 712100, China

Ying Zhao
Key Laboratory of Plant Nutrition and the Agri-Environment in Northwest China, Ministry of Agriculture, Northwest Agriculture and Forestry University, Yangling 712100, China
Department of Soil Science, University of Saskatchewan, Saskatoon, SK, S7N 5A8, Canada

Bingcheng Si
Department of Soil Science, University of Saskatchewan, Saskatoon, SK, S7N 5A8, Canada
College of Water Resources and Architecture Engineering, Northwest Agriculture and Forestry University, Yangling 712100, China

Robert Lee Hill
Department of Environmental Science and Technology, University of Maryland, College Park, MD 20742, USA

Lele Cui
Key Laboratory of Plant Nutrition and the Agri-Environment in Northwest China, Ministry of Agriculture, Northwest Agriculture and Forestry University, Yangling 712100, China
Suide Test Station of Water and Soil Conservation, Yellow River Conservancy Committee of the Ministry of Water Resources, Yulin, Shanxi 719000, China

Xiaoyu Liu
Key Laboratory of Plant Nutrition and the Agri-Environment in Northwest China, Ministry of Agriculture, Northwest Agriculture and Forestry University, Yangling 712100, China

E. Carina H. Keskitalo
Geography and Economic History, Umeå University, Umeå 901 87, Sweden

Qian Yu
State Key Laboratory of Hydroscience and Engineering, Tsinghua University, Beijing 100084, China

Yongcan Chen
State Key Laboratory of Hydroscience and Engineering, Tsinghua University, Beijing 100084, China

Zhaowei Liu
State Key Laboratory of Hydroscience and Engineering, Tsinghua University, Beijing 100084, China

Nick van de Giesen
Department of Water Management, Faculty of Civil Engineering and Geosciences, Delft University of Technology, PO Box 5048, 2600 GA Delft, The Netherlands

Dejun Zhu
State Key Laboratory of Hydroscience and Engineering, Tsinghua University, Beijing 100084, China

Terry Lucke
Stormwater Research Group, University of the Sunshine Coast, Sippy Downs, Queensland 4558, Australia

Richard White
Stormwater Research Group, University of the Sunshine Coast, Sippy Downs, Queensland 4558, Australia

Peter Nichols
Stormwater Research Group, University of the Sunshine Coast, Sippy Downs, Queensland 4558, Australia

Sönke Borgwardt
Büro BWB Norderstedt, Kattendorf 24568, Germany

Rayco Marrero-Diaz
Environmental Research Division, Institute of Technology and Renewable Energies (ITER), Granadilla de Abona, Tenerife, Canary Islands 38611, Spain
Instituto Volcanológico de Canarias (INVOLCAN), Puerto de la Cruz, Tenerife, Canary Islands 38400, Spain

Francisco J. Alcalá
Civil Engineering Research and Innovation for Sustainability (CERis), Instituto Superior Tecnico, Universtiy of Lisbon, Lisbon 1049-001, Portugal
Instituto de Ciencias Químicas Aplicadas, Facultad de Ingeniería, Universidad Autónoma de Chile, Santiago 7500138, Chile

Nemesio M. Pérez
Environmental Research Division, Institute of Technology and Renewable Energies (ITER), Granadilla de Abona, Tenerife, Canary Islands 38611, Spain
Instituto Volcanológico de Canarias (INVOLCAN), Puerto de la Cruz, Tenerife, Canary Islands 38400, Spain

Dina L. López
Instituto Volcanológico de Canarias (INVOLCAN), Puerto de la Cruz, Tenerife, Canary Islands 38400, Spain
316 Clippinger Laboratories, Department of Geological Sciences, Ohio University, Athens, OH 45701, USA

Gladys V. Melián
Environmental Research Division, Institute of Technology and Renewable Energies (ITER), Granadilla de Abona, Tenerife, Canary Islands 38611, Spain
Instituto Volcanológico de Canarias (INVOLCAN), Puerto de la Cruz, Tenerife, Canary Islands 38400, Spain

Eleazar Padrón
Environmental Research Division, Institute of Technology and Renewable Energies (ITER), Granadilla de Abona, Tenerife, Canary Islands 38611, Spain
Instituto Volcanológico de Canarias (INVOLCAN), Puerto de la Cruz, Tenerife, Canary Islands 38400, Spain

Germán D. Padilla
Environmental Research Division, Institute of Technology and Renewable Energies (ITER), Granadilla de Abona, Tenerife, Canary Islands 38611, Spain
Instituto Volcanológico de Canarias (INVOLCAN), Puerto de la Cruz, Tenerife, Canary Islands 38400, Spain

Marcos D. Mateus
MARETEC, Instituto Superior Técnico, Universidade de Lisboa, Av. Rovisco Pais, Lisboa 1049-001, Portugal

Guilherme Franz
MARETEC, Instituto Superior Técnico, Universidade de Lisboa, Av. Rovisco Pais, Lisboa 1049-001, Portugal